INTRODUCTORY FINITE
MATHEMATICS WITH COMPUTING

INTRODUCTORY FINITE MATHEMATICS WITH COMPUTING

William S. Dorn
University of Denver

Daniel D. McCracken

JOHN WILEY & SONS, Inc.
New York / London / Sydney / Toronto

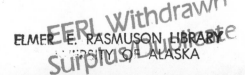

Library of Congress Cataloging in Publication Data:

Dorn, William S
  Introductory finite mathematics with computing.

  1. Mathematics—1961–   2. Mathematics—Data processing.
  3. Basic (Computer program language)
I. McCracken, Daniel D., joint author. II. Title.
QA39.2.D65   510   75–30647
ISBN 0-471-21917-7

Printed in the United States of America

10 9 8 7 6 5 4 3 2 1

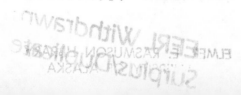

PREFACE

It is becoming clear that many important problems in the social, behavioral and business sciences—psychology, sociology, economics, finance, medicine, and education, for instance—cannot be solved through discussion and debate alone in the absence of careful analysis of factual data. Although we have no quarrel with philosophy and theology, for example, and indeed endorse their crucial position in establishing goals and objectives, in this book we emphasize the importance of a more formalized approach to some aspects of many significant contemporary problems.

With this in mind, this book has been designed for a first course in mathematics and computing at the university level. The mathematical topics have been selected with modern problems in the social, behavioral and business sciences in mind. Moreover, the selection of topics has been heavily influenced by the existence and almost universal availability of the digital computer. This last characteristic distinguishes this book from other books aimed at the same readers.

For example, difference equations play an important role in the first part of this text. Without the computer, difference equations could not be usefully introduced at this level. With the computer, however, they are readily accessible to the freshman, nonmathematics major. But, most important, many problems in the social, behavioral, and business sciences give rise naturally to difference equations. In this text, using difference equations, we treat population growth, the spread of epidemics, a nation's economy, a problem in learning theory, amortization of loans, and population movements.

In contrast to the physical sciences where differential equations play a prominent role, in the social, behavioral, and business sciences difference equations are more natural and more appropriate. The national income or the degree of learning usually is not measured at every point in time nor is its value available at any arbitrary point in time. (The federal government announces the gross national product at the end of each month, and a learner is tested for his knowledge or skill at fixed, distinct testing times.) To treat such phenomena as continuous processes using the calculus is, at best, a rough approximation.

The other mathematical topics included here are vectors and matrices, probability, linear systems of equations, and linear programming. Their presentation also has been

v

influenced by the availability of the computer and, indeed, it has been adjusted to exploit the computer. Again, as an example, although the inverse of a matrix is defined, described, and used in problem analysis, the numerical process of actually inverting a matrix is absent. Instead, we let the computer find the inverse of a matrix for the reader.

Another important facet of this book is its continual concern with the "mathematical approach" to a problem. That approach consists of:

1. Stating the problem and the assumptions underlying the problem quite precisely.

2. Expressing the problem in a concise and unambiguous language: mathematics.

3. Solving the mathematical equations (sometimes with the aid of a computer) to obtain some new information.

4. Examining the new information to see if it fits the problem we thought we were solving.

5. Modifying the statement of the problem or the underlying assumptions of step 1 based on these results.

6. Reanalyzing the problem, following the above steps, until satisfactory results are obtained.

7. Drawing conclusions of a qualitative nature from the quantitative results.

This approach is emphasized throughout the case studies. (There are 10 such case studies.) In a typical case study a problem, for example, from sociology, is stated in broad terms and then narrowed down until it satisfies the requirements of step 1. Learning how to carry out this narrowing-down process is an important part of this book. The problem is then written in mathematical terms, and the mathematical techniques necessary for the solution (step 3) are developed. The results are then examined and, in many cases, a new problem is formulated. In all cases we try to deduce some general principles from the observation of the results.

Often mathematics books and courses are concerned only with step 3. This is not true here. We consider each step as vital and necessary as any other.

The computer language we have chosen is BASIC, simply because it is widely available in the academic world, it is very easy to learn, and it is inexpensive to provide to students compared with the present implementations of most other languages. We are not in any way taking up the cudgels for BASIC versus FORTRAN, APL, PL/I, or whatever. It is just that BASIC seems to be the best choice at this time for our readers.

One mathematical topic is conspicuous by its absence: calculus. It is not that calculus is inappropriate for social-behavioral-business scientists. Indeed, we think that a brief, problem-oriented calculus course with computing would be an ideal sequel to this course. Another excellent sequel would be a statistics course with computing. With a course based on this text as background, the statistics course need not be a "cookbook" course nor too difficult for the students. It could be (and should be) a legitimate mathematics course with many practical applications.

In summary then, this book accomplishes these things.

1. It introduces the reader to the scientific method of analyzing problems, including the interpretation of numerical results, in order to draw quite general conclusions.
2. It shows areas in which mathematics and computing are applicable to the reader's own discipline and to allied disciplines.
3. It develops skills in both mathematics and computing.
4. It prepares the reader for a subsequent course in calculus or statistics.

All of this does not come cheaply. The reader is expected to have a rudimentary knowledge of algebra including inequalities and sets (see The Self-Assessment Exam that follows). He also is expected to do some honest-to-goodness mathematics and to write and run computer programs. A serious student of a social, behavioral, or business science cannot afford to do less.

(NOTE TO THE INSTRUCTOR: An Instructor's Guide to accompany this text is available. It contains:

1. Solutions to all exercises including computer printouts where appropriate.
2. Additional supplementary material that the instructor may use as a basis for classroom lectures.
3. Behavioral objectives for each chapter.
4. Sample examinations with their solutions for each chapter.
5. Sample final and/or mid-term examinations with their solutions.
6. A terminal practice session used to familiarize students with the use of a teletype and the BASIC language.
7. A two page summary of BASIC language statements.

It is a pleasure to acknowledge some of the many people who helped us in the preparation of this text. Patricia Bond typed the final manuscript and, in the process, discovered several significant blunders. Anita Johnson typed the printed version of the instructor's manual.

The staff of John Wiley and Sons, Inc.—the mathematics editor, Gary Ostedt; the production supervisor, Kenneth R. Ekkens; the editorial manager, Malcolm Easterlin; and the "real" editor, Vivian Kahane—all contributed in various ways to the contents and form of the book.

Three faculty members at the University of Denver, Herbert J. Greenberg, Jack K. Cohen, and Joel S. Cohen, all taught courses from notes that were early drafts of this book. These respected mathematicians and skilled teachers made valuable suggestions and criticisms that helped to shape the final manuscript. Gene Collins of the Jefferson County (Colorado) Schools kindly made that school district's computer (an HP 2000B) available to us; the computer printouts included in the text and in the instructor's guide were produced by that computer. Many people read the entire penultimate version of the manuscript, and their suggestions and criticisms had a strong influence on this final version. Among them were Professor Gora Bhaumik of the California State University at Fullerton, Professor Geoffrey Churchill of Georgia State University, Professor Harold S. Engelsohn of Kingsborough Community College, Professor Garret J. Etgen of the University of Houston, Dr. Peter Hilton of the Seattle Research Center, Professor Thomas E. Kurtz of Dartmouth College, Professor Ann D. Martin of Los Angeles Valley College, Professor Robert A. Mills of Eastern Michigan University, Professor Paul Purdom of Indiana University, and Professor Frank Scalzo of Queensborough Community College. But the bulk of our gratitude goes to the many University of Denver students who, over a period of five years, suffered through studying from the two early drafts—replete with errors—as their classroom notes. Their experiences and helpful comments resulted in substantive changes both in content and style.

Finally we wish to express our gratitude to the University of Denver and particularly to Dean Edward A. Lindell* for providing an atmosphere, a surrounding, and the physical facilities that encouraged and continues to encourage new courses such as the one on which this book is based. Such farsighted and enlightened leadership is, unfortunately, all too rare in higher education today.

William S. Dorn
Daniel D. McCracken

DENVER, COLORADO
OSSINING, NEW YORK
JULY, 1975

---

*New president of Gustavus Adolphus College, St. Peter, Minnesota.

CONTENTS

SELF-ASSESSMENT EXAMINATION

This examination is for the benefit of the prospective reader of this book. Such a reader should be able to handle all of the questions below. We do not mean that he should answer all of the questions without error. Instead, he should be reasonably confident that he understands the question and how to answer it. From time to time we all make blunders and silly mistakes, such as 2×3 is 5. Discount such slips when administering this examination to yourself. If you answer a question incorrectly but, after seeing the correct result say, "Of course. I knew that," then go on as if you had been correct all along. On the other hand, if there are several questions for which you have no idea where to begin, you should take a review course or read a review book such as *Intermediate Algebra*, 3d ed., by William Wooton and Irving Drooyan (Wadsworth, 1972).

The correct answers to this examination are found in Appendix A.

PART I EXPONENTS AND FACTORING

Fill in each blank in the following 10 exercises with either

a number such as 4 or -6

or

an arithmetic operator such as $+$ or $-$

or

a symbol such as y or a^3

For example, if an exercise were given as

$$3 + (t \; \Box \; 2) = 1 + t$$

then the correct entry is

$-$.

(*Note*. Questions 4, 9, and 10 require two blanks to be completed.)

1. $3(2 - \Box) = 3 \cdot 2 - 3 \cdot 6$

2. $5(a - b) = 5a \; \Box \; 5b$

3. $x(x - y) = \Box - xy$

4. $a - (b + 2) = a \; \Box \; b \; \Box \; 2$

5. $7 - (a + 3) = \Box - a$

6. $a - (4 \; \Box \; b) = a - 4 + b$

7. $3^2 \cdot 3^4 = 3^{\Box}$

8. $a \cdot a^{\Box} = a^5$

9. $x^2(x^3 - x) = x^{\Box} - x^{\Box}$

10. $3^3(3^{\Box} + a^2) = 3^7 + 3^{\Box} \cdot a^2$

PART II INEQUALITIES

11. Which of the following are true?

    (a) $3 > 1$

    (b) $-3 > -2$

    (c) $-2 < 0$

    (d) $-2 < 1$

12. Fill in the blanks with either $<$ or $>$.
    If $0 < x < 1$,

    (a) $x^2 \; \Box \; x$

    (b) $x^3 \; \Box \; x^2$

13. If $x > 1$, which of the following are true?

    (a) $x^2 > 1$

    (b) $x^2 > x$

(c) $x^2 < x$

(d) $x^3 > x^2$

(e) $x^3 < x$

14. Fill in the blank with + or −.
If $x > 0$,

$$1 \ \square \ x > 1$$

15. Fill in the blank with + or −.
If $0 < x < 1$,

$$0 < 1 \ \square \ x < 1$$

PART III GRAPHS

16. Draw a graph of

$$y = 3x - 2$$

What is the y-intercept? the x-intercept?

17. Plot on an x-y graph the following points.

(A) (2,3)   (B) (4,0)

(C) (0,6)   (D) (6,−7)

(E) (0,0)   (F) (0,−5)

PART IV   SETS

If set $A = \{2, 3, 4, 5, 6\}$ and set $B = \{3, 5, 7, 8\}$ then

18. Describe $A \cup B$ (i.e., the union of A and B).

19. Describe $A \cap B$ (i.e., the intersection of A and B).

If $A = \{$letters in the word "WINTER"$\}$
$B = \{$letters in the word "SPRING"$\}$
$C = \{$letters in the word "SUMMER"$\}$
$D = \{$letters in the word "AUTUMN"$\}$

20. Describe $A \cap B \cap D$.

21. Describe $B \cap C \cap D$.

PART V   VENN DIAGRAMS

If sets X and Y are represented by overlapping squares as follows:

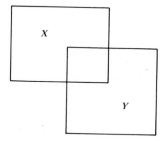

22. Draw a Venn diagram indicating $X \cup Y$.

23. Draw a Venn diagram indicating $X \cap Y$.

If sets A, B, and C are represented by three overlapping circles as follows:

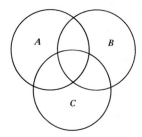

24. Draw a Venn diagram indicating $A \cap B \cap C$.

25. Draw a Venn diagram indicating $B \cap C$.

26. Draw a Venn diagram indicating $(A \cup B) \cap C$.

W. S. D.
D. D. M.

1

SOME PROBLEMS IN PSYCHOLOGY, ECONOMICS, AND BIOLOGY

1.1 INTRODUCTION

In this chapter we will discuss three seemingly different problems from three diverse subject areas: psychology, economics, and biology. In the end we will see that, at least mathematically, all three have a great deal in common.

We will be concerned throughout with building and analyzing "mathematical models" of real-life processes. That is, we will write down equations that mimic some process or some thing that has been described for us only through words and sentences. In what sense will our equations mimic a process? The solution of the equations, whether or not we can find the solution, will follow the same general pattern as the actual process does. Thus, if the process is a population that is exploding, then the equations must have a solution that explodes.

1

And what is the point in writing down such equations, such mathematical models? First, we can use a mathematical model to study what happens if we make certain changes in the process. For example, what happens to our exploding population if the death rate decreases? Second, but equally important, by studying the equations and their form, we hope to gain a better and clearer understanding of the process itself. Thus we would hope to gain some understanding of how and why populations grow (or decline) from our population models.

The practice of creating and analyzing mathematical models of real-life processes lies at the heart of modern science and engineering. The equations of physics and chemistry have been developed over hundreds of years and have helped scientists to understand the world around us better. We will concentrate our attention on modeling phenomena from the social-behaviorial sciences: psychology, biology, economics, business, medicine, law, sociology and geography, among others.

We will find that our first description (in words) of a process often is not sufficient to allow us to create a model. The description may be too vague or perhaps will be lacking in some essential detail. In such cases we will have to use our common sense to make some assumptions about the process. These assumptions are crucial and must continually be reexamined and questioned. It is quite possible for equally learned persons to disagree on what assumptions are to be made in any given case. You should acquire the habit of debating the validity of any and all assumptions. If you have a set of assumptions that you think better fit the real-life situation, then use them. In any case, from the assumptions, we create a model (equations) that leads us to conclusions. If we change the assumptions, the model and the conclusions also may change. Mathematics takes us from the assumptions to the conclusions, but the validity and reliability of the assumptions rests with us. The mathematics will accept whatever assumptions we provide with no questions asked.

1.2 CASE STUDY 1: A SIMPLE LEARNING MODEL

Consider an individual in a learning environment in which the task to be mastered is well defined and measurable. For example, think of a rat whose task is to run a particular maze. If the rat learns to run the maze, then it has mastered the task.[1] As another example consider a child whose task is to learn to ride a bicycle for several hundred yards without falling.

The important point about such tasks is that the total amount to be learned is well defined and fixed. This contrasts with learning mathematics or English literature where, no matter how much one has learned, there is always a substantial amount of material still to be mastered.

Now suppose that the individual is trained in the task during a succession of practice sessions. The rat is placed in the maze for a certain number of minutes and then

---

[1]Of course, the rat will never learn to run the maze without any missteps. One reasonable criterion would be that the rat successfully reaches the end of the maze with no more than three missed turns. Another reasonable criterion would be that the rat successfully completes the maze 90% of the time.

removed. At some later time he is again placed in the maze for the same number of minutes, and so on. We will number these practice sessions 1, 2, 3, ...[2] Similarly, the child practices riding the bicycle for, say, 30 minutes each day.

We now make a rather stringent assumption about what happens in each practice (or learning) session. We assume that the amount learned in each practice session is some fraction of the material that is unlearned at the start of the session. For example, suppose we assume that in each practice session the individual learns 30% of what is yet to be learned. Suppose also that at some point he has learned 40% of the task. Then there is still 60% of the task to be mastered. In the next practice session the individual will learn 30% of the unlearned 60%. Since 30% of 60% is 18%, he learns 18% of the task. At the end of the practice session, he has mastered 40% + 18% = 58% of the task. In the next practice session the individual will learn another 12.6% and thus increase his total learning to 70.6% of the total task.

Let us now try to write some equations that will help us understand this learning process. We let L_0 be that fraction of the total task that the individual has learned before any practice sessions at all.[3] We then let L_1 be the fraction of the total task mastered after one practice session. Similarly, L_2 is the fraction learned after two practice sessions and so on. Notice that L_0, L_1, and L_2 all represent fractions of the total task. Hence, none of L_0, L_1, and L_2 can exceed 1.

How much is learned *during* the second session? The individual has mastered a fraction L_1 at the end of the first session. Therefore he also knows a fraction L_1 just before the start of the second session. (Remember, no learning takes place between practice sessions.) At the end of the second session, he has mastered L_2. During the second session, he must have learned $L_2 - L_1$ (what he knows at the end less what he knew at the beginning).

We said earlier that we were assuming that the amount the individual learned was some fraction of what was yet to be learned. How much is yet to be learned at the start of the second session? Since the individual has learned a fraction L_1 at that time, he has yet to learn $1 - L_1$. Suppose we assume that the amount learned is 30% of what is to be learned. We will say in this case that the "learning rate" is 30%.

Then the amount learned, $L_2 - L_1$, is 30% of the amount yet to be learned, $1 - L_1$. In mathematical terms this becomes

$$L_2 - L_1 = .3(1 - L_1) \tag{1.1}$$

The .3 is the decimal equivalent of 30%.

In the third practice session the amount learned is $L_3 - L_2$. The amount yet to be

---

[2] The three dots . . . indicate that the pattern of numbers continues indefinitely. In this case 3 is followed by 4, 5, 6, and so on.

[3] Often L_0 will be zero (i.e., the individual knows nothing before training begins). This is not always so. It is possible that a rat has run other mazes and knows some patterns of movement. In that case, L_0 is positive. Similarly, just by watching other people ride bicycles, a child may know part of the task of learning to ride a bicycle without any formal practice sessions.

learned at the start of the third session is $1 - L_2$. By an argument entirely analogous to the one above,

$$L_3 - L_2 = .3(1 - L_2) \qquad (1.2)$$

Similarly, we can write

$$L_4 - L_3 = .3(1 - L_3) \qquad (1.3)$$

$$L_5 - L_4 = .3(1 - L_4) \qquad (1.4)$$

and so on where L_4 and L_5 are the fractions of the total task learned after four and five practice sessions, respectively. If we examine the amount learned in the first session, we obtain

$$L_1 - L_0 = .3(1 - L_0) \qquad (1.5)$$

Now suppose we wish to know what fraction of the task has been learned after five sessions (i.e., we wish the value of L_5). We could use (1.4) if we knew L_4. However, to find L_4, we must use (1.3), which requires that we know the value of L_3. Continuing in this way through (1.2), (1.1), and finally (1.5), we see that we must know L_0, the amount the individual knows before training starts. It is not unreasonable to expect to know the value of L_0. Suppose, for example, we are considering a rat that has never been in a maze. We could then assume that $L_0 = 0$ (i.e., there has been no learning prior to the first practice session). In this case (1.5) becomes

$$L_1 = 0.3$$

which says that after one practice session, 30% of the task has been learned. From (1.1),

$$L_2 = L_1 + .3(1 - L_1)$$

and, since $L_1 = .3$,

$$L_2 = .3 + .3(.7) = .51$$

We continue to (1.2):

$$L_3 = L_2 + .3(1 - L_2)$$
$$= .51 + .3(.49) = .657$$

Then, from (1.3),

$$L_4 = L_3 + .3(1 - L_3) = .657 + .3(.343) = .7599$$

and, from (1.4),

$$L_5 = L_4 + .3(1 - L_4) = .7599 + .3(.2401) = .83193$$

Thus less than 17% of the task remains to be learned after the fifth practice session.

We could, of course, continue to compute L_6, L_7, . . . indefinitely. If we did we would obtain

$$L_6 = .882351 \qquad L_{11} = .980227$$
$$L_7 = .917646 \qquad L_{12} = .986159$$
$$L_8 = .942352 \qquad L_{13} = .990311$$
$$L_9 = .959646 \qquad L_{14} = .993218$$
$$L_{10} = .971752 \qquad L_{15} = .995252$$

It appears, therefore, that succeeding values of the fraction learned come closer and closer to 1 without ever actually reaching the value of 1.

We might reasonably ask: Is such behavior characteristic of all learning? Is this behavior characteristic of learning that follows the assumption[4] we have made? Is this behavior unique to our choice of 30% for the ''learning rate'' and no initial knowledge ($L_0 = 0$)?

We set out now to answer these questions. In this section we will obtain only partial answers. Later, in Chapter 3, we will obtain more complete results.

First we note that we could abbreviate (1.1) to (1.5) by writing

$$L_{k+1} - L_k = .3(1 - L_k) \qquad (1.6)$$

where k takes on the values 0, 1, 2, 3, and 4, successively. Indeed, letting $k = 0$ in (1.6) we get

$$L_1 - L_0 = .3(1 - L_0)$$

which is (1.5). When $k = 1$ we get (1.1), and so on. Actually, we could let $k = 5, 6, 7,$ . . . in (1.6) and thereby obtain a sequence of equations that could be used to compute the fraction of the task learned after any given number of practice sessions. However, to compute L_9 we need the value of L_8. To get L_8, we need L_7, and so on, until we get back to L_0. Thus (1.6) must be ''solved'' first for $k = 0$. Then, with the results of that equation, we solve (1.6) for $k = 1$. It should be clear that (1.6) really represents not one equation but an infinite number of equations, and (1.6) is valid for

$$k = 0, 1, 2, 3, . . .$$

---

[4]Recall that we assumed that the fraction learned in any session was some percentage of what remained to be learned.

Next suppose that the learning rate is not 30% but some other percentage. Indeed, suppose that percentage when put into decimal form is A. Then (1.6) becomes

$$L_{k+1} - L_k = A(1 - L_k)$$

and this is valid for $k = 0, 1, 2, \ldots$. We can rewrite this last equation (it really is not one equation but an infinite number of equations) as

$$L_{k+1} = (1 - A)L_k + A \qquad (1.7)$$

It turns out that this last equation can be "solved." The solution is given by

$$L_k = (1 - A)^k(L_0 - 1) + 1 \qquad (1.8)$$

We will see how to obtain this solution in Chapter 3. However, even at this point it is important to understand what this solution means.

Recall that previously to find L_5 we first needed L_0, L_1, L_2, L_3, and L_4, in that order. Now, however, (1.8) provides the value of L_5 directly without any intervening calculations. Indeed, letting $k = 5$ in (1.8), we get

$$L_5 = (1 - A)^5(L_0 - 1) + 1$$

In the example we used earlier, $A = .3$ and $L_0 = 0$, so

$$L_5 = -(.7)^5 + 1$$

$$L_5 = .83193$$

which agrees with the previously calculated value. In this sense (1.8) is the "solution" of (1.7), $\left[$i.e., (1.8) produces the same values for L_k (where $k = 0, 1, 2, \ldots$) as does (1.7)$\right]$. The reason we consider (1.8) to be the solution is that (1.8) provides the value of L_k directly without calculating all the previous values $L_0, L_1, L_2, \ldots, L_{k-1}$.

Now, assuming[5] that (1.8) is indeed the solution of (1.7), we can make some significant observations. First, if the learning rate is a percentage less than 100% and greater than 0%,

$$0 < A < 1$$

since A is the decimal equivalent of that percentage. Thus

$$0 < 1 - A < 1$$

In (1.8) we raise $1 - A$ to successively larger powers as k (the number of practice sessions) increases. Since $1 - A$ is less than 1, the successive powers of $1 - A$ become

[5]As we mentioned earlier, we will show that (1.8) is the solution of (1.7) in Chapter 3.

smaller and smaller.[6] However, these powers of $1 - A$ are never zero. Thus the first term,

$$(1 - A)^k(L_0 - 1)$$

becomes smaller and smaller no matter what the value of L_0, but it never becomes zero. [We are excluding the case where $L_0 = 1$ (i.e., the task is completely mastered before any training.)] It follows that as k becomes larger and larger, L_k becomes closer and closer to 1 but never reaches 1. Notice that this conclusion is valid provided

$$0 < A < 1 \tag{1.9}$$

and

$$L_0 \neq 1 \tag{1.10}$$

We can now answer two of the questions posed earlier. The "behavior" referred to in those questions was: learning steadily increased, getting arbitrarily close to 1, but never reaching 1. We now know that this behavior is characteristic of the assumption we made regarding the learning rate (see footnote 4), since that assumption leads directly to (1.7). The only other assumptions made were (1.9) and (1.10). Thus the behavior is not unique to the choice of $A = .3$ and $L_0 = 0$.

We cannot say that all learning processes exhibit this behavior. Of course, we cannot say that other learning processes do not have this type of behavior, either. However, we should suspect just from our own experience that many learning situations will have other types of behavior.

In any case, if the assumption regarding the amount learned in one session [i.e., (1.7)] is false, or if one or both of the assumptions regarding the values of A and L_0 are false, our discussion of the behavior of the solution is meaningless. In any real situation the underlying assumptions should be reexamined after the mathematical equations have been analysed. Such assumptions must be questioned continually, and any results of a mathematical analysis can be interpreted only in light of them. This reevaluation of the basic assumptions is an often overlooked but important part of any problem analysis.

We close this section by raising some additional questions about the learning process that we have been discussing. We will not answer these questions. However, some clues to the answers can be found in the solutions to the exercises at the end of the section.

In all cases learning steadily increased, getting arbitrarily close to 1, but never reaching 1.

If the learning rate (value of A) is increased (or decreased), does the amount learned approach 1 faster or slower or at the same rate?

If the initial learning, L_0, is increased (or decreased), does the amount learned approach 1 faster or slower or at the same rate?

---

[6] ½ squared is ¼, and ½ cubed is ⅛. Each successive power of ½ is only one half as large as its predecessor.

EXERCISES FOR SECTION 1.2

(Note. Solutions are given for those exercises marked with an asterisk.)

*1. (a) Compute $L_1, L_2, \ldots, L_{10}$ from (1.7) for $A = 0.7$ and $L_0 = 0$. (The subject learns 70% of the unlearned material in each session.)
(b) Compare your results with those in this section. Does an increase in A change the way in which the amount learned approaches 1?

2. (a) Compute $L_1, L_2, \ldots, L_{10}$ from (1.7) for $A = 0.1$ and $L_0 = 0$.
(b) Compare your results with those in this section. Does a decrease in A change the way in which the amount learned approaches 1?

*3. (a) Compute $L_1, L_2, \ldots, L_{10}$ from (1.7) for $A = 0.3$ and $L_0 = 0.5$.
(b) Compare your results with those in this section. Does an increase in the initial learning, L_0, change the way in which the amount learned approaches 1?

4. (a) Compute $L_1, L_2, \ldots, L_{10}$ from (1.7) for $A = 0.7$ and $L_0 = 0.5$.
(b) Compare your results with those in this section. Does an increase in both A and L_0 change the way in which the amount learned approaches 1?

*5. (a) Compute $L_1, L_2, \ldots, L_{10}$ from (1.7) for $A = 0.1$ and $L_0 = 0.5$.
(b) Compare your results with those in this section. Does a decrease in A and a simultaneous increase in L_0 change the way in which the amount learned approaches 1?

6. Compute $L_1, L_2, \ldots, L_{10}$ for $A = 1.2$ and $L_0 = 0$. Discuss the results in terms of the amount learned and the learning rate.

1.3 CASE STUDY 2: A MODEL OF THE NATIONAL ECONOMY

In this case study we will discuss a simple model of a nation's economy. We will derive an equation—actually an infinite number of equations—that describes the behavior of the total national income. These equations will bear a striking resemblance to (1.7), which dealt with the learning model. However, we will not be able to write down a solution of these equations in the same sense that (1.8) was the solution of (1.7). On the other hand, we will be able to obtain some numerical results and to draw some general conclusions about both the short- and long-term behavior of the economy.

Suppose that the national income can be separated into three parts or components: (1) consumer expenditures for goods and services; (2) private investment in plants and equipment; and (3) government expenditures. It is not clear that such a division of the national income can always be achieved. For example, given any one expenditure of money, it may not be obvious into which of these three categories it falls. Nevertheless, we make this assumption and see where it leads us. After obtaining some results, we will reexamine this assumption. If the results are unrealistic, then this, or some of the other assumptions we will make shortly, must be rejected.

Let T denote the total national income. We will assign variables (or names) to each of the three components of this income as follows: let C be consumer expenditures; let I be private investment; and let G be government expenditures.

Our assumption regarding the division of the total income into three components leads us to the equation

$$T = C + I + G \qquad (1.11)$$

That is, the total income is the sum of consumer expenditures, private investment, and government expenditures.

All four quantities in (1.11) vary with time, so we need to examine all of these for various times. Usually the values of T, C, I, and G are known only for specific times. For example, the Commerce Department may announce the total national income, T, for each quarter of a year or perhaps for each year. Similarly, government expenditures may be known for four different three-month periods (quarters) during the year. It is natural and essential that we discuss the income only for certain discrete periods of time and not for any arbitrary time.[7] We will let T_k be the total national income in the kth time period. (You may think of a time period as being a quarter of a year or a whole year of any other convenient period of time.) In a like fashion we let C_k, I_k, and G_k be, respectively, the consumer expenditures, private investment, and government expenditures in the kth time period. Equation 1.11 is assumed to be valid for all time periods. Thus we write

$$T_k = C_k + I_k + G_k \qquad k = 0, 1, 2, \ldots \qquad (1.12)$$

Notice that k starts at zero, so we designate the first time period as the zeroth one. This might be, for example, the year 1910. If so, and if the time period is a year, then $k = 1$ is the year 1911, $k = 2$ is 1912, and so on. Notice also that (1.12) is not one equation but an infinite number of equations.

We now make some further assumptions about the economy. First, we assume that consumers' buying habits are affected favorably by the total national income. That is, if a government press release indicates a large national income, consumers will spend more money. Of course, the consumer only knows the national income in the years prior to the current one. Therefore consumer expenditures in the second period can be affected only by the national income in the zeroth or first period. Indeed, we will assume that consumers have a short memory, so that their buying habits in the second period are affected only by the national income in the first period. Similarly, consumer expenditures in the third period are affected only by national income in the second period and so on.

Just how are consumer expenditures affected by the national income? We will assume that these consumer expenditures are some percentage of the previous period's

---

[7]It is not possible to discuss the national income on say March 2 since there is no way to know what that income is on that specific day. On the other hand, we may very well know the national income for the first three months of the year.

national income. Suppose, for the moment, that consumer expenditures are ½ (or 50%) of the previous period's income. Then C_1, the consumer expenditures in the first period, is given by

$$C_1 = 0.5T_0$$

Similarly,

$$C_2 = 0.5T_1$$
$$C_3 = 0.5T_2$$

and so on. We can abbreviate these equations and the corresponding ones for all time periods by

$$C_k = 0.5T_{k-1} \qquad k = 1, 2, 3, \ldots \tag{1.13}$$

Notice that k here starts at 1. In (1.12), k started at 0. If, instead of (1.13), we wrote

$$C_{k+1} = 0.5T_k \tag{1.14}$$

then k would start at zero (i.e., $k = 0, 1, 2, \ldots$). It is important to see that both (1.13) and (1.14) represent the same infinite set of equations. That infinite set of equations is

$$C_1 = 0.5T_0$$
$$C_2 = 0.5T_1$$
$$C_3 = 0.5T_2$$

and so on.

We now make a second assumption. We will assume that private investment depends not on the *level* of consumer expenditures, but on the *change* in consumer expenditures. If consumers spend more in one time period than in the previous time period, then private investment increases. For example, if industry sees that consumers are spending more this year than they did last year, then industry will invest more money in new plants, additional trucks, and so forth. In this same vein, if consumer spending has gone down, then industry will liquidate some of its plant (i.e., sell trucks, curtail building programs, etc.).

For definiteness suppose that the investment is equal to the change in consumer spending. Now consumer expenditures in the first time period are C_1. Similarly, C_0 represents the consumer expenditures in the zero time period. The change in consumer expenditures during the first time period then is $C_1 - C_0$. Because investment in the first period is equal to this change,

$$I_1 = C_1 - C_0$$

If consumer expenditures are increasing, then $C_1 > C_0$, and I_1 will be positive. In this case, private capital is invested in new plant to meet the increasing consumer demand. If consumer expenditures are decreasing, then $C_1 < C_0$, and I_1 will be negative. In this latter case corporations will sell trucks, not replace worn-out equipment, because demand is dropping.

Of course, in the second time period, private investment is also equal to the change in consumer spending. Now, however, we are concerned with the change in consumer spending that occurs in the second time period in relation to the first year; that is,

$$I_2 = C_2 - C_1$$

Similarly,

$$I_3 = C_3 - C_2$$

Again we can abbreviate these equations and their successors by

$$I_k = C_k - C_{k-1} \qquad k = 1, 2, 3, \ldots \qquad (1.15)$$

Once again k starts at 1, as it did in (1.13). The reader should be able to rewrite (1.15) so that k starts at zero.

We make one final assumption: that government expenditures are constant and do not vary from one time period to the next. We will choose the unit of money so that government expenditures are 1; that is,

$$G_k = 1 \qquad k = 0, 1, 2, 3, \ldots \qquad (1.16)$$

Keep in mind that if government expenditures are \$1 billion per period, all of our results must be multiplied by \$1 billion at the end. Similarly, if government expenditures are \$2.5 billion per period, then the final results must be multiplied by \$2.5 billion. The crucial assumption is that the government expenditures are constant. The particular value assigned to G_k merely establishes the units in which money is measured.

Using (1.15) to replace I_k in (1.12), we get

$$T_k = C_k + C_k - C_{k-1} + G_k$$

or

$$T_k = 2C_k - C_{k-1} + G_k \qquad (1.17)$$

For what values of k is (1.17) valid? Since (1.17) resulted from (1.12) and (1.15), it is valid only for those values of k for which *both* (1.12) and (1.15) are true.

Now (1.12) is valid for $k = 0, 1, 2, 3, \ldots$, but (1.15) is valid only for $k = 1, 2, 3, \ldots$. Thus (1.17) is only valid for $k = 1, 2, 3, \ldots$.

From (1.13),

$$C_k = 0.5 T_{k-1}$$

and we can use this to replace C_k in (1.17). But what are we to do about C_{k-1}? We would like to write C_{k-1} in terms of some expression in the Ts; that is

$$C_{k-1} = \ldots$$

But what expression involving the Ts should appear on the right?

Recall that (1.13) and (1.14) represented the same set of infinite equations; that is,

$$C_1 = 0.5 T_0$$

$$C_2 = 0.5 T_1$$

$$C_3 = 0.5 T_2$$

and so on. By letting k take on the values $1, 2, 3, \ldots$, successively, we wrote these equations as

$$C_k = 0.5 T_{k-1} \qquad k = 1, 2, 3, \ldots$$

Since we wish to write

$$C_{k-1} = \ldots$$

then $k - 1$ must take on the values $1, 2, 3, \ldots$. To accomplish this, k must take on the values $2, 3, 4, \ldots$. Thus k starts at 2. The first T that appears is T_0 and, since the first k is 2, it must be that the right side is T_{k-2} (if $k = 2$ then $k - 2 = 0$). Thus,

$$C_{k-1} = 0.5 T_{k-2} \qquad k = 2, 3, 4, \ldots \tag{1.18}$$

represents the same infinite set of equations as does (1.13) and as does (1.14).

Using (1.18) and (1.13) to replace C_k and C_{k-1} in (1.17) we obtain

$$T_k = 2(0.5 T_{k-1}) - 0.5 T_{k-2} + G_k$$

But $G_k = 1$, so

$$T_k = T_{k-1} - 0.5 T_{k-2} + 1 \qquad k = 2, 3, 4, \ldots \tag{1.19}$$

Notice that k starts at 2 and increases in steps of 1. That is because we used (1.18), which is only valid for $k = 2, 3, 4, \ldots$. The other equations used included $k = 1$, and some also included $k = 0$. In (1.19) we can only use values of k that are valid for *all* equations used in the analysis leading up to that equation.

If we write out the first three equations represented by (1.19), we get,

$$T_2 = T_1 - 0.5T_0 + 1 \qquad (1.20)$$

$$T_3 = T_2 - 0.5T_1 + 1 \qquad (1.21)$$

$$T_4 = T_3 - 0.5T_2 + 1 \qquad (1.22)$$

If we know the values of T_0 and T_1, we can calculate T_2 from (1.20). Once we have the value of T_2, we can calculate T_3 from (1.21). Next we compute T_4 from (1.22). Continuing in this way, we can calculate the national income in any period *if* we have the national income in periods 0 and 1.

Suppose, for example, that the national income in two consecutive time periods is 2 and 3. (Recall that the unit of money is 1.) Then we denote these two periods as the 0 and 1 periods, so that

$$T_0 = 2 \quad \text{and} \quad T_1 = 3$$

Using (1.20),

$$T_2 = 3 - 0.5(2) + 1 = 3$$

Then, using (1.21),

$$T_3 = 3 - 0.5(3) + 1 = 2.5$$

From (1.22),

$$T_4 = 2.5 - 0.5(3) + 1 = 2$$

If we continue in this way using six digits in all computations we will obtain the table shown on page 14.

In this case the national income oscillates above and below 2 and eventually stabilizes at 2. As time progresses, the oscillations become smaller. In such a case, we say that the national income "approaches" 2 or that "in the long run" the national income is 2. Recall that in the learning model of Section 1.2 the fraction of the material learned "approached" 1.

Before going on with this analysis of the economy, we ask the following questions.

1. If government expenditures were increased for a few years, would it change the nature of the economy?
2. When is the best time for government expenditures to be increased or decreased?
3. Should incentives be given to industry to increase their dollar investment?

Time Period	National Income	Time Period	National Income
0	2.		
1	3.	21	1.99902
2	3.	22	1.99902
3	2.5	23	1.99951
4	2.	24	2.
5	1.75	25	2.00024
6	1.75	26	2.00024
7	1.875	27	2.00012
8	2.	28	2.
9	2.0625	29	1.99994
10	2.0625	30	1.99994
11	2.03125	31	1.99997
12	2.	32	2.
13	1.98438	33	2.00002
14	1.98438	34	2.00002
15	1.99219	35	2.00001
16	2.	36	2.
17	2.00391	37	2.
18	2.00391	38	2.
19	2.00195	39	2.
20	2.	40	2.

4. Should credit regulations be changed to encourage or discourage consumer spending?
5. Would some changes in the assumptions lead to a different kind of economy?
6. Is it possible to obtain an expanding economy using the assumptions we have made?

See how many questions of this type you can formulate. Examine the above questions critically. What do we mean by "expanding," "different kind of economy," and "best time"? On what basis can we answer "yes" or "no" to such questions?

Questions of the type given in questions 1 to 6 and their critical evaluation can, and often do, form the basis for policy decisions at the national, local, industry, and personal levels. In what follows we will broaden our discussion to help us in answering some of these questions. The exercises will discuss some specific cases, which should give you some insight into the answers to the questions. Keep such questions in mind as you go through the remainder of this section. However, do not expect explicit or definitive answers to them.

We will maintain our four basic assumptions about the economy.

1. The national income is the sum of consumer expenditures, private investment, and government expenditures.

$$T_k = C_k + I_k + G_k \qquad k = 0, 1, 2, \ldots \qquad (1.12)$$

2. Consumer expenditures in any time period depend on the national income in the previous period. Moreover, C_k is some percentage of T_{k-1}, or

$$C_k = AT_{k-1} \qquad k = 1, 2, 3, \ldots \qquad (1.23)$$

where A is the percentage in decimal. In the numerical example used earlier, $A = .5$, so that consumer expenditures were 50% of the previous period's national income. The percentage A is called the *marginal propensity to consume* by economists.

3. Private investment in any time period is some percentage of the change in consumer expenditures in that period over those expenditures in the previous time period. If the percentage, in decimal, is B,

$$I_k = B(C_k - C_{k-1}) \qquad k = 1, 2, 3, \ldots \qquad (1.24)$$

In the numerical example above, $B = 1$ (i.e., the percentage is 100%). In principle there is no reason why the percentage cannot exceed 100%. In that case, B would exceed 1.

4. Government expenditures are constant (do not change from one time period to the next).

$$G_k = 1 \qquad k = 0, 1, 2, \ldots \qquad (1.25)$$

Again, 1 merely establishes the units in which money are measured.

As before we use (1.24) to eliminate I_k from (1.12):

$$T_k = C_k + B(C_k - C_{k-1}) + G_k \qquad k = 1, 2, 3, \ldots$$

where again k starts at 1, since (1.24) is only valid for $k = 1, 2, 3, \ldots$. Simplifying this last equation, we get

$$T_k = (1 + B)C_k - BC_{k-1} + G_k \qquad (1.26)$$

Now, from (1.23),

$$C_k = AT_{k-1} \qquad k = 1, 2, 3, \ldots$$

and

$$C_{k-1} = AT_{k-2} \qquad k = 2, 3, 4, \ldots \qquad (1.27)$$

Using these in (1.26) to eliminate the terms in C_k and C_{k-1},

$$T_k = (1 + B)AT_{k-1} - BAT_{k-2} + G_k$$

Finally, using (1.25) and rearranging,

$$T_k = A(1 + B)T_{k-1} - ABT_{k-2} + 1 \qquad k = 2, 3, 4, \ldots \qquad (1.28)$$

Notice that k starts at 2, since (1.27) is only valid for $k = 2, 3, 4, \ldots$. All of the other equations that we have used are valid for these values of k as well.

In our previous numerical example, A was 0.5 and B was 1. Using these values in (1.28), we get

$$T_k = T_{k-1} - 0.5T_{k-2} + 1 \qquad k = 2, 3, 4, \ldots$$

which is precisely (1.19). Why did we bother to derive this last equation (1.19) a second time? It is good practice when a general equation has been derived to check that it works for a special case where the equation already is known. Thus we use $A = 0.5$ and $B = 1$ in (1.28) in order to "check" that it reproduces (1.19) as it should.

We now have a general equation (1.28) that is based on our four assumptions. Given the values of T_0 and T_1, we can calculate successively T_2, T_3, T_4, and so on. Of course, in any given case, we must know the values of A and B. Indeed, A and B characterize the economy. A large value of A implies that consumers react very favorably to a large national income. Similarly, a large value of B indicates that investments will rise sharply if consumer expenditures increase. Of course, if consumer spending decreases, then a large value of B means that investment will drop just as sharply. It is, therefore, far from obvious that a large B is desirable.

Let us try another example of a specific economy. Recall that we have examined an economy in which $A = 0.5$ and $B = 1$. Starting with $T_0 = 2$ and $T_1 = 3$, that economy oscillated but settled down to 2. We now increase both consumer and investment reaction. We will suppose that consumer expenditures are 80% of the previous period's national income; that is,

$$A = 0.8$$

We will also assume that private investment is twice the change in consumer expenditures so that

$$B = 2$$

Then (1.28) becomes

$$T_k = 2.4T_{k-1} - 1.6T_{k-2} + 1 \qquad k = 2, 3, 4, \ldots \qquad (1.29)$$

Writing out the first three of these for $k = 2$, $k = 3$, and $k = 4$, we get

$$T_2 = 2.4T_1 - 1.6T_0 + 1 \qquad (1.30)$$

$$T_3 = 2.4T_2 - 1.6T_1 + 1 \qquad (1.31)$$

$$T_4 = 2.4T_3 - 1.6T_2 + 1 \qquad (1.32)$$

We again start the economy with $T_0 = 2$ and $T_1 = 3$. Then, from (1.30),

$$T_2 = 2.4(3) - 1.6(2) + 1 = 5$$

From (1.31),

$$T_2 = 2.4(5) - 1.6(3) + 1 = 8.2$$

Then, from (1.32),

$$T_4 = 2.4(8.2) - 1.6(5) + 1 = 12.68$$

It appears that we may have an expanding economy—one in which national income increases each time period. But do we? How can we be sure?

At the moment the only way we have to answer these questions is to continue computing successive national incomes: $T_5, T_6, T_7, \ldots$. We can never be sure that the pattern that develops will continue, but the more time periods for which we have computed the national income, the more confidence we will have that the pattern will continue. The three values (T_2, T_3, and T_4) computed above are hardly enough to inspire much confidence. If we continue to compute the national incomes from (1.29), we get the following values.

$$T_5 = 18.312$$
$$T_6 = 24.6608$$
$$T_7 = 30.88672$$
$$T_8 = 35.67085$$
$$T_9 = 37.19128$$

The economy continues to expand, but the rate of expansion seems to be decreasing. After all, from periods 7 to 8, there was an increase of about 5 while, from periods 8 to 9, the increase was less than 2. It looks as though the expansion may be losing some momentum.

Continuing our computations, we get

$$T_{10} = 33.18572$$

Our fears were well founded. The economy is no longer expanding. If we go one step further,

$$T_{11} = 21.13968$$

The drop in income from period 10 to period 11 is dramatic—more than 12. Does disaster lie ahead? The next time period produces

$$T_{12} = -1.36192$$

—a negative national income! Disaster has arrived. There is no practical value in continuing the computations, since the economy has collapsed. However, from a mathematical point of view, we can continue the computations if we choose to do so. If we do continue, we will see that the economy oscillates below and above 5. However, the oscillations become larger and larger as time progresses (see Exercise 1). Contrast this with the previous case ($A = 0.5$ and $B = 1$) where the oscillations became smaller and smaller. Other examples that exhibit quite different types of behavior are given in Exercises 2 and 3.

The fact that we can get at least four different types of behavior—oscillations that get smaller, oscillations that get larger, steadily decreasing to a fixed value, expanding economy[8]—leads us to the following questions. Are there other types of behavior besides these four? How can we tell which type of behavior will occur? Does the behavior depend on the values of A or B or both? Does the behavior depend on the national economy in the first two years (i.e., on the values of T_0 and T_1)?

Some of the exercises at the end of the section are designed to help you try to answer some of these questions. Keep in mind, however, that any conclusions drawn from numerical experiments are "guesses." To be certain of any conclusions, we must find a mathematical solution of the equation in question, in this case (1.28). We will not be able to solve that equation, so we will have to content ourselves with numerical results and guesses.

In order to facilitate the computation of numerical results, we will turn to the computer. In Chapter 2 we will develop enough knowledge of computer programming to help us to perform numerical experiments not only with this economic problem but also with more complicated problems.

We close this section by making three observations. First, we should now be able to answer at least partially some of the questions posed earlier, especially if we have done some of the exercises at the close of this chapter. Second, it should be clear that the assumptions could be changed to provide a more realistic representation of a nation's economy. The equations would become more complicated, and their solution (even a numerical solution) would be more difficult to obtain. Given enough patience and perseverance, however, we should have some confidence that we could handle a more realistic example.

Finally, we point out that if a given economy does not behave in accordance with these numerical results, it would be tempting to conclude that either the equations are in error or that the economy is misbehaving. Neither conclusion is valid. All we can say is that the equation (1.28), and the economy itself do not agree. In particular, in the case of such a disparity, one or more of the four assumptions must not be true for the economy being considered. We should, therefore, reconsider the particular economy to see which of the assumptions is violated. By suitably altering the offending assumption, we may be able to obtain a useful description of the economy. Even if we do not alter any assumptions, just by discovering which assumption does not apply to the economy, we will have learned a great deal about that particular economy.

[8]The last two of these are exhibited in Exercises 2 and 3.

EXERCISES FOR SECTION 1.3

1. Using (1.29), compute T_{13}, T_{14}, and so on, until you can see the oscillations occur and start to get larger.

*2. Use (1.28) with an economy that has $A = 0.5$ and $B = 6.0$. Start with $T_0 = 2$ and $T_1 = 3$. Compute T_2, T_3, . . . , T_{20}. What can you say about the behavior of this economy?

3. Use (1.28) with an economy that has $A = 0.5$ and $B = 0.1$. Start with $T_0 = 2$ and $T_1 = 3$. Compute T_2, T_3, . . . , T_{20}. Describe the behavior of the economy in words.

*4. (a) Use (1.28) with $A = 0.5$ and $B = 1.0$. Start with $T_0 = 4$ and $T_1 = 5$.
   (b) Compare your results with those given in Section 1.3, where $T_0 = 2$ and $T_1 = 3$.
   (c) Describe the long-run behavior of the solution.
   (d) Do you think that the long-run behavior depends on the choice of T_0 and/or T_1?

5. (a) Use (1.28) with $A = 0.8$ and $B = 2.0$. Start with $T_0 = 4$ and $T_1 = 5$.
   (b) Compare your results with those given in Section 1.3 where $T_0 = 2$ and $T_1 = 3$.
   (c) Describe the long-run behavior of the solution.
   (d) Do you think the long-run behavior depends on the choice of T_0 and/or T_1?

*6. Use (1.29) and start with $T_0 = 2$ and $T_1 = 3$. When the national income becomes negative at T_{12}, change government spending so that the national income is zero. That is, in period 12, increase government spending by 1.36192 to 2.36192. Continue to compute T_{13}, T_{14}, and so on. Each time the national income is about to become negative, increase government spending just enough to keep the national income at zero. Otherwise keep government spending at 1. How often do you have to increase government spending above 1? By how much must it be increased? Do the increases continue to grow in time?

7. Suppose government spending is zero. Then (1.28) becomes

$$T_k = A(1 + B)T_{k-1} - ABT_{k-2} \qquad k = 2, 3, 4, \ldots$$

For $A = 0.8$ and $B = 2.0$, this becomes

$$T_k = 2.4T_{k-1} - 1.6T_{k-2} \qquad k = 2, 3, 4, \ldots$$

Compare this with (1.29). Let $T_0 = 2$ and $T_1 = 3$ and compute the national income for the next 14 years (i.e., compute T_2, T_3, . . . , T_{15}). Describe the behavior of the solution and compare your results with the solution of (1.29) given in Section 1.3. Do you think the long-run behavior of the national income depends on the amount of government spending?

*8. Use

$$T_k = 2.4T_{k-1} - 1.6T_{k-2} \qquad k = 2, 3, 4, \ldots$$

as described in Exercise 7. Start with $T_0 = 2$ and $T_1 = 3$ and compute T_2, T_3, ... When the national income becomes negative, introduce enough government spending to keep the national income at zero. Otherwise keep government spending at zero.

*9. Equation 1.24 assumed that in any given year private investment I_k depended on the difference in the consumer expenditure for that same year, C_k, and consumer expenditure for the previous year, C_{k-1}. But, since the investment I_k occurs in the same year as does the consumer expenditure C_k, the value of C_k is not known to the investor who must decide on the value of I_k. Hence a more reasonable assumption might be that investment in a given year depends on the difference between consumer expenditures for the previous year and consumer expenditures for the year two years previous; that is,

$$I_k = B(C_{k-1} - C_{k-2})$$

Using this to replace (1.24), find an equation similar to (1.28) for the total national income, T_k.

*10. Assume that the total national income of a nation can be separated into two categories: private spending and government spending. Private spending in any given year is 50% of the increase in total income in that year over the total income in the previous year. Government spending in any given year is 40% of the private spending in that same year. Let T_k be the total national income in the kth year. Let P_k and G_k be the private and government spending respectively. Write an equation relating the total national income in the kth year, T_k, to the total national income in the $(k - 1)$st year, T_{k-1}.

11. Same as Exercise 10 above except that private spending is 65% of the increase in total income and government spending is 30% of private spending.

1.4 CASE STUDY 3: A POPULATION MODEL

We now turn to a biology problem. Consider a single species of life living in a closed environment. We can think of a culture of bacteria in a test tube or humans living on the face of the earth. We start by raising some questions similar in spirit to those we posed in Section 1.3 regarding the economy. Of course, we should keep in mind that we will have to make some assumptions in order to obtain any quantitative information about this problem. The information that we obtain will have to be interpreted in light of the assumptions we make, and the assumptions will need to be reevaluated. As we will see, our first set of assumptions will lead to unacceptable results. However, the analysis leading to these undesirable results will guide us to a more acceptable set of assumptions. In any case, during the entire development, we would do well to keep the following qualitative questions in the back of our minds. Our hope is to be able to provide answers to them or at least to some of them.

1. How does the population change as time goes on?
2. Is there a limit to the size of the population?
3. If some catastrophe suddenly kills a portion of the population, will the population grow back to its previous level?
4. Should measures be taken to increase or decrease the birth rate? the death rate?
5. Should attempts be made to change the average life span of the species?
6. If the species were wildlife, should some of the species be killed at times by, for example, permitting them to be hunted? When? How many should be killed?
7. Should the species be protected from attack by predators, human or otherwise?

Try to think of other questions that would help someone make policy decisions on birth control, hunting laws, health measures, and the like.

We now turn to a mathematical discussion of the growth of a single species of life in a closed environment. We will assume that the population can be measured at various points in time. For example, the population of the United States is counted every 10 years through a census. Wildlife censuses are taken at the beginning or close of a mating season (or perhaps of a hunting season). We will consider a time period to be the interval between censuses. For the population of the United States a period is 10 years. For wildlife a period may be a year (from the start of one hunting season to the start of another) or six months (from the start of one mating season to the start of the next). Some periods are natural ones, as in the case of mating seasons, and some are quite arbitrary, as with the United States census. The salient point is that there is a fixed amount of time between censuses and that the population is only known at the time the census is taken. There is no way to know precisely the number of people in the United States in 1972, although some estimate may be given. Similarly, we do not know a wildlife population in the middle of a mating season. With all of this in mind, we will let N_k be the number of individuals alive at the *end* of the kth period. Once again we will let $k = 0, 1, 2, \ldots$. By way of example, $k = 0$ could represent the first census. We will assume that the beginning of the $(k + 1)$st period coincides with the end of the kth period. Thus N_k is also the number of individuals alive at the start of the $(k + 1)$st period.

In any given period the population changes. Individuals are born and individuals die. We now make an assumption regarding the birth and death rates. We will assume that the number of births in any period is some percentage of the number of living individuals at the beginning of that period. In other words, if we start with 200 individuals, there will be twice as many births as if we started with 100 individuals. Similarly, we assume that the death rate in any period is some percentage of the number of individuals alive at the period's start. That is, the more people there are, the more deaths there will be. What we are interested in is the population growth (i.e., births less deaths). (If deaths exceed births, the population declines.)

If the number of births is 15% of the population and if the number of deaths is 7% of the population, then the population growth is 15% − 7%, or 8%. In general we will not be concerned with births and deaths themselves, but with the difference between them. We will use the excess of births over deaths as the growth rate, realizing that this growth rate may be negative if deaths exceed births.

Suppose that the increase in the number of individuals in any period is $A\%$ of the number of individuals alive at the start of the period. (Recall that the increase is the number of births less the number of deaths.) If A is given in decimal (.7 for 70%) then the population increase in the $(k + 1)$st period is

$$AN_k$$

since there are N_k individuals alive at the end of the kth period. Recall that the kth period's end coincides with the beginning of the $(k + 1)$st period. But the change in population in the $(k + 1)$st period is also given by the difference between the population at the end of the period, N_{k+1}, and the population at the start of the kth period, N_k. That is, the population change is

$$N_{k+1} - N_k$$

Equating this to the other expression for the population change in the $(k + 1)$st period,

$$N_{k+1} - N_k = AN_k$$

Rearranging terms,

$$N_{k+1} = (1 + A)N_k \qquad k = 0, 1, 2, \ldots \tag{1.33}$$

If we know the value of A, the rate of growth, and if we know N_0, we can calculate successively N_1, N_2, N_3, and so on. (See Exercises 1 to 4.)

Even without knowing the specific values of A or N_0, however, we can draw some conclusions regarding the behavior of the population.

For $k = 0$, (1.33) is

$$N_1 = (1 + A)N_0 \tag{1.34}$$

for $k = 1$,

$$N_2 = (1 + A)N_1$$

Using (1.34) to replace N_1 in this last equation,

$$N_2 = (1 + A)(1 + A)N_0 = (1 + A)^2 N_0 \tag{1.35}$$

Next let $k = 2$ in (1.33).

$$N_3 = (1 + A)N_2$$

Using (1.35) in this last equation,

$$N_3 = (1 + A)^3 N_0$$

We could continue in this way to obtain

$$N_4 = (1 + A)^4 N_0$$
$$N_5 = (1 + A)^5 N_0$$

Having done so it would be quite reasonable to "guess" that

$$N_k = (1 + A)^k N_0 \qquad k = 0, 1, 2, \ldots \tag{1.36}$$

This guess is, in fact, correct. However, we cannot be sure at this point that it is correct. In Chapter 3 we will develop the necessary tools to verify this statement. For now we accept the statement on faith.

We call (1.36) the "solution" of (1.33). Both (1.33) and (1.36) will produce the same values of N_k for $k = 0, 1, 2, \ldots$. If we use (1.33) to compute N_{10} we must first compute, in order, $N_1, N_2, \ldots, N_9$. On the other hand, N_{10} can be computed directly from (1.36) without computing any other N_k. It is in this sense that (1.36) is the solution of (1.33) instead of the other way around. To repeat, (1.36) provides a more direct way of obtaining any particular value of N_k than does (1.33).

Let us now examine (1.36) to see what conclusions we can draw from it. Suppose first that $A > 0$. This implies that births exceed deaths. Then

$$1 + A > 1$$

and as k gets large, $(1 + A)^k$ gets large. It follows that N_k also gets large. In fact, N_k gets larger and larger without bound. No population in a closed environment grows in such an unbounded way. Bacteria in a test tube eventually fill the test tube completely. Similarly, people on the earth would eventually be shoulder to shoulder with no place left to put any additional people. Our conclusion from this observation is that something is wrong with our assumptions.

One possibility is that A is not positive, but negative. In that case

$$1 + A < 1$$

Since A is a percentage in decimal, we will assume[9] that A cannot be less than -1. (This says that the decrease in population cannot exceed 100%.) Thus

$$1 + A > 0$$

Since $1 + A$ is between 0 and 1, if we raise $1 + A$ to the kth power, we get smaller and smaller numbers as k gets larger.[10] Therefore, for large k, the population N_k as given in (1.36) becomes very small and approaches zero (see Exercises 3 and 4). This means that the species eventually will be exterminated. This is certainly possible in some cases—species have vanished. Fortunately, most species survive. However, even in this latter case, where the population does not vanish, there is still some fallacy in our discussion. The decision so far has led to either extermination or unrestricted growth. If we choose $A = 0$, the only other possibility, then the population is constant and equal to N_0. Indeed, (1.36) becomes simply

---

[9]For $A \leq -1$, see Exercise 12.

[10]½ squared is ¼, and ½ cubed is ⅛. Each successive power of ½ is only one half as large as its predecessor.

$$N_k = N_0 \qquad k = 0, 1, 2, \ldots$$

Since we have not achieved satisfactory results from our development, we pause to reexamine the assumptions with which we started. They are:

1. The population can be, and is, measured at certain times called periods.
2. The number of individuals that are born (or die) in any given period is some fixed percentage of the individuals alive at the start of the period.

Apparently at least one of these is not valid for a species living in a closed, confined environment. What is the offending assumption? How can we use the results already obtained to achieve a better, more realistic assumption?

One reasonable conclusion is that assumption 2 is not correct. Let us look at an alternative to assumption 2. Keep in mind that there are other alternatives as well. First, we note that as long as there is ample room to place newly born individuals, assumption 2 may be quite all right. However, when the population becomes large (the test tube is nearly filled or the earth is extremely heavily populated), the individuals start to consume all of the food supply, pollute the environment and, in other ways, make it difficult to maintain life. The result of this overcrowding is to reduce the birth rate and increase the death rate. Either of these will decrease A. If we accept the argument just presented, then we cannot assume A is a constant for all periods. Indeed, A should depend on the population itself. One way to exhibit such a dependence is to replace A by

$$A - BN_k \tag{1.37}$$

where A and B are positive numbers. Suppose, for example, that $A = 0.4$ and $B = .0002$. Then, if N_k, the population, is small, say 10,

$$A - BN_k = .4 - .0002(10) = .398$$

In this case the rate of growth is very close to 40%, which is equivalent to a value of $A = .4$ in the earlier case. Thus, for a small population, this assumption regarding the rate of growth should produce results quite similar to those produced by assumption 2.

On the other hand, for larger N_k such as 1000,

$$A - BN_k = .4 - .0002(1000) = .2$$

which is only a 20% growth rate. As the population grows, the growth rate decreases. Indeed, when $N_k = 2000$,

$$A - BN_k = .4 - .0002(2000) = 0$$

and the growth rate is zero. In other words, when the population reaches 2000, there is no change in the population. If N_k exceeds 2000, the birth rate becomes negative, and the population decreases (see Exercises 5 and 6).

We therefore replace assumption 2 with the following assumption:

3. The number of individuals that are born (or die) in any given period is some variable percentage of the number of individuals alive at the start of the period. The variable percentage is largest when the population is small and becomes smaller as the population increases. Eventually the variable percentage becomes zero and even negative for sufficiently large populations.

Using this last assumption, the population change during the $(k + 1)$st period is

$$(A - BN_k)N_k$$

This compares with

$$AN_k$$

used earlier The population change is also given by

$$N_{k+1} - N_k$$

so

$$N_{k+1} - N_k = (A - BN_k)N_k$$

or

$$N_{k+1} = (1 + A - BN_k)N_k \qquad k = 0, 1, 2, \ldots \qquad (1.38)$$

B will generally be much smaller than A because B is multiplied by the population N_k before the result is subtracted from A. Only when N_k is large do we want the product BN_k to have a significant effect on A.

It remains to be seen whether assumption 3, which led to (1.38), produces realistic and satisfactory results. We investigate that now. We will look at a specific example.

Again, let $A = .4$ and $B = .0002$ so that (1.38) becomes

$$N_{k+1} = (1.4 - .0002N_k)N_k \qquad k = 0, 1, 2, \ldots \qquad (1.39)$$

Suppose the initial population is 1000. Then, letting $k = 0$ in (1.39),

$$N_1 = (1.4 - .0002 \times 1000)1000 = 1200$$

Then, letting $k = 1$ in (1.39),

$$N_2 = (1.4 - .0002 \times 1200)1200 = 1392$$

Continuing with (1.39) by letting $k = 2, 3, 4, \ldots$, we find that

$$N_3 = 1561.2672 \qquad N_7 = 1920.2546$$

$$N_4 = 1698.2630 \qquad N_8 = 1950.8809$$

$$N_5 = 1800.7488 \qquad N_9 = 1970.0460$$

$$N_6 = 1872.5091 \qquad N_{10} = 1981.8482$$

One might ask what meaning fractions such as .8482 (see N_{10}) can have when discussing populations. Several interpretations are possible. The interpretation we will use is: our computations, and indeed our entire analysis, is approximate at best. Therefore the numerical results are to be taken as approximate. If the fraction is less than ½ we will ignore it. If the fraction is ½ or greater we will add one individual. Thus N_3 is 1561, since the fraction (.2672) is less than ½. However, N_{10} is 1982 since the fraction is greater than ½, so we add 1 to 1981.

We now will observe the population figures above and will try to discern a pattern in the successive populations. First we observe that the population continues to grow with each succeeding time period. However, the rate of increase is decreasing since, for example, the increase in period nine is $1970.0460 - 1950.8809 = 19.1651$, while the increase in period ten is $1981.8482 - 1970.0460 = 11.8022$. This should not be too surprising, since the term $-BN_k$ in (1.38) has the effect of retarding population growth as the population itself gets larger.

It turns out that the population is leveling off (see Exercises 7 and 8) and approaches a fixed value that it will never exceed. The behavior is similar to that of the learning model in Section 1.2, where the fraction of the amount learned approached 1 but never quite reached that value. Here N_k approaches 2000 but never quite reaches that value.

This type of behavior appears to be acceptable. Just how realistic this behavior is remains to be seen. The test will be, Do any real populations behave this way? The answer is "yes," and we will demonstrate this in Chapter 5, where we examine the population of the United States.

Based on assumptions 1 and 3, we can now attempt to answer some of the questions posed earlier.

1. How does the population change as time goes on? It increases rapidly at first and then, although it continues to increase, does so at a slower rate when the population becomes large.
2. Is there a limit to the size of the population? Yes.
3. If some catastrophe suddenly kills a portion of the population, will the population grow back to its previous level? Yes, since starting at 1000 the population increased to 2000, it follows that starting at some other population even below 1000 it will grow back to 1000 and hence eventually grow to 2000.

The remaining questions are not readily answered by the results we have obtained. Most of them require a study of more than one species.

EXERCISES FOR SECTION 1.4

*1. (a) Use (1.36) with $A = 0.08$ (8% rate of growth) and $N_0 = 1000$. In how many periods will the population exceed 7500?
(b) In how many periods will the population exceed 100,000?

2. (a) Use (1.36) with $A = 0.1$ (10% rate of growth) and $N_0 = 1000$. In how many periods will the population exceed 5000?
(b) In how many periods will the population exceed 10,000?

*3. (a) Use (1.36) with $A = -0.1$ (10% decrease in population in each period) and $N_0 = 1000$. In how many periods will the population go below 500?
(b) In how many periods will the population go below 100?

4. (a) Use (1.36) with $A = -0.05$ (5% decrease in population in each period) and $N_0 = 1000$. In how many periods will the population go below 500?
(b) In how many periods will the population go below 200?

*5. If the rate of population growth is given by (1.37) and if $A = 0.3$ and $B = 0.0001$, for what values of the population, N_k, will the rate of growth be:
a. Increasing?
b. Zero?
c. Decreasing?

6. If the rate of population growth is given by (1.37) and if $A = 0.08$ and $B = 0.00001$, for what values of the population, N_k, will the rate of growth be:
a. Increasing?
b. Zero?
c. Decreasing?

*7. Using (1.38) with $A = 0.4$, $B = 0.0002$, and $N_0 = 1000$, we computed $N_{10} = 1981.8482$. For what time period will the population exceed 1990?

8. Using (1.38) with $A = 0.4$, $B = .0002$, and $N_0 = 1000$, for what time period will the population exceed 1999?

*9. Using (1.38) with $A = 0.4$ and $B = 0.0002$, let $N_0 = 500$ and compute $N_1, N_2, \ldots, N_{15}$. Discuss the behavior of the solution.

10. Using (1.38) with $A = 0.4$ and $B = 0.0002$, let $N_0 = 2000$ and compute $N_1, N_2, \ldots, N_{10}$. Discuss the behavior of the solution.

*11. Using (1.38) with $A = 0.4$ and $B = 0.0002$, let $N_0 = 3000$ and compute $N_1, N_2, \ldots, N_{20}$. Discuss the behavior of the solution.

*12. Discuss the behavior of (1.36)—the solution of (1.33)—for the case where $A \leq -1$.

13. The solutions of both (1.7) and (1.38) approach a fixed value as time progresses. The solution of (1.7) approaches 1, and the solution of (1.38) approaches some fixed population. Discuss the differences in the way in which the approaches take place.

1.5 ORGANIZING THE COMPUTATIONS

The arithmetic computations in the three previous sections, while straightforward, are extremely tedious and hence quite likely to lead to numerical errors. Moreover, if you wanted to get someone, say a little brother who is in the sixth grade, to help you with the arithmetic, you probably would have considerable difficulty explaining to him just exactly what you wanted him to do. In this section we will describe a method of organizing such calculations so that you, or an assistant who doesn't understand the problem, can correctly carry them out.

Consider, for example, the population growth problem in Section 1.4. Equation 1.38 prescribes the calculation of the population, N_k. In order to compute successive populations from this equation, we need the values of A and B and also the value of the beginning population, N_0. In the numerical example in the text $A = 0.4$, $B = 0.0002$, and $N_0 = 1000$. From (1.38),

$$N_1 = (1.4 - .0002 \times 1000) \times 1000 \qquad (1.40)$$

or

$$N_1 = 1200$$

and

$$N_2 = (1.4 - .0002 \times 1200) \times 1200 \qquad (1.41)$$

that is,

$$N_2 = 1392$$

These arithmetic expressions for N_1 and N_2 are quite similar. The only change from the right side of (1.40) to the right side of (1.41) is the replacement of 1000 by 1200. Each of these populations and each succeeding population is computed from

$$(1.4 - .0002 \times N) \times N \qquad (1.42)$$

where N represents the value of the immediately preceding population. Thus N_3 will be

$$(1.4 - .0002 \times 1392) \times 1392$$

since 1392 is the population just prior to N_3.

Therefore, if we think of the successive populations as being written in a column, row by row, such as

$$1000$$

$$1200$$

$$1392$$

and so on, in order to obtain the next row in the column we use the number recorded in the row above and insert its value for N in (1.42).

With this as background we can write down a procedure for computing a sequence of population values.

1. Obtain a starting value of the population and record it in the first row. Call it N.
2. Compute $(1.4 - .0002 \times N) \times N$.
3. Record this value in the next row.
4. Call the latest value recorded N.
5. Return to step 2.

This process also can be represented by the diagram in Figure 1.1 where the numbers at the upper right corner of each box correspond to the step numbers above. The diagram is to be read as follows.

Find the word "Start." Follow the arrow leaving "Start" until a box is encountered. Carry out the instructions contained in the box. Follow the arrow leaving that box until another box is encountered. Follow the arrows from box to box and do what you are instructed to do in each box as you encounter the box.

For example, Box 1 in Figure 1.1 instructs us to obtain a value of the starting population. Suppose, as before, we use 1000. Then $N = 1000$. In Box 2 we use 1000 for N and compute

$$X = 1200$$

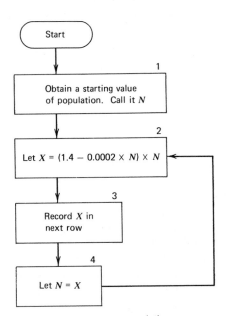

FIGURE 1.1   Beginning flowchart for computing populations.

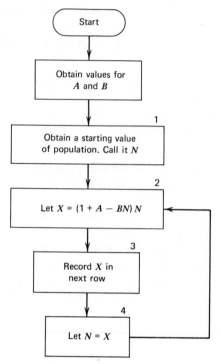

FIGURE 1.2 Flowchart for computing populations with various A and B values.

We record this value as instructed in Box 3. Next, we change the value of N to 1200 and then return to Box 2, where we calculate

$$(1.4 - .0002 \times 1200) \times 1200$$

so that

$$X = 1320$$

The process then continues.

A diagram such as the one in Figure 1.1 is called a *flowchart* because it describes the "flow" of the procedure. All flowcharts are to be read the same way as this one. From a box containing "Start," follow the arrows from one box to another. As each box is reached, carry out the actions contained within the box.

If we follow the flowchart in Figure 1.1 slavishly, we can carry out the desired population computations even if we know nothing at all about the underlying problem that the flowchart represents. For example, you could give this flowchart to a sixth-grade student, tell him to follow the arrows and do what each box tells him to do, and

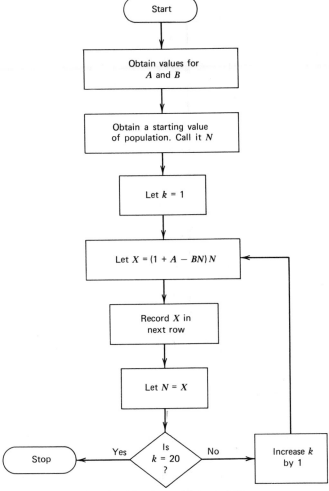

FIGURE 1.3 Flowchart for computing 20 values of the population.

he would be able to produce a list of populations for you in a column. He would not need to know that they were populations. All he would need to know is how to do arithmetic and how to follow orders. As you might have guessed, computers have the ability to do both of these things, and flowcharts are useful in getting computers to help us in carrying out computations.

Let us return to our flowchart in Figure 1.1 and observe some of its shortcomings. First, it would be a much more useful flowchart if we could use it for populations with different values for *A* and *B* than the values 0.4 and 0.0002 used so far. This is easily accomplished by adding one box and changing box 2, as shown in Figure 1.2.

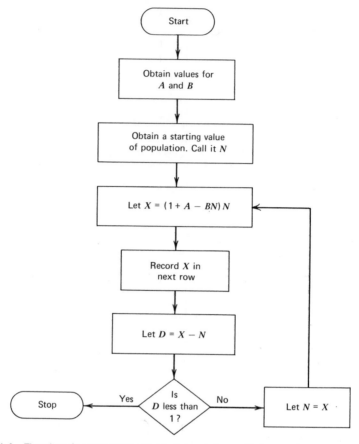

FIGURE 1.4 Flowchart for computing values of population until changes become small.

But an even more serious problem is: When do the calculations stop? If we follow the instructions as given so far, we will compute and record values of the population endlessly. A sixth-grade assistant or a computer, if left alone, would continue to compute and record until it became exhausted or ran out of paper.

Suppose we wish to compute exactly 20 successive population values and then stop the calculations. We need some way, in our flowchart, to count to 20. To this end we will let the value of a new variable, say k, increase by one each time we calculate a new population. When k reaches 20, we will stop. The flowchart in Figure 1.3 accomplishes this. Notice the diamond-shaped box at the bottom of the chart. In our previous flowcharts, each box had only one arrow leaving it, so we had no difficulty in following the arrows from box to box. But this diamond-shaped box has two arrows leaving it. Moreover, the box contains no action for us to take. Instead, a question is asked (Is $k = 20$?). Instead of taking any overt action, we merely answer the question with "yes" or "no." Depending on the answer to that question, we follow one or the other of the two arrows leaving the box. Note that one arrow is labeled "yes" and the other is labeled "no." If $k = 20$ (answer of "yes"), the left pointing arrow is followed. On the other hand, if $k \neq 20$ (answer of "no"), the right pointing arrow is used. The remainder of the flowchart should be self-explanatory.

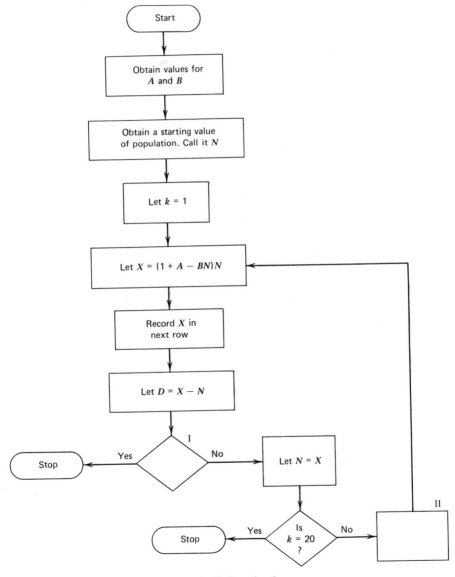

FIGURE 1.5 Partial flowchart to be used with Exercise 3.

In flowcharts *rectangular* boxes contain the description of an action to be taken (e.g., "Record X in the next row"); *diamond-shaped* boxes ask a question that has two possible answers, "yes" or "no," and two arrows leading outward, one for each response; *oval-shaped* boxes designate a "start" or "stop." We will continue to

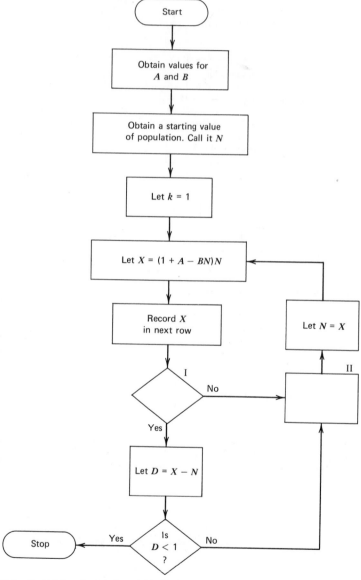

FIGURE 1.6 Partial flowchart to be used with Exercise 4.

follow these conventions regarding the shapes of the boxes in flowcharts throughout the text.

Of course, we could use some other way to stop the computations of the population values. In Figure 1.3 we counted to 20. But we might prefer to continue to compute the values of the population until those values are changing by very little. For example, we might wish to stop if two succeeding populations differ by less than one. Prior to commencing the calculations, however, we have no idea when this will occur, so we cannot count to 20 or to any preassigned number. Instead, we compute the difference between X, the new population value, and N, its predecessor, as each new population

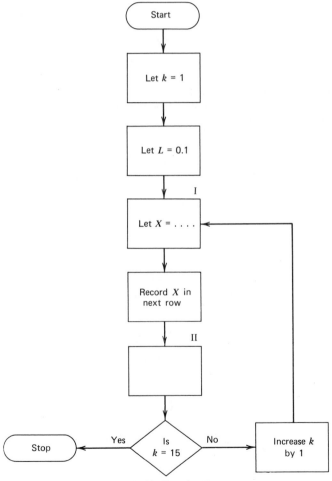

FIGURE 1.7 Partial flowchart to be used with Exercise 5.

is calculated. If this difference is less than 1, we stop. A flowchart to do this is shown in Figure 1.4. The reader should follow it through step by step and assure himself of each box's purpose.

EXERCISES FOR SECTION 1.5

*1. Modify the flowchart in Figure 1.3 so that 37 populations are computed and recorded.

2. Modify the flowchart in Figure 1.3 so that the number of the period is recorded in a column preceding the value of the population. In other words, for $A = 0.4$, $B = 0.0002$, and $N_0 = 1000$, the following results should be recorded.

1	1200
2	1392
3	1561.2672

and so on.

*3. Complete the flowchart in Figure 1.5 so that it results in the computation of successive populations until *either* (a) two succeeding populations differ by less than 1 *or* (b) at most 20 populations have been recorded.

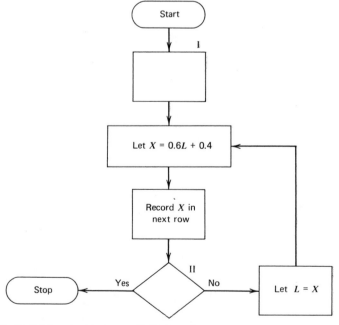

FIGURE 1.8 Partial flowchart to be used with Exercise 6.

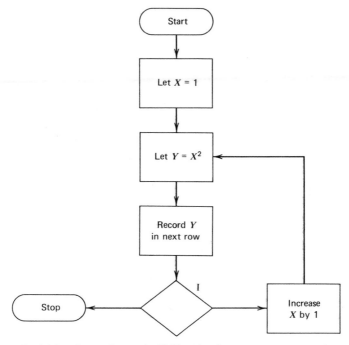

FIGURE 1.9 Partial flowchart to be used with Exercise 7.

4. Complete the flowchart in Figure 1.6 on page 34 so that it results in the computation of successive populations until *both* (a) two succeeding populations differ by less than 1 *and* (b) at least 20 populations have been recorded.

*5. Complete the flowchart in Figure 1.7 on page 35 so that it computes and records the values of $L_1, L_2, \ldots$ up to L_{15} as given in (1.7) for $A = 0.3$ and $L_0 = 0.1$.

6. Complete the flowchart in Figure 1.8 so that it computes and records the values of $L_1, L_2, \ldots$ and so on as given in (1.7) until a value greater than 0.999 has been recorded. The value of $A = 0.4$ and the value of $L_0 = 0.2$.

*7. Complete the flowchart in Figure 1.9 so that is computes and records in a column the squares of the integers from 1 to 30. Notice that the labels "yes" and "no" are missing from the diamond-shaped box and should be completed.

8. Complete the flowchart in Figure 1.10 on page 38 so that it computes and records the cubes of the *odd* integers from 1 to 27.

1.6 RECAPITULATION AND A PREVIEW

We close this chapter with a summary of the previous sections and a brief discussion of some of the problems that we will encounter in succeeding chapters.

In Sections 1.2, 1.3, and 1.4 we arrived at the following equations.

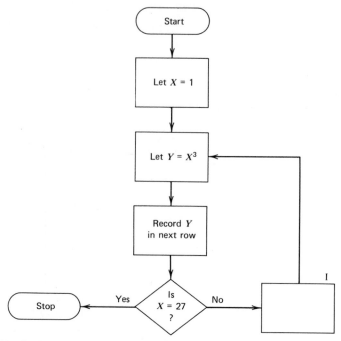

FIGURE 1.10 Partial flowchart to be used with Exercise 8.

$$L_{k+1} = (1 - A)L_k + A \qquad\qquad k = 0, 1, 2, \ldots \qquad (1.7)$$

$$T_k = A(1 + B)T_{k-2} - ABT_{k-1} + 1 \qquad k = 2, 3, 4, \ldots \qquad (1.28)$$

$$N_{k+1} = (1 + A - BN_k)N_k \qquad\qquad k = 0, 1, 2, \ldots \qquad (1.38)$$

The first described a learning process, the second described a national economy, and the third described a population. However, there is a certain similarity in the three equations. Each of the equations relates the value of something at one point in time to its value at some other time or times. For example, (1.7) relates the amount learned at one time to the amount learned at one period later. Equations of this type are called *difference equations* or *recurrence relations*. We will use the former term. We reemphasize that a difference equation is really an infinite number of equations. Equation 1.7 is shorthand for

$$L_1 = (1 - A)L_0 + A$$

$$L_2 = (1 - A)L_1 + A$$

$$L_3 = (1 - A)L_2 + A$$

and so on, indefinitely.

There also are some factors that distinguish each of the above difference equations from each other. Notice that (1.7) and (1.38) only involve L and N, the quantities whose value we seek, at two different times. Notice also that (1.7) and (1.28) contain the unknown quantity, L or T, linearly (i.e., no squares, square roots, or the like appear in the equations). On the other hand, (1.38) contains N_k^2 if the parentheses are removed and is, therefore, nonlinear.

In Chapter 3 we will investigate solutions of difference equations in detail. We will only find solutions for equations of the type shown in (1.7): two different times and no nonlinear expressions. Equations of the type (1.28) can be solved, but their solution is beyond the scope of this text. Nonlinear equations such as (1.38) generally cannot be solved except by numerical techniques such as the one demonstrated in Section 1.4.

As we progress through the development of mathematics and computing, we will pause frequently to discuss applications of the techniques we develop to subjects such as sociology, biology, geography, business, medicine, and law. We will try to give you some idea of the nature of these applications in the following paragraphs.

Suppose we have collected data on land use in some city. We may know, for instance, that over the past several years 10% of the vacant land has been changed to parking lots, while 5% of what were parking lots is now vacant land. Similarly, we know the change of land use back and forth between residential use, commercial use, parking lots, and vacant land. If this pattern of change continues, what will the city look like in 100 years? Will it be one huge parking lot? Will it be half commercial and half residential? The point being that if the long-run land use is one that we like, then we should strive for policies that will maintain the present pattern of change. If we do not like the eventual land distribution, then we should work to change the patterns of change that exist.

Along these same lines, suppose we know patterns of population movement over some period. We might know that 10% of the people living in the eastern part of a country move to the middle, 15% move to the west, and 75% stay in the east. We also know the movement out of the middle and out of the west. Where will the people be located in the long run?

Turning to the field of business, suppose you borrow money that is to be paid back in fixed monthly payments over a certain number of months. What is the true rate of interest? Would you be better off to get a conventional bank loan even at a higher interest rate?

Suppose we know the incidence of a disease in a country. We also know that if we administer a diagnostic test to people with the disease, a certain percentage will be diagnosed and the others will not. (An infallible and inexpensive test is hard to find.) We also know the success of the test with people who do not have the disease. Then suppose the test is administered to someone who may or may not have contracted the disease. Suppose also that the test indicates the person is well (i.e., does not have the disease in question). What are the chances that the test has predicted correctly? An erroneous prediction can, of course, lead to the death of the patient, so we would like the chances of an erroneous prediction to be very small.

Hopefully, these examples will give you an idea of what is to come. All of these and many more problems will be encountered in the remainder of this book. We will develop the necessary mathematics to examine these problems and to answer, at least in part, the questions posed.

2

PROGRAMMING IN THE BASIC LANGUAGE

2.1 WHAT IS PROGRAMMING ALL ABOUT?

Computer programming is a human activity. A person who has a problem that he wants to use a computer to help him solve must develop a procedure consisting of a sequence of the elementary operations that a computer is capable of carrying out. The procedure must then be expressed in a language that the computer can "understand." BASIC is such a language, and it is one of the most widely used for instructional purposes. (The letters stand for "Beginner's All-purpose Symbolic Instruction Code.")

People have problems, whereas computers follow *procedures*. A computer cannot "solve a problem," no matter how many Sunday supplements say so. When we have a problem that we want to use a computer to help solve, we must first devise a precise method of solving it. The method chosen must, in principle, be something that a person could do, if given enough time. In other words, it must be absolutely clear at every stage what is to be done and what the sequence of actions is. Such a rigorously specified sequence of actions for solving a problem is called an *algorithm*. An algorithm can be expressed in many ways: in English, in the graphical form of a flowchart, or as a computer program, among others. If the algorithm is not expressed as a computer program in the first place, and usually it is not, the next step is to write a computer program that carries out the processing actions required by the algorithm.

In the mathematical applications in this book we need to talk both about devising algorithms and about writing computer programs. At the beginning, it is necessary to focus on the BASIC language, so that we will have a way of "expressing ourselves" to the computer when we turn to problems that require more time to be spent on devising and expressing appropriate algorithms.

2.2 A SIMPLE BASIC PROGRAM

Let us begin our study of the BASIC programming language and how to use it by considering a simple but complete program, one to compute the area of a rectangle from the product of its height and width.

```
100   LET H=2.2
200   LET W=4
300   LET A=H*W
400   PRINT H,W,A
500   END
```

We see that a BASIC program consists of a set of *statements,* each of which is preceded by a *line number,* and each of which generally specifies some action to be carried out. Every statement begins with an English word that indicates what the action is, and is usually followed by other information.

The first three statements are LET statements, which give values to the *variables* named H, W, and A. BASIC permits us to choose any letters we like for variable names, but we generally try to pick letters that remind us of the meaning of the variable if possible. The first statement gives the variable named H the value 2.2, which it will keep through the rest of the program. Likewise, the next statement gives W the value 4. The quantities 2.2 and 4 are called *constants* in BASIC, just as in mathematics. The third statement is a bit different: it calls for a calculation to be performed. The asterisk is the symbol that calls for multiplication, so the values that we gave to the variables named H and W are multiplied, and their product assigned to the variable named A. The PRINT statement says to print these three values, and the END specifies that there is nothing more to the program. The last statement in every BASIC program must be an END.

The way BASIC is almost always used, it is an *interactive* language. This means that we use it with a typewriterlike *terminal* that is connected to a computer by wires, which is sometimes done with a telephone line. When we type a command to the computer at our terminal, the computer immediately responds—or "interacts"—appropriately. Our interaction with the computer in BASIC takes place in two stages in a simple program like this one. First, we type the program into the computer. No calculations are carried out during this phase. The statements can be typed in any order, if we wish; they will automatically be put into ascending sequence for us. After we have typed in the program, we can issue a command called "LIST" to get a copy of it. If errors are discovered in inspecting the program, the incorrect statements can be retyped, or any necessary additional statements inserted. (It is to leave room for such additional statements that we initially number statements by tens or hundreds.) When we are satisfied that the program is correct, we type in the command "RUN," which causes the program to be *executed,* or carried out, following the instructions we wrote. Here is the output when the above program was run.

```
    2.2             4           8.8
```

The values that were assigned to H and W have been printed correctly, and the computed value of the area of the rectangle has been printed.

BASIC is able to distinguish between program statements like LET H = 2.2 and commands like LIST and RUN because statements always begin with statement numbers, whereas commands never do.

It is often desirable in producing computer output to identify what the various numbers mean, and we will commonly do so in this book. This is easy to do in BASIC, since anything enclosed in quotation marks in a PRINT statement is simply printed as it appears. Quoted and nonquoted material can be mixed any way we like, leading to considerable flexibility in the manner of printing results. Here is the area of a rectangle program with one form of identification added.

```
100   LET H=2.2
200   LET W=4
300   LET A=H*W
400   PRINT "HEIGHT =",H,"WIDTH =",W
450   PRINT "AREA =",A
500   END
```

This is the output that was produced.

```
HEIGHT =        2.2        WIDTH =          4
AREA =          8.8
```

One might think it would be more natural to have printed the answers all on one line instead of on two lines as we have done. Notice, however, that each item of information is printed in 15 printing positions. If we had attempted to print all six items (three numbers and three identifying phrases) on one line, we would have needed $6 \times 15 = 90$ printing positions. But the page of most terminals is only 72 printing positions wide. Thus we can print at most five items on a line. We chose to print four items on the first line and two items on the second line.

Notice the above output is aligned in columns. This is automatic in BASIC as long as the items in the list of the PRINT are separated by commas.

Now we can get the six items above all on one line if we are willing to give up the alignment in columns. To do so, we separate the PRINT entries with semicolons.

```
100   LET H=2.2
200   LET W=4
300   LET A=H*W
400   PRINT "HEIGHT =";H;"WIDTH =";W;"AREA =";A
500   END
```

The output is:

```
HEIGHT = 2.2          WIDTH = 4     AREA = 8.8
```

Notice that we also have reduced the blank space between items in the output. To emphasize that the use of semicolons destroys the column alignment, we will rewrite the last program with the output divided between two rows.

```
100   LET H=2.2
200   LET W=4
300   LET A=H*W
400   PRINT "HEIGHT =";H;"WIDTH =";W
450   PRINT "AREA =";A
500   END
```

The output is:

```
HEIGHT = 2.2          WIDTH = 4
AREA = 8.8
```

It is almost never true that there is only one correct way to write a computer program. Sometimes one way is better than another for reasons that we occasionally point out, but there are almost always various possibilities. We can illustrate another feature of BASIC by changing this program slightly.

The items that are listed in a PRINT statement are not restricted to quoted identifications and variable names. Actually, we can put ordinary constants there, and we can even call for calculations. Consider this version.

```
100   LET H=2.2
200   LET W=4
400   PRINT "HEIGHT =";H;"WIDTH =";W;"AREA =";H*W
500   END
```

We see that the previous statement 300 has been eliminated; no variable named A is ever evaluated. Instead, in the PRINT itself, the product H*W appears. The output is exactly the same as with the previous version.

The extreme of this process would be to put everything in the PRINT statement and, in this simple example, that is actually possible.

```
100   PRINT "HEIGHT =";2.2;"WIDTH =";4;"AREA =";2.2*4
200   END
```

Once again the output is unchanged.

It will be uncommon to find situations where something this rudimentary is a really legitimate computer problem. After all, the multiplication of two numbers can be done more simply and cheaply with simpler tools than an expensive computer.

Even in simple tasks, however, it often is desirable to write programs that have a bit more flexibility than our examples so far. One of the ways that BASIC provides for doing this is the combination of the READ and DATA statements. Using this facility, we can group some or all of our numeric data into one statement, where it is easier to change if the need should arise, then use the READ statement to make the data available to program variables. Here is our little task as the program might be rewritten to use this feature.

```
10   READ H,W
20   LET A1=H*W
30   PRINT "HEIGHT =";H;"WIDTH =";W;"AREA =";A1
40   DATA 33,23.1
50   END
```

Observe that the line numbers run from 10 in steps of 10 this time; the choice of line numbers is largely a matter of personal preference. The READ statement lists the variables that are to receive values from the DATA statement, in the same order in which the values appear. That is, the 33, since it appears first, will be assigned to H, and so forth. The DATA statement may be placed anywhere in the program, either before or after the READ. Notice that the variable name for the area this time is A1. A variable name in BASIC may be either a single letter or a single letter followed by a single digit.

```
HEIGHT = 33     WIDTH = 23.1      AREA =  762.3
```

Throughout the discussion so far it has been assumed that statements are executed in the order they are listed, starting with the lowest-numbered statement. This is true, unless we explicitly specify otherwise, using additional features of the BASIC language. One of these features is the GOTO statement, with which we state that the next statement to be executed is *not* the next one in ordinary sequence, but the one having the line number given. In the program below, for instance, the statement GOTO 10 is followed by a statement with the line number 60—but it is statement 10 that will be executed after the GOTO, not statement 60.

```
10    READ H,W
20    LET A1=H*W
30    PRINT "HEIGHT =";H;"WIDTH =";W;"AREA =";A1
40    PRINT
50    GOTO 10
60    DATA 33, 23.1, 2,3, 60.96,98.1
70    DATA 1,1,0.5,3.2
80    END
```

Let us look at the program in question a little more closely. Its first three statements are just as before, but then we come to a PRINT with nothing else on the line. This will produce a blank line in the output, to provide easier readability for the more extensive output that we are going to produce this time. Next is the GOTO 10 statement, which returns us to the READ. In other words, after computing the area of one rectangle, we are going to return to the beginning of the program to pick up data (from the DATA statement again) on another rectangle, and so on, until we run out of data. Finally, in the way of new features, observe that there are two DATA statements here. The principle is simply that values are picked up from the DATA statements in the order in which they appear within each DATA statement, and then values are picked up from the next DATA, for as many DATA statements as there are.

We also notice in passing a minor item of some interest: there is no space between the comma and the last constant in the first DATA statement. This is a small example of a feature of most BASIC systems: the spacing of the statements is of almost no importance. Whether to place spaces around the equal sign in a LET statement, for instance, is a matter of personal preference. This remark does not apply, naturally, to material within quote marks in a PRINT; all spaces here are reproduced just as written.

Here is the output when the program was run.

```
HEIGHT = 33     WIDTH = 23.1        AREA = 762.3

HEIGHT = 2      WIDTH = 3      AREA = 6

HEIGHT = 60.96       WIDTH = 98.1         AREA = 5980.18

HEIGHT = 1      WIDTH = 1      AREA = 1

HEIGHT = .5          WIDTH = 3.2        AREA = 1.6


OUT OF DATA    IN LINE 10
```

The last line is the notification that we ran out of data in the DATA statement, which was deliberate. The notice of this fact would differ in details in other BASIC systems. It might be noted as an error in some cases, but that is true only if the condition is not deliberate, since it might not be in some applications.

In order to see more of the fundamentals of BASIC in action, let us change the requirements of our illustrative program so that it will produce the perimeter of a rectangle instead of its area. This, of course, is given by two times the sum of the height and width of the figure. One possible program and its output are:

```
100   READ H,W
200   DATA 20,30
300   LET P=2*(H+W)
400   PRINT "HEIGHT =";H;"WIDTH =";W;"PERIMETER =";P
500   END
```

```
HEIGHT = 20    WIDTH = 30    PERIMETER = 100
```

We see the plus sign used in the ordinary way to indicate addition, together with parentheses to indicate the desired grouping of quantities. It would not do to write

```
2*H + W
```

because, in BASIC just as in usual mathematical notation, this would mean for the multiplication to be done first, followed by the addition, which would give an incorrect result. On the other hand, it would be acceptable to write

```
2*H + 2*W
```

In other situations we have no choice but to use parentheses. In the expression

$$\frac{A}{B + C}$$

for example, we must write

```
A/(B+C)
```

where the slash(/) stands for division. There is no other way to force the sum of B and C to be in the denominator. Another example, and one that beginners frequently make mistakes with, is

$$\frac{A}{B*C}$$

If we write

```
A/B*C
```

we will have an unintended result. In the absence of parentheses, a series of multiplications and divisions are done in sequence from left to right. Therefore A will first be divided by B, then the quotient multiplied by C, giving the same result as

$$\frac{A*C}{B}$$

The solution, once again, is to use parentheses.

A/(B*C)

to force the product of *B* and *C* to be in the denominator.

There are occasions when extreme cleverness can avoid a few parentheses. We do not recommend the exercise of such cleverness, which has no real benefits and often causes trouble. The use of extra parentheses to remove any possible doubt of the intended meaning is generally good programming practice.

We have now seen three of the four usual *arithmetic operators,* as they are called. The fourth, the minus sign for subtraction, presents no new concepts. Besides these four operations, a fifth is commonly included in describing programming languages: *exponentiation.* In BASIC exponentiation is denoted by an upward pointing arrow or, in a few systems, by two asterisks. Here is a brief program to show how this operation is written and how it operates.

```
100   LET X1=2↑3
200   LET X2=(5-2)↑2
300   LET X3=(5+2*2)↑.5
400   LET Y5=2↑(1/4)
500   PRINT "X1","X2","X3","Y5"
600   PRINT X1,X2,X3,Y5
700   END
```

X1	X2	X3	Y5
8	9	3.	1.18921

In line 100, 2 is raised to the third power, which is 8. In line 200, the parentheses force the subtraction to be done first, so the exponentiation is 3 raised to the second power, which is 9. In line 300 the operations within parentheses are done first, and there the multiplication is done first, so the quantity within parentheses is 9. This is raised to the one-half power, which is the same as taking the square root. The square root of 9 is 3. In line 400 the parentheses force the division to be done first, so we have 2 raised to the one-fourth power, which is the same as taking the fourth root. There are two PRINT statements this time, the first of which produces only identifications, which are in the form of headings written over the numerical information. This is the way we would commonly want identification for columns of output. Then the second PRINT calls for the values themselves.

This example was deliberately devised to emphasize the distinction between variable names, written without quotation marks in the second PRINT, and identifications, written with quotation marks in the first. The variable names, without quotes, mean whatever the earlier parts of the program say they mean. If we had failed to assign values to them, there would be no values to print, and most BASIC systems would indicate an error. The same names written in quotes are simply printed exactly as written and have no "meaning" in the sense of a relation to anything else in the program. We could scramble the names or use names different from the variable names, and the program would run without any indication of the problem. The results might or might not mean much to us, but the program would run just fine.

After discussing one more small matter, we turn to some exercises to let the reader test and develop his skills in writing simple BASIC programs himself.

The question is: what happens when numbers are too large or too small to be represented conveniently in the ordinary form used so far? The answer is the exponent notation, in which a number is followed by the letter E and a positive or negative number that is a power of 10 by which the number is to be multiplied. Consider the examples in this program.

```
100 LET A=1.2/1000000
200 LET B=12345†2
300 LET C = 1E2 + 1E-2
400 PRINT A, B, C, 0.1234E9
500 END
```

```
1.20000E-06     1.52399E+08     100.01     1.23400E+08
```

In line 100 we carry out a division that results in the quotient 0.0000012. Looking at the printed results, we see that instead of printing it in that form, it has been printed as 1.2E-06, which means $1.2 \cdot 10^{-6}$, which is the same quantity written in a different form. Some BASIC systems would print this as .12E-05, which is also the same number in a different form. In line 200 we have a number that can be written readily enough in nonexponent form, which is then squared. The result, with all its digits, is 152399025. In the printed output we see that this quantity has been written in exponent form as .152399E+09, which is the same number except that the last three digits have been lost. Actually, in many computers, those last three digits don't even exist. There is some limit on the number of digits that can be held in a number in any computer, and numbers requiring more than that number of digits have to be approximated. This applies not only to large integers, but to most fractions. The simple fraction ⅓, for example, cannot be represented exactly as a decimal fraction regardless of how many digits are used: the representation is an infinitely repeating decimal fraction.

In line 300 we see that it is quite permissible to write numbers in exponent form within statements if we wish. The numbers, in ordinary form, are simply 100 and 0.01 and, in the output, the result is presented as 100.01. The BASIC system makes a decision on each number regarding whether or not it can be represented conveniently without exponents. Finally, in the PRINT, we see, first, that variables and constants can appear in the same list if we wish it and, second, that a number that is in exactly the same form as BASIC would choose for it will not be changed in the output.

EXERCISES FOR SECTION 2.2

(Note. Solutions are given for those exercises marked with an asterisk.)

1. Write BASIC LET statements to do the following.
   *(a) Assign to the variable named K the value 12.
   (b) Assign to the variable named Q the value 56.88.
   *(c) Assign to the variable named F9 the value of the expression $2.1 + 3 \times 12.3$. Do not do the arithmetic by hand in advance; write an appropriate expression in the LET statement.
   (d) Assign to the variable named Z2 the value of the expression

$$88 - \frac{12.3}{45.9}$$

   *(e) Assume that a previous LET statement has already given a value to the variable named L; you do not know what that value is, but you can safely assume that it has a value. Then write a statement to multiply that value by 6 and assign the product to the variable named X.
   (f) Assume that a previous LET statement has given a value to the variable named A. Subtract 123 from that value and assign the result to the variable named A1.
   *(g) Assume that a previous statement has given a value to the variable named Z. Subtract 1 from that value and assign the result to the variable named Z. (The correct answer will convince you that a BASIC LET statement is not an equation!) This kind of operation is commonly required, and we will see many examples of applications of it.
   (h) Assume that a previous statement has given a value to a variable named M. Multiply that value by 1.1 and assign the product back to M, much as in the previous exercise.

*2. Identify three errors in the following BASIC program.

```
100 A = 1.2
200 LET B = 2A + 4
300 PRINT A,B
```

3. Identify three errors in the following BASIC program.

```
100 LET B12 = 3
200 LET Q = B1
300 PRINT "B12 =" B12
400 END
```

*4. Modify the following BASIC program (which is correct) so that it uses the READ and DATA statements instead of the first two LET statements.

```
100 LET R = 2
200 LET S = -3.4
300 LET T = R/S
400 PRINT R, S, T
500 END
```

5. Modify the following BASIC program so that it uses the READ and DATA statements instead of the first LET statements.

```
10 LET B1 = 12
20 LET B2 = 400.1
30 LET W = (B1 + B2)↑2
40 PRINT B1, B2, W
50 END
```

*6. Add a PRINT statement to print the words REGULAR, OVERTIME, and TOTAL over (i.e., on the line before) the three numbers printed by the following program.

```
100 READ R, Ø
200 DATA 83.50, 12.39
300 LET T = R + Ø
400 PRINT R, Ø, T
500 END
```

7. In the program of the previous exercise, modify the PRINT statement appropriately so that the output is

```
REGULAR = 83.50     ØVERTIME = 12.39     TØTAL = 95.89
```

8. Write a complete BASIC program to do the following. Assign to the variables named A, B, and C the values 1.2, 7.3, and 3.4, respectively, then compute the value of the expression

$$A + \frac{B}{C}$$

and assign that value to the variable named D. Finally, print the values of all four variables without identifications.

9. (a) Write a complete BASIC program to do the following. Assign to the variables named F, G1, and G2 the values 12, 34, and -67, respectively, then compute the value of the expressions

$$X = F + G1$$

and

$$Y = G1 / G2 + 6$$

Finally, print the values of F, G1, G2, X, and Y without identifications.

(b) Now write a program to produce the same output, but consisting of only two statements (PRINT and END) and no variables.

2.3 SOME BASIC CONTROL STATEMENTS

In this section we will learn some elementary but fundamental ideas about how to write programs that contain parts that repeat. Since it is not practical to set up a program to repeat indefinitely, we will explore ways to instruct a program to make decisions about its own actions, based on data values or computed results. We will also begin to see that these techniques are of great importance in many ways in writing most computer programs.

Let us begin with a simple program that introduces a new and useful BASIC statement: INPUT. The computational situation is that we wish to be able to read a value of the radius of a circle from our terminal, then have the program compute and print the area of a circle having that radius.

The INPUT statement is somewhat similar to the READ in that it gives a value to each variable listed in it; it is different in that with the READ, the value comes from the program itself (in the DATA statement), whereas with the INPUT the value is typed in as the program is run. When an INPUT statement is executed, the BASIC system prints a question mark and waits for us to type in a number (or numbers, if the INPUT has more than one variable name), followed by a carriage return. Here is the simplest version of the program for computing the area of a circle.

```
LIST

100    INPUT R
200    LET A=3.14159*R↑2
300    PRINT "RADIUS =";R;"AREA =";A
400    END

RUN

?12
RADIUS = 12    AREA = 452.389
```

Here we have shown the complete printout, including the LIST and RUN commands, as the interaction proceeded. This version of the program calls for no repetition, so after it has called for one value of the radius and printed the line of output, everything is finished with this run. We could, of course, issue the run command again to call for another execution of the complete program but, if we really want to compute the areas of a number of circles, that is not an attractive way to proceed. Here is a second version of the program, which takes a step in the direction of something more usable.

```
100    INPUT R
200    LET A=3.14159*R↑2
300    PRINT "RADIUS =";R;"AREA =";A
400    REM  **  THIS EXTRA PRINT IS TO GET A BLANK LINE  **
500    PRINT
600    GOTO 100
700    END

RUN

?12
RADIUS = 12    AREA = 452.389
```

```
? 10
RADIUS = 10    AREA = 314.159

? 1
RADIUS = 1    AREA = 3.14159

? C
DONE
```

An extra PRINT statement has been inserted at line 500 to get a blank line after each result; this is preceded by a remark line, which begins with REM, to explain the feature to anyone reading the program. Remarks may be used freely in this way. Any line that begins REM is simply passed over in executing the program. In other words, REM statements are for our use in reading and understanding the program. The computer ignores them.

At line 600 we have a GOTO that takes program execution back to the INPUT statement at line 100. This means that we can compute the area of as many circles as we like . . . but how do we ever get it to stop? The answer—not a very satisfactory one—is that every terminal has some kind of *interrupt* button on it with which we can inform the system to break out of its normal mode of operation. Exactly what is done as a result varies from one BASIC system to the next, but we may safely generalize that *somehow* it will stop program execution in a case like this with or without an explanatory comment.

This is not really a very satisfying way to arrange things. We should much prefer to have a way to inform the program more explicitly of what our intentions are. Fortunately, BASIC provides such a technique, which we will see has much wider applicability as well. The facility is the IF-THEN statement, exhibited in statement 200 of this final version of the program.

```
100    INPUT R
200    IF R=0 THEN 800
300    LET A=3.14159*R↑2
400    PRINT "RADIUS =";R;"AREA =";A
500    REM  **  THIS EXTRA PRINT IS TO GET A BLANK LINE  **
600    PRINT
700    GOTO 100
800    END


RUN


? 23
RADIUS = 23    AREA = 1661.9

? 0.1
RADIUS = .1              AREA = 3.14159E-02

? 10.01
RADIUS = 10.01          AREA = 314.788

?
DONE
```

Line number 200 contains the new feature: it says that if the value of R is equal to zero, then the statement executed next should not be line 300, but line 800, the END statement. This gives us a simple way to inform the program that we are ready to stop, since we would never need to know the area of a circle of radius zero. When, in response to the question mark printed from the INPUT statement, we enter a zero, the system simply returns us to the command mode. That is, instead of expecting input values, it is ready for commands such as RUN or LIST or for the typing of new BASIC statements.

BASIC provides six relational tests, of which equality is one.

BASIC Symbol	Mathematical Symbol	Meaning
$=$	$=$	Equal to
$<$	$<$	Less than
$>$	$>$	Greater than
$<=$	$\leq$	Less than or equal to
$>=$	$\geq$	Greater than or equal to
$<>$	$\neq$	Not equal to

For a second and rather different example of the usefulness of the IF statement, let us turn to a payroll calculation. The program will use an INPUT statement to call for the number of hours worked and the hourly pay rate. If the number of hours worked is 40 or less, the pay for the week is simply the product of the hours worked and the pay rate. If the number of hours worked is greater than 40, however, the worker is to be paid at the normal rate for the first 40 hours and at time-and-a-half for the hours over 40. Here is the program.

```
100    INPUT H,R
200    REM  **  CHECK FOR ZERO SENTINEL  **
300    IF H=0 THEN 1100
400    LET P=H*R
500    REM  **  CHECK WHETHER MORE THAN 40 HOURS WORKED **
600    IF H <= 40 THEN 800
700    LET P=R*(40+1.5*(H-40))
800    PRINT "HOURS =";H;"RATE =";R;"PAY =";P
900    PRINT
1000    GOTO 100
1100    END

RUN


?32,2.56
HOURS = 32    RATE = 2.56       PAY = 81.92

?35.7,2.56
HOURS = 35.7        RATE = 2.56        PAY = 91.392

?40,3.00
HOURS = 40    RATE = 3    PAY = 120
```

```
? 41.,3.00
HØURS = 41    RATE = 3    PAY = 124.5

? 50,3.00
HØURS = 50    RATE = 3    PAY = 165

? 0,0

DØNE
```

At line 300 we have an IF used in the same way as in the previous program: if the number of hours worked as entered in response to the INPUT is zero, that will be taken as a signal that we wish to stop program execution. The value zero is called a "sentinel," since it signals to the computer that the program is to end. At line 400 we compute the pay according to the formula for hours of no more than 40. Naturally, we don't know at that point whether the hours worked is over or under 40, but we compute the pay anyway, knowing that if the hours are over 40 the situation will be correctly handled by the following statements.

At line 600 we have an IF that asks whether the hours worked (H) were less than or equal to ($<=$) 40. If so, we jump to line 800, skipping around the computations in line 700. At line 700 we recompute the pay altogether, which will wipe out the previously computed value. Reading the expression in the statement, we see that it pays at the rate of R dollars per hour for 40 hours, plus 1.5 times R for H-40 hours, that is, the hours in excess of 40. This gets the "time-and-a-half" that was specified. If it is any easier to follow, observe that the statement could also have been written

```
LET P = R*40 + 1.5*R*(H-40)
```

with the same result.

The results shown demonstrate that the program operated as desired. Thirty-two hours at $2.56 per hour is indeed $81.92. (Dollar signs could have been provided within the quotes if desired, although there is no way in BASIC to get them to print immediately to the left of the first digit of the number.) The second line shows 35.7 hours at $2.56 per hour; we will see later in the chapter how to round off dollars-and-cents figures to the nearest penny. The third line shows nothing exceptional, and the fourth line proves that the overtime calculation does work correctly. Forty-one hours at $3 per hour would be $123 straight time but, with time-and-a-half for the 1 hour in excess of 40, the proper result is $124.50, as shown. The last line gives another example of overtime pay, after which we enter two zeros to signal the end of the run.

For a final example in this section we take up a computation that was used in Chapter 1 and show how it might be programmed. Equation 1.8 stated that the solution to the learning model as formulated in the previous chapter was

$$L_k = (1 - A)^k(L_0 - 1) + 1 \tag{1.8}$$

The problem we set ourselves is to prepare a table of the values of L_k for values of k from 1 to 20, given values of A and L_0.

```
100   READ LO,A
200   DATA 0,.3
300   PRINT "K","L"
400   LET K=1
500   LET L=(1-A)↑K*(LO-1)+1
600   PRINT K,L
700   LET K=K+1
800   IF K <= 20 THEN 500
900   END

RUN
```

K	L
1	.3
2	.51
3	.657
4	.7599
5	.83193
6	.882351
7	.917646
8	.942352
9	.959646
10	.971752
11	.980227
12	.986159
13	.990311
14	.993218
15	.995252
16	.996677
17	.997674
18	.998372
19	.99886
20	.999202

```
DØNE
```

The program begins by establishing values for L_0 and A and printing column headings. Then it assigns the value 1 to K. The value of K will change as the program is executed, as we will see; this particular step (statement 400) will not be repeated. In line 500 we have the actual calculation according to the formula of (1.8), with a change that may need some explanation. In the formula, the variable on the left was written with a subscript, L_k, to show that the value of L depends on the value of k. Within the program, however, this dependence does not have to be written in this explicit fashion—and, indeed, cannot be. It is up to us in writing the program to make sure that when we print a value of K and a value of L, they are the ones that are correctly associated with each other. After L has been computed it is printed.

Now we come to a new feature. We need to increase, or *increment,* the value of K by one. That is, we need to take the old value of the variable named K, add one to it, and assign the sum to be the new value of the same variable named K. The old value is destroyed by this action, but we have no further need for the old value. Next we need to know if the table has been completed (i.e., have we yet printed the line for which

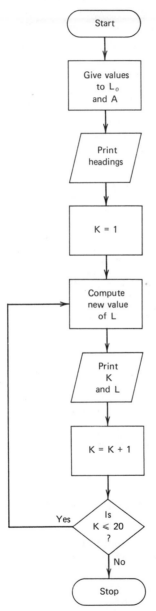

FIGURE 2.1 Flowchart describing computations for learning model (Case Study 1).

$K = 20$?). The way to ask that question most simply, in this case, is to specify going back to line 500 if K is less than or equal to 20. If it is, we have another line to compute and print; if it is not, we are finished.

When there are choices to be made in a program, as there are here, it can be helpful in understanding the sequence of the actions—and especially the effects of decisions—to have a graphical representation of the process. That is, we turn to the flowchart notation introduced in Chapter 1. A flowchart of the procedure used in this program is shown in Figure 2.1.

We see that the general appearance is the same as when the flowchart was not tied to a computer program, with one addition: a box in the shape of a parallelogram is used to denote input and output operations. Notice also that the "language" of BASIC is used freely in writing this flowchart, such as by writing simply $K = 1$ instead of some English equivalent like "Give K the value 1."

EXERCISES FOR SECTION 2.3

*1. What will this program do?

```
100   INPUT A,B
200   LET G=A
300   IF A>B THEN 500
400   LET G=B
500   PRINT G
600   END
```

2. What will this program do?

```
100   INPUT G,D
200   LET T=G-675*D
300   IF T>0 THEN 500
400   LET T=0
500   PRINT G,D,T
600   END
```

*3. State what would be done by the program and corresponding flowchart shown.

```
100   INPUT N
200   IF N=0 THEN 600
300   IF N<0 THEN 100
400   PRINT N
500   GOTO 100
600   END
```

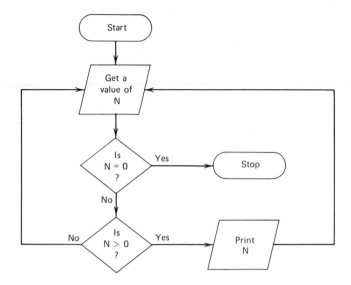

4. State what would be done by the program and corresponding flowchart shown at end of exercises on page 60.

```
100   LET N=1
200   PRINT N
300   LET N=N+1
500   IF N <= 25 THEN 200
600   END
```

*5. Draw a flowchart and write a program to carry out the following actions. Using INPUT, get a value for D. If D is less than or equal to 8, set E to zero; but if D is greater than 8, set E to D − 8. Print the value of E.

6. Draw a flowchart and write a program to carry out the following actions. Using INPUT, get values for S and Q. If S is less than Q, make S2 equal to the amount by which S is less than Q; however, if S is greater than or equal to Q, set S2 to zero. Print S, Q, and S2.

*7. Draw a flowchart and write a program to carry out the following actions. Get a value for A (annual earnings). Compute T (tax) as zero if A is less than or equal to $2000, and as 2% of the amount over $2000 otherwise. Print A and T, then stop.

8. Draw a flowchart and write a program to carry out the following actions. Get values for B (base), S (sales), and Q (quota). Compute BO (bonus) as zero if S is not greater than Q, and as 8% of the amount by which S exceeds Q otherwise. Compute G (gross) as the sum of B and BO. Print B, S, Q, and G and then stop.

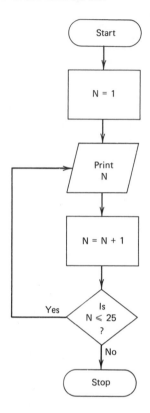

2.4 THE FOR-NEXT LOOP

Situations in which we wish to have a variable run through a series of values arise so frequently that most programming languages provide features for simple specification of the actions desired. In BASIC the language element is the combination of the FOR and NEXT statements. The simplest way to introduce the subject is with an example, and we will simply rewrite the program for the calculation of the 20 values of L.

```
100   READ LO,A
200   DATA 0,.3
300   PRINT "K","L"
400   FØR K=1 TØ 20
500   LET L=(1-A)↑K*(LO-1)+1
600   PRINT K,L
700   NEXT K
800   END

RUN
```

K	L
1	.3
2	.51
3	.657
4	.7599
5	.83193
6	.882351
7	.917646
8	.942352
9	.959646
10	.971752
11	.980227
12	.986159
13	.990311
14	.993218
15	.995252
16	.996677
17	.997674
18	.998372
19	.99886
20	.999202

DØNE

The key element is in line 400, where we specify that K is to run through all the integer values from 1 to 20 inclusive, while the part of the program from the FOR statement down to the NEXT K statement is executed repeatedly. More explicitly, here is what the FOR-NEXT means here. "Assign the value 1 to the variable named K, and carry out the statements between the FOR and the NEXT. Add 1 to the value of K and carry out these statements again. Continue this process until the steps between the FOR and the NEXT have been carried out with K equal to 20." We see that the results are identical to those computed earlier.

Let us carry out this task in a different way, which will reemphasize a fundamental fact about BASIC variables and the assignment of values to them. Recall from Chapter 1 that we found a relationship between a current value of L, denoted by the subscript k, and the next value of L, denoted by the subscript $k + 1$.

$$L_{k+1} = (1 - A)L_k + A$$

If we wish, we may calculate the successive values of L directly from this relationship, according to the following procedure.

1. Give L a starting value.
2. Using the current value of L, evaluate the formula $(1 - A)L + A$; this becomes the new current value of L. Repeat this step as many times as desired.

Here is a program that does exactly that.

```
10   LET A=.3
20   LET L=0
30   PRINT "K","L"
40   FØR K=1 TØ 20
50   LET L=(1-A)*L+A
60   PRINT K,L
70   NEXT K
80   END
```

The first statement simply gives a value to A. The second statement is a literal translation of the first rule stated above. The FOR-NEXT combination causes the evaluation and printing to be done 20 times; the value of K does not appear in the formula. Statement 50 is a literal translation of the second rule above. It says to take the current value of L and use it to evaluate the formula; when this has been done, the newly computed value of L becomes the new value of L. It is this "new" value that will be used in the next evaluation of the formula and that will, in turn, be replaced, and so forth.

The important concept is that the variable that appears on the left of the equal sign in a LET statement has a value assigned to it as a result of the execution of the statement, regardless of whether it had a value before and regardless of whether any previous value is used in the computation specified on the right side of the equal sign.

We will not show the output, since it is identical to that displayed earlier.

Let us take a similar task but do it a different way to show that to a large measure the interpretation of what comes out of a computer is up to the user.

We saw in Chapter 1 that the growth of a population can be represented by the formula

$$N_{k+1} = (1 + A - BN_k)N_k$$

Suppose that we try to use this formula to predict the population of the United States between 1890 and 1970 using

$$A = 0.232912$$

$$B = 0.000671071$$

and starting with $N_0 = 62.948$. This last means that we are taking the unit of measurement to be a million people, and that the population in 1890 was therefore 62,948,000.

Here is one way the program could be written.

```
100   LET A=.232912
200   LET B=6.71071E-04
300   LET N=62.948
400   PRINT "YEAR","PØPULATIØN"
500   FØR K=0 TØ 8
600   PRINT 1890+10*K,N
700   LET N=(1+A-B*N)*N
800   NEXT K
900   END
```

Notice that the variable in the FOR statement starts with zero this time and, very important, we print the current value of N *before* computing a new one. This means, in effect, that what we are printing is the old value of N, not the new one. Notice also that the FOR index is converted to a value for the year by an appropriate formula. Here is the output of the program.

YEAR	POPULATION
1890	62.948
1900	74.9503
1910	88.6373
1920	104.01
1930	120.975
1940	139.331
1950	158.755
1960	178.818
1970	199.009

DONE

Since the FOR index is used for nothing except producing the values for the year in the output, we might as well set it up to produce the years directly, using the STEP option to get increments of 10 years.

```
100   LET A=.232912
200   LET B=6.71071E-04
300   LET N=62.948
400   PRINT "YEAR","POPULATION"
500   FOR Y=1890 TO 1970 STEP 10
600   PRINT Y,N
700   LET N=(1+A-B*N)*N
800   NEXT Y
900   END
```

The output is the same as before.

For another example we will suppose that an American student planning a bicycle trip in Europe wants to prepare a short table converting distances in miles to kilometers, according to the formula

$$\text{Kilometers} = 1.6093 \times \text{Miles}$$

He wants the table to run from 1 to 10 in steps of 1. Here is the program.

```
100   PRINT "MILES","KILOMETERS"
200   PRINT
300   FOR M=1 TO 10
400   PRINT M,1.6093*M
500   NEXT M
600   END
```

MILES	KILØMETERS
1	1.6093
2	3.2186
3	4.8279
4	6.4372
5	8.0465
6	9.6558
7	11.2651
8	12.8744
9	14.4837
10	16.093

After looking at this table, our imaginary student decides that he wants it extended to 100 miles, but with the additional portion to run in steps of five miles instead of one. That is, the table should run from 1 to 10 in steps of 1, and from 15 to 100 in steps of 5.

```
100   PRINT "MILES","KILØMETERS"
200   PRINT
300   FØR M=1 TØ 10
400   PRINT M,1.6093*M
500   NEXT M
550   FØR M=15 TØ 100 STEP 5
560   PRINT M,1.6093*M
570   NEXT M
600   END
```

We see that three statements have been added, line numbers 550, 560, and 570, defining another FOR-NEXT loop. But this time the FOR statement has the STEP parameter that adds 5 instead of 1 to M between repetitions of the loop. The statement in line 300 is as though it had been written

```
300 FØR M=1 TØ 10 STEP 1
```

The output is:

MILES	KILØMETERS
1	1.6093
2	3.2186
3	4.8279
4	6.4372
5	8.0465
6	9.6558
7	11.2651
8	12.8744
9	14.4837
10	16.093
15	24.1395
20	32.186
25	40.2325
30	48.279
35	56.3255
40	64.372
45	72.4185
50	80.465
55	88.5115
60	96.558
65	104.604

MILES	KILØMETERS
70	112.651
75	120.697
80	128.744
85	136.79
90	144.837
95	152.883
100	160.93

The parameters for the FOR statement (the starting, ending, and step values) are not restricted to positive integers, as the following examples show.

```
100    PRINT "MILES","KILØMETERS"
200    PRINT
300    FØR M=.5 TØ 5 STEP .5
400    PRINT M,1.6093*M
500    NEXT M
600    END

RUN
```

MILES	KILØMETERS
.5	.80465
1	1.6093
1.5	2.41395
2	3.2186
2.5	4.02325
3	4.8279
3.5	5.63255
4	6.4372
4.5	7.24185
5	8.0465

```
100    PRINT "MILES","KILØMETERS"
200    PRINT
300    FØR M=5 TØ .5 STEP -.5
400    PRINT M,1.6093*M
500    NEXT M
600    END

RUN
```

MILES	KILØMETERS
5	8.0465
4.5	7.24185
4	6.4372
3.5	5.63255
3	4.8279
2.5	4.02325
2	3.2186
1.5	2.41395
1	1.6093
.5	.80465

```
DØNE
```

It is essentially the programmer's responsibility to be sure that the parameters of the FOR statement make sense. If we write something meaningless, such as

$$\text{F\O R X = 1 T\O 5 STEP -1}$$

most BASIC systems will just ignore that statement and skip over the statements between the FOR and the NEXT, but sometimes strange and unintended things will happen.

Considering the fundamental importance of this feature of the BASIC language, we will show one more example of its use, one that will be needed later in the book.

The factorial of a positive integer N is defined as the product of all integers from 1 up to N. Thus the factorial of 2 is $1 \cdot 2 = 2$, the factorial of 4 is $1 \cdot 2 \cdot 3 \cdot 4 = 24$, and so on. Here is a program to compute and print the factorial of a number read as input.

```
100    PRINT "TYPE N";
200    INPUT N
300    IF N<0 THEN 1100
400    LET F=1
500    FØR X=1 TØ N
600    LET F=F*X
700    NEXT X
800    PRINT N;"FACTØRIAL =";F
900    PRINT
1000   GØTØ 100
1100   END
```

The program begins by typing an instruction to the user, which is commonly done. Note that the statement ends with a semicolon, which is intentional. When a PRINT statement ends with a comma or semicolon, the terminal carriage does *not* space a line after printing. This means that the number typed in response to the INPUT in the next statement will be on the same line as the instruction "TYPE N." This feature is often useful and can be used in many other ways. The program is designed with an IF statement to use a negative value as a signal that we wish to terminate execution of the program.

Now comes the actual computation. We set the variable F equal to 1; this will become the value of the factorial when the FOR-NEXT loop is completed. This loop runs the variable X through all values from 1 to N, multiplying F by X and setting F equal to the product. F is thus repeatedly replaced by larger numbers as the loop proceeds. This phrasing reads as though N has to be larger than 1, but actually the program will operate correctly for N = 1: F starts at 1, is multiplied by 1, and the FOR loop is completed. After printing N, N factorial, and a blank line for readability, the program goes back to get another value of N. Here is the output for some representative values.

```
TYPE N?1
 1    FACTØRIAL = 1

TYPE N?3
 3    FACTØRIAL = 6

TYPE N?5
 5    FACTØRIAL = 120

TYPE N?10
10    FACTØRIAL = 3.62880E+06

TYPE N?20
20    FACTØRIAL = 2.43290E+18

TYPE N?34

ØVERFLØW - WARNING ØNLY  IN LINE 600
34    FACTØRIAL = 1.70141E+38
```

It may be seen that the factorials become quite large for even moderate values of N. In fact, the last line shows that it is possible to ask for results that cannot be computed. The error comment refers to the way numbers are represented in the computer and need not concern us in detail. What it means is simply that the computer could not represent a number of the size that would have been needed. Program execution was stopped without going back to the INPUT.

EXERCISES FOR SECTION 2.4

*1. What will this program do?

```
100    LET S=0
200    FØR J=1 TØ 15
300    LET S=S+J
400    PRINT J,S
500    NEXT J
600    END
```

2. What will this program do?

```
100    LET S=0
200    FØR N=1 TØ 21 STEP 2
300    LET S=S+N
400    PRINT N,S
500    NEXT N
600    END
```

*3. Write a program that will print the integers from 1 to 40, one to a line. There will be no input.

4. Write a program that will read an integer. If the integer is greater than 50, stop program execution. Otherwise print all the integers from 1 up to and including the number read.

*5. Write a program that will produce a conversion table from temperatures in Centigrade degrees (C) to Fahrenheit (F), according to the formula

$$F = 1.8C + 32$$

for all Centigrade temperatures from 0 to 100 in steps of 1 degree. There will be no input.

6. Write a program that obtains two numbers from the terminal; these are the beginning and ending temperatures for a conversion table like that of the previous exercise. Make a test to see that the first number is actually smaller than the second before proceeding, and determine that the difference between the two is not greater than 200; stop program execution without producing the table if the data fail either test.

*7. Write a program to print 20 lines, each line giving one of the numbers from 1 to 20, together with the square of the number.

8. Write a program to print 20 lines, each line giving one of the integers from 50 to 69, together with the cube of the number.

*9. Write a program corresponding to the flowchart of Figure 1.3 for computing 20 values of a population. Do not use the FOR-NEXT technique, but program the computation exactly as shown in the flowchart. When you run the program, use $A = 0.4$, $B = 0.0002$, and $N_0 = 1000$.

10. Same as Exercise 9, but use the FOR-NEXT capability.

2.5 FUNCTIONS

It often is necessary in writing programs to compute values of various functions, such as finding the square root of a number. We will need only three functions in this book, but BASIC provides a dozen or so, including things such as logarithms and trigonometric functions, which do not appear in this book.

Every function in BASIC has a three-letter name. We call for the computation of a function by writing the name of the function followed by parentheses enclosing an expression. For instance, the name of the square root function is SQR; if we write

```
LET X = SQR(5)
```

the program will compute the square root of 5 and assign that value to X. If we write

```
PRINT X, SQR(2*X)
```

the program will evaluate the expression 2*X, take its square root, and print that value following the value of X. If we write

```
PRINT X, X + SQR(X)
```

the square root of X will be added to X and the sum will be printed following the value of X.

In other words, the *argument* of a function (what appears in parentheses—the quantity of which we want the square root) can be any BASIC expression. It can be a constant, a single variable, or any expression, including other function references. We can write things like

```
LET B = 2 + SQR(X + SQR(Y))
```

Notice that parentheses are used to enclose the arguments of a function, which is quite different from the earlier use to specify the order of arithmetic operations. As a matter of fact, both uses can occur in the same statement; BASIC has no trouble in distinguishing. For example, we can write

```
LET A = SQR(B/(C+D))
```

or

```
LET P = 1/(1+SQR(X))
```

or even

```
LET R = SQR(1/(SQR(B/(C+D))))
```

and everything will be kept straight.

Here is an illustrative program that demonstrates a simple square root operation. Nine values are supplied to the function through a FOR-NEXT loop. The last one is a deliberate error, to show what a BASIC system may do when asked to find the square root of a negative number. (Other systems would give a response that would differ in details, and some would simply take the square root of the positive value, with no error indication.) We have placed the square root in the PRINT statement in this program, since we have nothing else to do with the result, but a function can be written anywhere in a program that a variable name can.

```
10    DATA 4,2,3.5,.25,100,1000,10000,1.E+20,-9
20    FØR I=1 TØ 9
30    READ X
40    PRINT "SQUARE RØØT ØF ";X;" = ";SQR(X)
50    PRINT
60    NEXT I
70    END

RUN


SQUARE RØØT ØF   4      =   2
SQUARE RØØT ØF   2      =   1.41421
```

```
SQUARE ROOT OF   3.5            =   1.87083

SQUARE ROOT OF   .25            =   .5

SQUARE ROOT OF   100    =   10

SQUARE ROOT OF   1000           =   31.6228

SQUARE ROOT OF   10000          =   100

SQUARE ROOT OF   1.00000E+20      =   1.00000E+10

SQUARE ROOT OF  -9      =
SQR OF NEGATIVE ARGUMENT IN LINE 40
```

The absolute value function simply gives the positive value of its argument, regardless of whether the argument is positive or negative. Putting it another way, the absolute value function discards the minus sign of a negative argument and does nothing to a positive argument. Here is an illustrative program showing how simple the operation of this function is.

```
10   DATA 4,-4,6.88,-6.88,0
20   FOR I=1 TO 5
30   READ X
40   PRINT "ABSOLUTE VALUE OF ";X;" = ";ABS(X)
50   PRINT
60   NEXT I
70   END


RUN


ABSOLUTE VALUE OF   4      =   4

ABSOLUTE VALUE OF  -4      =   4

ABSOLUTE VALUE OF   6.88            =   6.88

ABSOLUTE VALUE OF  -6.88            =   6.88

ABSOLUTE VALUE OF   0      =   0
```

The integer function (INT) takes as argument any number, and returns as function value the greatest integer less than or equal to the argument. This slightly complicated way of phrasing things is required to correctly define the action for negative numbers. For positive numbers we can simply say that the integer function discards any fractional part of the argument, so that INT(4.1) = 4, INT(4.9) = 4, INT(4) = 4, and so on. But, with negative numbers, the action of returning the next smaller integer (unless the argument is itself an integer) means that the fractional part is not just dropped. For example, INT(−4.1) = −5, since −5 is the greatest integer not exceeding −4.1. Here is a program, similar to the previous two, that illustrates these actions.

```
10    DATA 4.1,4,4.9999,-1,-1.1,-1.999,0
20    FØR I=1 TØ 7
30    READ X
40    PRINT "INTEGER PART ØF ";X;" = ";INT(X)
50    PRINT
60    NEXT I
70    END

RUN


INTEGER PART ØF    4.1            =    4

INTEGER PART ØF    4        =    4

INTEGER PART ØF    4.9999         =    4

INTEGER PART ØF   -1        =   -1

INTEGER PART ØF   -1.1           =   -2

INTEGER PART ØF   -1.999         =   -2

INTEGER PART ØF    0        =    0
```

One common application of the INT function is in *rounding off* a number to an integer, that is, finding the integer nearest to the given value. The INT function itself will not do the job, because it returns the greatest integer less than or equal to the given number, but all that has to be done is to add 0.5 to the number before using the INT function. Consider some examples.

$$INT(2.34 + 0.5) = INT(2.84) = 2$$

This is the correct answer, since the integer nearest to 2.34 is 2.

$$INT(2.67 + 0.5) = INT(3.17) = 3$$

This is also correct, since the nearest integer to 2.67 is 3. The method, perhaps surprisingly, also works for negative numbers.

$$INT(-2.34 + 0.5) = INT(-1.84) = -2$$

$$INT(-2.67 + 0.5) = INT(-2.17) = -3$$

The INT function takes any arithmetic expression as its argument, just as the other functions do, so we can write things like INT(X + 0.5), or INT(100*D + 0.5), or whatever.

If we wish to round off a value to something other than an integer, we have a bit more work to do. Take, for instance, the common operation of rounding off a value in dollars and cents to the nearest cent. The INT function can still be used, if we convert everything to cents first. We do this by multiplying the dollars-and-cents value by 100. After adding 0.5 and using the INT function, we will have rounded off to the nearest penny. If we then divide by 100, we get back to dollars and cents.

For an example, consider an amount such as $3.0168, which might have been obtained by multiplying a price per part of .1257 by 24 parts. If we multiply 3.0168 by 100, we get 301.68 cents; adding 0.5 gives 302.18; dropping the fractional part gives 302 cents; dividing by 100 gives 3.02, which is the original dollars-and-cents amount rounded to the nearest penny.

Here is a program that shows this operation in action for a set of representative quantities.

```
SCR


10 DATA 23.4567, 87.654, 1.23456, 1.111, 5.5555, 6.001, 6.009
20 DATA 9.7531, 1.3579, 13.579, 135.791, 4.9999
30 FOR I = 1 TO 12
40 READ D
50 PRINT D; INT(100*D + 0.5)/100
60 NEXT I
70 END

RUN


   23.4567      23.46
   87.654       87.65
   1.23456      1.23
   1.111        1.11
   5.5555       5.56
   6.001        6
   6.009        6.01
   9.7531       9.75
   1.3579       1.36
   13.579       13.58
   135.791      135.79
   4.9999       5
```

EXERCISES FOR SECTION 2.5

*1. Write a statement, using the square root function, that will give to S the value of the square root of 2. (Why do you suppose the proviso "using the square root function" was added? Can you do this without the square root function?)

2. Write a statement, using the square root function, that will give to T the value of the square root of X + 12.

*3. Write a program that will produce a table of the first 40 integers and their square roots.

4. Write a program that will produce a table of the first 40 integers, their squares, and their square roots.

*5. Write a statement that will leave X unchanged if its value is positive, but reverse its sign if it is negative.

6. Write a statement that will give to D the value of A − B if A is larger than B, the value of B − A if B is larger than A, and the value zero if A and B are equal.

*7. Write a statement that will transfer execution to statement 800 if the difference between M and N is less than 0.0001. M and N are both positive, but you do not know which is larger.

8. Write a few statements that will give to G the absolute value of E or the absolute value of F, whichever absolute value is larger.

*9. BASIC has no function that returns the fractional part of a number, in comparison with the INT function, which returns its integer portion. This is partly because accomplishing the result is relatively simple for positive numbers. Write a statement that gives to F the fractional part of a positive number X.

10. Write a statement that will transfer to statement 300 if X has an integer value.

*11. Write a statement that will give to A the value of the number B rounded to the nearest one tenth.

12. Write a statement that will round the value of Q to the nearest thousand and assign that value to P.

*13. Write a BASIC program corresponding to the flowchart in Figure 1.4 for computing populations until the changes become small.

14. Write a BASIC program that computes and prints successive populations using

$$N_{k+1} = (1 + A - BN_k)N_k$$

until either (a) two successive populations differ by less than 1 or (b) at most 20 populations have been computed. (See Exercise 3 of Section 1.5.)

3
DIFFERENCE EQUATIONS

3.1 WHAT IS A DIFFERENCE EQUATION?

We have already encountered difference equations in our earlier discussions. In Chapter 1 we wrote equations describing a learning process (1.7), a model of a national economy (1.28), and the growth of a population (1.38). Each of these equations was a difference equation. We solved those equations in a straightforward, although tedious, way. Later we used a computer and a BASIC program to reduce the tedium. In this chapter we will look at difference equations from a different point of view. We will find that in many cases of interest we can "solve" such equations (i.e., find a formula that precisely describes the behavior of the difference equation). Moreover, even in the case of difference equations that we cannot solve, we often will be able to give a qualitative description of the behavior of the solution even though we cannot write down a formula.

Before attempting to solve a difference equation, however, we must know precisely what we mean by the terms "difference equation" and "solution." In this section, therefore, we will concentrate on the ideas and terminology behind difference equations. In succeeding sections we will exploit these ideas.

An algebraic equation deals with ordinary variables. For example,

$$x + y = 3$$

is an algebraic equation. A difference equation deals with subscripted variables. Again, by way of example,

$$y_{k+1} - 3y_k = 0 \qquad (3.1)$$

where k takes on each and every one of the values 0, 1, 2, 3, . . . , is a difference equation. The fact that k does take on all of these values is important. Equally important is that the variable, y, appears twice and with different subscripts—$k + 1$ and k. With this as background we now turn to a definition of a difference equation.

We define a *difference equation* as an equation that relates a subscripted variable—where the subscript itself is a variable—to one or more of its "neighbors." The term "neighbors," although somewhat vague at this point, can be clarified through some examples.

Returning to (3.1), the variable is y. The subscript, which is a variable, is k. When we say a subscript is a variable, we mean that it takes on any one of the values 0, 1, 2, 3, ... Equation 3.1 relates y_k to one of its neighbors. The "neighbor" is y_{k+1}, since $k + 1$ is a neighbor to k. In general, a neighbor to a subscripted variable, y_k, has a subscript that is the original subscript, k, *plus or minus an integer*. In (3.1), y_k is related to only one of its neighbors, y_{k+1}.

As another example of a difference equation consider

$$2y_{k+3} - 4y_{k+2} + y_k = 3 \qquad (3.2)$$

which relates y_k to two other values, y_{k+2} and y_{k+3}. Our definition required that y_k be related to one or more of its neighbors. Both (3.1) and (3.2) satisfy this requirement.

Yet another example of a difference equation is

$$y_k y_{k+2} - 2y_k^2 = -1 \qquad (3.3)$$

In this case there are two arbitrary values, y_k and y_{k+2}. However,

$$y_k = 6$$

is *not* a difference equation, since no neighbors of y_k appear. Similarly,

$$y_3 - y_4 = 7$$

is not a difference equation, since no arbitrary subscripts appear. The following two equations also are not difference equations.

$$y_{k+1} + 2y_5 = 0$$
$$y_k + 3y_q = 6$$

The first has one arbitrary subscript, $k + 1$, but no neighbor to y_{k+1}. The second has two arbitrary subscripts that are unrelated to each other.

If we wish to decide whether or not an equation is a difference equation, there is a simple two-step test that we can apply.

1. In all references to the variable there should be one, and only one, unknown quantity in the subscripts.
2. At least two subscripts must be different.

The reader should apply this test to the equations above. For those that are not difference equations, determine whether it is step 1 or 2 that fails.

Throughout our discussion we have assumed that the variable subscript was any one of the nonnegative integers. Thus, when k appeared, it could take on *any* one of the values $0, 1, 2, 3, \ldots$. Moreover, k takes on *all* of those values in succession. When $k + 1$ appears, it can and does take on the values $1, 2, 3, 4, \ldots$, and $k + 2$ takes on $2, 3, 4, 5, \ldots$. At times we may find it convenient to restrict the allowable values for k to be all the positive integers, starting with an integer greater than zero or perhaps all positive integers greater than 6 or some such number. In such cases we still have a difference equation, provided conditions 1 and 2 above are satisfied. Unless we specify otherwise, however, we will assume that the subscript can be any one of the nonnegative integers: $0, 1, 2, \ldots$.

It should be clear that what we have defined as a difference equation represents an infinite number of algebraic equations. For example, (3.1) represents

$$\left.\begin{aligned} y_1 - 3y_0 &= 0 \\ y_2 - 3y_1 &= 0 \\ y_3 - 3y_2 &= 0 \\ \cdots \end{aligned}\right\} \tag{3.4}$$

and so on as k takes on the values $0, 1, 2, \ldots$ in succession.

Two difference equations may appear to be different and yet be the same. For example, consider

$$y_k - 3y_{k-1} = 0 \tag{3.5}$$

where k takes on the values 1, 2, 3, ... If we write out the first three algebraic equations of this last difference equation, we get again (3.4). Therefore (3.5) is the same difference equation as (3.1). Moreover,

$$y_{k-3} - 3y_{k-4} = 0$$

where $k = 4, 5, 6, \ldots$ is also the same equation as either (3.1) or (3.5).

Two difference equations are said to be *identical* if the infinite set of algebraic equations that they represent are identical.

A difference equation such as (3.1) is said to be *linear* because it involves the function y only linearly (i.e., no squares, square roots, products, etc.). On the other hand, (3.3) is nonlinear because there is a term in y_k^2 and also the product of y_k and y_{k+2}. Either of these would be sufficient to classify the equation as nonlinear.

Not all linear difference equations can be treated alike. Equation 3.1, which is linear, can be solved for all y_k if y_0 is specified. For example, if $y_0 = 2$, then setting $k = 0$ in (3.1),

$$y_1 = 3y_0 = 6$$

and setting $k = 1$ in (3.1),

$$y_2 = 3y_1 = 18$$

and so on. But consider the linear difference equation:

$$y_{k+2} + y_{k+1} - 2y_k = 0 \tag{3.6}$$

and suppose that once again we are given that $y_0 = 2$. Can we solve this equation for all y_k? We proceed as before and let $k = 0$; that is,

$$y_2 + y_1 - 2y_0 = 0$$

or

$$y_2 + y_1 - 4 = 0 \tag{3.7}$$

But now we are stymied. We have two unknown quantities, y_2 and y_1, in this one equation. If we go on to let $k = 1$ in (3.6) in an attempt to correct this deficiency, we get

$$y_3 + y_2 - 2y_1 = 0 \tag{3.8}$$

and we introduce a third unknown, y_3. We have made no progress toward finding a solution.

The only way to resolve this dilemma is to specify the value of y_1 as well as the value of y_0. For example, if we set $y_1 = -1$, from (3.7),

$$y_2 = 5$$

and from (3.8),

$$y_3 + 5 - 2(-1) = 0$$

or

$$y_3 = -7$$

We can continue to compute y_4, y_5, and so on. The key point is that we needed to specify two consecutive values of y_k (i.e., y_0 and y_1), while to solve (3.1) we needed only to specify one value, y_0.

Of course, sometimes we may need to specify three or even more values of y_k. For the difference equation

$$y_{k+3} + 3y_{k+2} - 4y_{k+1} - 12y_k = 0 \tag{3.9}$$

we need y_0, y_1, and y_2 before we can solve the equation.

Notice that (3.1), (3.6), and (3.9) all were linear, but each required a different number of starting values. We will classify linear difference equations according to the number of starting values that are necessary to obtain a solution.

Equations such as (3.1), which require one starting value, are called *first-order* equations. ("First" corresponds to "one" starting value). Equations such as (3.6), which require two starting values, are called *second-order* equations. In general, if N starting values are required to find a solution, then the difference equation is said to be "of Nth order" or "of order N." Equation 3.9 is of third order.

We always can write a linear, first-order difference equation as

$$y_{k+1} = My_k + C \tag{3.10}$$

where $M \neq 0$. The restriction that M is not zero is crucial. If M were zero there would be only one subscript, and step 2 of our test for a difference equation would fail (i.e., (3.10) would not even be a difference equation). Similarly, it is essential that no other terms appear in (3.10). If there were other terms with different subscripts (e.g., y_{k+2} or y_{k-1}), then we would need more than one starting value, and the equation would not be first order.

A linear, second-order difference equation always can be expressed:

$$y_{k+2} = Ny_{k+1} + My_k + C \tag{3.11}$$

where again $M \neq 0$. Notice that N may be zero. We turn now to some examples that we encountered earlier in Chapter 1.

The learning model given in (1.7) is

$$L_{k+1} = (1 - A)L_k + A$$

The coefficient of L_k is $1 - A$ and, if $A \neq 1$, this does not vanish, so this difference equation is a linear, first-order equation.

As another example, consider (1.28), which described the national economy.

$$T_k = A(1 + B)T_{k-1} - ABT_{k-2} + 1 \qquad k = 2, 3, \ldots$$

We rewrite this

$$T_{k+2} = A(1 + B)T_{k+1} - ABT_k + 1 \qquad k = 0, 1, 2, \ldots$$

This is in the form (3.11), where $N = A(1 + B)$, $M = -AB$, and $C = 1$. Moreover, $M \neq 0$ unless either A or B is zero, so this is a linear, second-order difference equation.

We will confine our discussion in this chapter to the solution of linear, first-order difference equations. We hasten to point out, however, that the results we obtain will be useful in our investigations of nonlinear equations such as (1.38). We also note that much of what we say can be extended to second- and higher-order equations. However, these latter types of equations require complex numbers and are beyond the scope of this text.

EXERCISES FOR SECTION 3.1

(Note. Solutions are given for those exercises marked with an asterisk.)

1. Which of the following are difference equations? For those that are not difference equations, explain why they are not. In each case the allowable values for k are shown at the right.
   *(a) $f_{k+2} - 2f_{k+1} + f_k = 3$ $k = 0, 1, 2, \ldots$
    (b) $y_2 - y_1 = 4$ $k = 0, 1, 2, \ldots$
   *(c) $f_{k+1}f_k = 2$ $k = 0, 1, 2, \ldots$
    (d) $x_k^2 = 0$ $k = 0, 1, 2, \ldots$
   *(e) $(k + 1)f_k = 4$ $k = 0, 1, 2, \ldots$
    (f) $g_{k-1} = g_k$ $k = 1, 2, 3, \ldots$
   *(g) $g_k^2 = k + 1$ $k = 0, 1, 2, \ldots$
    (h) $H_{k+2} + (k + 3)H_k = k$ $k = 0, 1, 2, \ldots$

*2. For each difference equation on the left there is one on the right that is identical to it. The allowable values for k in each equation are shown below the appropriate equation. Match identical equations.

(a) $y_{k+3} - y_{k+1} = 2$
$\quad k = -1, 0, 1, 2, \ldots$

(i) $-2 - y_k + y_{k+2} = 0$
$\quad k = 2, 3, 4, \ldots$

(b) $y_k - y_{k+2} = 2$
$\quad k = 0, 1, 2, \ldots$

(ii) $2 - y_{k-1} + y_{k+1} = 0$
$\quad k = 1, 2, 3, \ldots$

(c) $y_{k+4} - y_{k+2} = 2$
$\quad k = 0, 1, 2, \ldots$

(iii) $y_k - y_{k+2} = 2$
$\quad k = 1, 2, 3, \ldots$

(d) $2 - y_{k+1} + y_{k+3} = 0$
$\quad k = 0, 1, 2, \ldots$

(iv) $y_{k+2} - y_k = 2$
$\quad k = 0, 1, 2, \ldots$

3. For each difference equation on the left there is one on the right that is identical to it. The allowable values for k in each equation are shown below the appropriate equation. Match identical equations.

(a) $2f_{k+1} - f_k = 0$
$\quad k = 1, 2, 3, \ldots$

(i) $f_{k+1} = 2f_{k+2}$
$\quad k = 2, 3, 4, \ldots$

(b) $f_{k+1} = 2f_{k+2}$
$\quad k = -1, 0, 1, \ldots$

(ii) $2f_{k+1} - f_k = 0$
$\quad k = 2, 3, 4, \ldots$

(c) $2f_{k+3} - f_{k+2} = 0$
$\quad k = 1, 2, 3, \ldots$

(iii) $f_{k+2} - 2f_{k+3} = 0$
$\quad k = -1, 0, 1, 2, \ldots$

(d) $f_{k+2} - 2f_{k+3} = 0$
$\quad k = 0, 1, 2, \ldots$

(iv) $2f_{k+3} - f_{k+2} = 0$
$\quad k = -2, -1, 0, 1, \ldots$

4. Put each of the following in the form of (3.10) or (3.11) and identify M, N, and C.

*(a) $3y_{k+1} - y_k + 2 = y_k - 3$

(b) $x_{k+2} - 2x_{k+1} = 6$

*(c) $2r_{k+1} - 3r_{k-1} = 0$

(d) $5G_{k+4} - 2G_{k+2} = 6$

(e) $3f_{k+1} = 2f_k - f_{k+1} + 3$

(f) $6y_k - y_{k+1} = y_{k+1} - 2$

(g) $3y_{k+2} - y_k = 2y_{k+1} - 1$

(h) $f_k - f_{k+1} + 2 = 2f_{k+2}$

5. Which of the following difference equations are identical to (1.28)

(a) $T_k = A(1 + B)T_{k-1} - ABT_{k-2} + 1$ $\quad k = 0, 1, 2, 3, \ldots$

(b) $T_{k+1} = A(1 + B)T_k - ABT_{k-1} + 1$ $\quad k = 1, 2, 3, \ldots$

(c) $T_{k+2} = A(1 + B)T_{k+1} - ABT_k + 1$ $\quad k = 0, 1, 2, 3, \ldots$

(d) $T_{k+2} = A(1 + B)T_k - ABT_{k-2} + 1$ $\quad k = 0, 1, 2, 3, \ldots$

3.2 WHAT IS A SOLUTION OF A DIFFERENCE EQUATION?

We turn now to a discussion of what we mean by a solution of a difference equation. We also will discuss how we can verify whether or not a given formula is or is not a solution of a specific difference equation. In Section 3.3 we will see how to find the solution of some difference equations. Before we can do that, however, we need to know what a solution is and how to test whether or not we have a solution.

The solution of an algebraic equation consists of all values of the variables that made the equation true. Thus the solution of

$$x^3 + 3x^2 - 13x - 15 = 0 \qquad (3.12)$$

consists of -5, -1, and 3, because when (and only when) x takes on any one of these values, (3.12) is true. Thus the solution of an algebraic equation is, in general, a set of numbers.

The solution of a difference equation is not a set of numbers. Instead, it is a formula or expression usually involving the variable subscript, k.

Such a formula is said to be a *solution of a difference equation* if it makes the difference equation true for *each and every* value of k. As has been our custom, we examine some examples. Consider (3.1), which was

$$y_{k+1} - 3y_k = 0 \qquad (3.1)$$

where $k = 0, 1, 2, 3, \ldots$. Consider the function

$$y_k = 3^k \qquad (3.13)$$

with $k = 0, 1, 2, 3, \ldots$. We ask: Is (3.13) a solution of (3.1)?

How are we to answer such a question? To do so we "substitute" the proposed solution (3.13) in the difference equation (3.1) and test whether the equation is valid for all k (i.e., is valid for $k = 0$, $k = 1$, $k = 2$, and so on).

From (3.13), with k replaced by $k + 1$,

$$y_{k+1} = 3^{k+1}$$

Using this and (3.13) in (3.1), the latter becomes

$$y_{k+1} - 3y_k = 3^{k+1} - 3(3^k) = 3^{k+1} - 3^{k+1} = 0$$

Notice that all steps are valid for *any* value of k; therefore (3.13) is a solution of (3.1). It is essential that we be certain that each and every step be valid for an arbitrary value of k.

Let us consider another sequence and check if it is a solution of (3.1).

$$y_k = 2^k \qquad (3.14)$$

Then

$$y_{k+1} = 2^{k+1}$$

and, using these in (3.1),

$$y_{k+1} - 3y_k = 2^{k+1} - 3 \cdot 2^k = 2^{k+1} - (2 + 1)2^k = 2^{k+1} - 2^{k+1} - 2^k = -2^k$$

But $-2^k \neq 0$ for any value of k, so (3.14) is not a solution of (3.1).

Have we found in (3.13) the only solution of (3.1)? Let us try

$$y_k = 2 \cdot 3^k \qquad (3.15)$$

Substituting this into (3.1),

$$y_{k+1} - 3y_k = 2 \cdot 3^{k+1} - 3(2 \cdot 3^k) = 0$$

for any choice of k. Thus (3.15) is also a solution of (3.1). If we try

$$y_k = -4 \cdot 3^k \qquad (3.16)$$

then

$$y_{k+1} - 3y_k = -4 \cdot 3^{k+1} - 3(-4 \cdot 3^k) = 0$$

and we have found a third solution. In summary,

$$y_k = 3^k$$
$$y_k = 2 \cdot 3^k$$
$$y_k = -4 \cdot 3^k$$

are all solutions of (3.1).

It appears, therefore, that there might be an infinite number of solutions to (3.1). Indeed this is so, but this fact will not cause us any difficulty as we shall see. To verify that there are indeed an infinite number of solutions consider

$$y_k = Q \cdot 3^k \qquad (3.17)$$

where Q is any constant: positive, negative, zero, integer, noninteger, and so on. Substituting (3.17) in the difference equation (3.1)

$$y_{k+1} - 3y_k = Q \cdot 3^{k+1} - 3 \cdot Q \cdot 3^k = Q \cdot 3^{k+1} - Q \cdot 3^{k+1} = 0$$

Thus (3.17) is a solution of (3.1) for any choice of Q. Since there are an infinite number of choices for Q, there are an infinite number of solutions.

Each of the solutions (3.13), (3.15), and (3.16) is called a *particular solution* of (3.1). The solution (3.17) is called *the general solution* of (3.1). Note the use of the article "the" preceding the words "general solution," implying that there is only one such general solution. This is, in fact, so, and all possible solutions of the difference equation (3.1) can be written in the form of (3.17). We note that each of (3.13), (3.15), and (3.16) is a special (or particular) case of (3.17). It is in this sense that the first three of these are called "particular" and the last, (3.17), is termed "general."

Suppose now that we are given the value of y_0 in addition to being given the difference equation (3.1). For definiteness suppose

$$y_0 = 6$$

From the solution (3.17), when $k = 0$,

$$y_0 = Q \cdot 3^0 = Q$$

If both of these last two equations are to be true, then it must be that $Q = 6$. The solution then becomes

$$y_k = 6 \cdot 3^k$$

and this is the *only* solution of (3.1) for the case when $y_0 = 6$. Therefore we no longer have an infinite number of solutions, and this is a much happier situation than before.

The fact that $y_0 = 6$ is called an *initial condition*. It specifies the "initial" value of the subscripted variable. The initial condition determines the value of Q in the general solution. That is, the initial condition selects a particular solution from among the infinite number of solutions that make up the general solution.

For linear, first-order difference equations such as (3.1), the general solution will contain one arbitrary constant such as Q. One initial condition is needed to determine the value of that constant. Therefore, given a linear, first-order difference equation and an initial condition, there is one and only one particular solution.[1]

We look at one more example before summarizing this section.

$$y_{k+1} + y_k = 2 \tag{3.18}$$

where $k = 0, 1, 2, \ldots$. A particular solution is given by

$$y_k = 1 + (-1)^k \cdot 5 \tag{3.19}$$

To verify this, we substitute (3.19) into (3.18).

$$
\begin{aligned}
y_{k+1} + y_k &= 1 + (-1)^{k+1} \cdot 5 + 1 + (-1)^k \cdot 5 \\
&= 2 + 5 \cdot (-1)^k \big[(-1) + 1 \big] \\
&= 2 + 0 \\
&= 2
\end{aligned}
$$

---

[1] The crucial point is that the equation is of first order. This should come as no surprise, since we defined first-order equations as those that required only one starting value. The statement given here is still valid when the difference equation has certain types of nonlinearity. However, if the difference equation is of second order, then the statement is false.

Thus (3.18) is true for all values of k. The general solution of (3.18) is

$$y_k = 1 + (-1)^k \cdot Q \qquad (3.20)$$

The reader should verify that this is a solution of (3.18) for any choice of Q whatsoever.

We close by summarizing some of the results that we have collected about solutions of difference equations.

1. A solution of a difference equation is an expression (or formula) that makes the difference equation true for all values of the variable subscript, k.
2. To check whether or not a given formula is a solution of a difference equation, we "substitute" the formula into the difference equation and check whether it makes the equation true for all values of k.
3. There are an infinite number of solutions to a difference equation.
4. The infinite number of solutions often can be reduced to one solution if an initial condition is specified.

EXERCISES FOR SECTION 3.2

1. Verify whether or not the sequences on the left are solutions of the difference equations on the right.
   (a) $y_k = k(k - 1)/2$ $y_{k+1} - y_k = k$
   (b) $y_k = 3^k + 3$ $-2y_{k+1} + 6y_k = 3$
   (c) $x_k = -1 - 2^k$ $x_{k+1} = 2x_k + 1$
   (d) $T_k = \frac{1}{2} + (-1)^k$ $T_{k+1} = -T_k + 1$

*2. Which of the following are solutions of $y_{k+1} - 4y_k = 3$?
   (a) $y_k = 4^k - 1$ (b) $y_k = -1$
   (c) $y_k = 4^k + 1$ (d) $y_k = 3 \cdot 4^k$
   (e) $y_k = 4 \cdot 3^k - 1$ (f) $y_k = 2^{2k+1} - 1$

3. Which of the following are solutions of $x_{k+1} - 8x_k = 14$?
   (a) $x_k = 2$ (b) $x_k = 8^k - 2$
   (c) $x_k = 8^k$ (d) $x_k = 8^k + 2$
   (e) $x_k = 2^{3k} - 2$ (f) $x_k = 2^k - 2$

*4. Match the difference equations on the left with the solutions on the right. In all cases, $k = 1$, 2, 3, ...
   (a) $(k + 1)y_{k+1} + ky_k = 2k - 3$ (i) $y_k = -4 + 4 \cdot 2^k$
   (b) $y_{k+1} - y_k = k$ (ii) $y_k = k(k - 1)/2$
   (c) $y_{k+2} - 4y_{k+1} + 4y_k = 0$ (iii) $y_k = 1 - (2/k)$
   (d) $y_{k+2} - 3y_{k+1} + 2y_k = 0$ (iv) $y_k = (k + 1) \cdot 2^k$

5. Match the difference equations on the left with the solutions on the right. In all cases, $k = 0$, 1, 2, 3, ...
   (a) $x_{k+1} - x_k = 1$ (i) $x_k = k + 1$
   (b) $x_{k+1} - x_k = k$ (ii) $x_k = 1/(k + 1)$

(c) $x_{k+2} - x_k = 0$

(d) $x_{k+1}(x_k + 1) = x_k$

(iii) $x_k = 1 + (-1)^k$

(iv) $x_k = k(k - 1)/2$

*6. If the general solution of a difference equation is

$$y_k = -2 + A \cdot 3^k$$

which of the following are particular solutions?

(a) $y_k = -2 + 9^k$

(b) $y_k = -2$

(c) $y_k = -2 + 3^{k+1}$

(d) $y_k = -2 + (-3)^k$

7. If the general solution of a difference equation is

$$x_k = 2 + B \cdot 2^k$$

which of the following are particular solutions?

(a) $x_k = 2$

(b) $x_k = 2 + 2^{k+2}$

(c) $x_k = 2^k$

(d) $x_k = 2 + (-2)^k$

*8. By successively letting $k = 0, 1, 2, 3, \ldots$, guess at a formula for the solution of

$$y_{k+1} - y_k = k + 1$$

where $y_0 = 0$.

$\left[Hint.\ 1 + 2 + \ldots + k = k(k + 1)/2 \right]$

9. By successively letting $k = 0, 1, 2, 3, \ldots$, guess at a formula for the solution of

$$y_{k+1} - y_k = k^2$$

where $y_0 = 0$.

$\left[Hint.\ 1^2 + 2^2 + \ldots + k^2 = k(k + 1)(2k + 1)/6 \right]$

3.3 SOLUTION OF LINEAR, FIRST-ORDER DIFFERENCE EQUATIONS

We now know how to verify whether or not a given formula is a solution of a given difference equation. However, when faced with a difference equation, we still do not know how to go about finding a solution. In this section, we turn our attention to this latter task.

Recall that the general, linear, first-order difference equation is given by (3.10):

$$y_{k+1} = My_k + C \tag{3.25}$$

where $M \neq 0$. We assume that M and C are constants (i.e., do not involve k at all). For $k = 0$, (3.25) is

$$y_1 = My_0 + C \tag{3.26}$$

Then, for $k = 1$,

$$y_2 = My_1 + C$$

But y_1 is given by (3.26), so substituting the value of y_1 from (3.26) in this last equation for y_2,

$$y_2 = M(My_0 + C) + C$$

or

$$y_2 = M^2y_0 + C(1 + M) \tag{3.27}$$

Returning to (3.25) and letting $k = 2$,

$$y_3 = My_2 + C$$

Using (3.27) to replace y_2 and rearranging,

$$y_3 = M^3y_0 + C(1 + M + M^2) \tag{3.28}$$

From this and (3.27) it would be reasonable to "guess" that the value of y_k is

$$y_k = M^ky_0 + C(1 + M + M^2 + \ldots + M^{k-1}) \tag{3.29}$$

We could verify this guess by substituting the expression on the right into the difference equation (3.25). However, we will delay any verification until we have rearranged (3.29) into the final form in which we wish to use it.

It would be more convenient, particularly from a computational point of view, if we could write

$$1 + M + M^2 + \ldots + M^{k-1}$$

without the three dots. To this end, note that

$$(1 - M)(1 + M) = 1 \cdot (1 + M) - M(1 + M) = (1 + M) - (M + M^2)$$

Rewriting the right side so that it appears on two lines, the second of which is to be subtracted from the first, we obtain

$$(1 - M)(1 + M) = 1 + M \\ - M - M^2$$

so

$$(1 - M)(1 + M) = 1 \qquad - M^2$$

Next consider

$$(1 - M)(1 + M + M^2) = 1 \cdot (1 + M + M^2) - M(1 + M + M^2)$$
$$= 1 + M + M^2$$
$$\quad - M - M^2 - M^3$$
$$= 1 \qquad\qquad - M^3$$

Notice that except for the first term in the first line (i.e., 1) and the last term in the second line $(-M^3)$, each term is matched by a similar term in the other line. Moreover, the matching terms have opposite signs and therefore cancel each other. Finally, we consider a more general case.

$$(1 - M)(1 + M + M^2 + \ldots + M^{k-1}) = 1 \cdot (1 + M + M^2 + \ldots + M^{k-1})$$
$$- M(1 + M + M^2 + \ldots + M^{k-1})$$
$$= 1 + M + M^2 + \ldots + M^{k-1}$$
$$- M - M^2 - \ldots - M^{k-1} - M^k$$

Once again, with the exception of 1 and $-M^k$, each term in each line has a matching term in the other line. Once again the matching terms have opposite signs and cancel. Therefore,

$$(1 - M)(1 + M + M^2 + \ldots + M^{k-1}) = 1 - M^k$$

Now, if $M \neq 1$, we can divide both sides of this last equation by $1 - M$ so that

$$1 + M + M^2 + \ldots + M^{k-1} = \frac{1 - M^k}{1 - M}$$

On the other hand, if $M = 1$,

$$1 + M + M^2 + \ldots + M^{k-1}$$

is just

$$1 + 1 + 1 + \ldots + 1$$

That is, it is the sum of k 1's or just k. We can summarize these results as

$$1 + M + M^2 + \ldots + M^{k-1} = \begin{cases} \dfrac{1 - M^k}{1 - M} & M \neq 1 \\[2mm] k & M = 1 \end{cases} \tag{3.30}$$

Thus (3.29) becomes, after some rearranging,

$$y_k = \begin{cases} M^k \left(y_0 - \dfrac{C}{1 - M} \right) + \dfrac{C}{1 - M} & \text{if } M \neq 1 \\[4mm] y_0 + kC & \text{if } M = 1 \end{cases} \tag{3.31}$$

We propose this last formula as *the* solution of (3.25). At this point it is only a "proposed" solution, since we made a guess when we jumped from (3.28) to (3.29) and another guess in obtaining (3.30). To verify that (3.31) is the solution of (3.25), we substitute the former in the latter.

Consider first $M \neq 1$. Then

$$y_k = M^k \left(y_0 - \frac{C}{1 - M} \right) + \frac{C}{1 - M}$$

and

$$y_{k+1} = M^{k+1} \left(y_0 - \frac{C}{1 - M} \right) + \frac{C}{1 - M}$$

Substituting these into (3.25),

$$M^{k+1} \left(y_0 - \frac{C}{1 - M} \right) + \frac{C}{1 - M} = M \cdot \left\{ M^k \left(y_0 - \frac{C}{1 - M} \right) + \frac{C}{1 - M} \right\} + C$$

or

$$M^{k+1} \left(y_0 - \frac{C}{1 - M} \right) + \frac{C}{1 - M} = M^{k+1} \left(y_0 - \frac{C}{1 - M} \right) + \frac{C}{1 - M}$$

The two sides of this equation are identical; hence the equation is satisfied for all values of k, and (3.31) is the solution of (3.25).

Next consider $M = 1$. Then (3.25) is

$$y_{k+1} = y_k + C \tag{3.32}$$

From (3.31) for $M = 1$,

$$y_k = y_0 + kC$$

and

$$y_{k+1} = y_0 + (k + 1)C$$

Using these last two equations in (3.32),

$$y_0 + (k + 1)\ C = y_0 + kC + C$$
$$= y_0 + (k + 1)C$$

Again this equation is satisfied for all values of k, so (3.31) is the solution of (3.25).

In summary, (3.31) is the desired solution for any value of M (recall that $M \neq 0$).

As an example of the use of (3.31), consider the linear, first-order difference equation

$$2y_{k+1} - 3y_k = 6 \tag{3.33}$$

where $y_0 = 2$. Dividing both sides of the difference equation by 2 and rearranging, we get

$$y_{k+1} = \frac{3}{2}y_k + 3$$

Thus $M = 3/2$ and $C = 3$. From (3.31),

$$y_k = \left(\frac{3}{2}\right)^k \left(2 - \frac{3}{1 - \frac{3}{2}}\right) + \frac{3}{1 - \frac{3}{2}}$$

or

$$y_k = 8\left(\frac{3}{2}\right)^k - 6 \tag{3.34}$$

The reader should verify, using the techniques of Section 3.2, that (3.34) is the solution of (3.33) and, moreover, that for $k = 0$, (3.34) produces the correct initial value, 2.

As another example, the same difference equation (3.33) but with the initial condition $y_0 = -2$ has the solution

$$y_k = 4\left(\frac{3}{2}\right)^k - 6$$

Notice that in general *the solution depends not only on the difference equation but on the initial condition* as well.

As one final example of the use of (3.31), we return to the learning model (1.7).

$$L_{k+1} = (1 - A)L_k + A$$

Recall that A is that fraction of the as yet unlearned material that is learned in one training session. Thus

$$0 < A < 1 \qquad (3.35)$$

Using (3.31),

$$L_k = (1 - A)^k \left(L_0 - \frac{A}{1 - {}^{\prime}(1 - A)} \right) + \frac{A}{1 - (1 - A)}$$

which can be rewritten

$$L_k = (L_0 - 1)(1 - A)^k + 1 \qquad (3.36)$$

for $k = 0, 1, 2, \ldots$. This solution also depends on both the difference equation (1.7) and the initial condition, L_0.

Notice that $1 - A$ is less than 1. Therefore $(1 - A)^k$ gets smaller and smaller as k gets larger and larger.[2] More precisely,

$$0 < (1 - A)^{k+1} < (1 - A)^k$$

If we take k to be sufficiently large, then $(1 - A)^k$ can be made as close to zero as we please (see Exercise 1.) The requirement (3.35) is sufficient to guarantee this. In any case the first term on the right in (3.36) can be made as close to zero as we like. In other words, L_k can be made as close to 1 as we wish, provided k is large enough. We call the solution for large values of k the "long-run" solution of (1.7), because it is the solution after a long time or, in other words, a large number of training sessions. In this case, the learning model, the long-run solution, is *independent* of the initial condition L_0. Thus, while the solution depends on the difference equation and the initial condition, the long-run solution depends *only on the difference equation.* This is not always so. Nevertheless, there are many interesting difference equations whose long-run solutions do not depend on the initial conditions. For these difference equations, no matter where we start, we always end in the same place. Difference equations that exhibit this behavior are called *stable*.

---

[2]For example, if $A = 3/4$, then $1 - A = 1/4$ and $(1/4)^2 = 1/16$, $(1/4)^3 = 1/64$, and so on. Each succeeding power of $1 - A$ is one fourth its predecessor.

EXERCISES FOR SECTION 3.3

1. (a) Write a BASIC program that takes as its input a value for A and another positive number named T. The program should check that $0 < A < 1$ and $T > 0$. The program should then compute the *smallest* value of k such that

$$0 < (1 - A)^k < T$$

(b) Test your program by letting $A = .9$. For this value of A and $T = 10$, the result should be $k = 0$. For $T = .01$, the program should print 3, and for $T = 1$, $k = 1$.

(c) Run the program for arbitrary values of A and quite small values of T, say $T = .00001$.

*2. (a) Use (3.31) to find the solution of

$$6S_{k+1} = 3S_k + 2$$

where $S_0 = -\frac{1}{3}$

(b) Discuss the long-run solution.

3. (a) Use (3.31) to find the solution of

$$3R_{k+1} = 6R_k - 2$$

where $R_0 = -\frac{1}{3}$.

(b) Discuss the long-run solution.

*4. (a) Use (3.31) to find the solution of

$$3R_{k+1} = 6R_k - 2$$

where $R_0 = \frac{2}{3}$.

(b) Discuss the long-run solution.

5. (a) Use (3.31) to find the solution of

$$6S_{k+1} = 3S_k + 2$$

where $S_0 = \frac{2}{3}$.

(b) Discuss the long-run solution.

6. Find the solution of the following difference equations with the initial condition as indicated.

*(a) $y_{k+1} - y_k - 2 = 0$ $y_0 = 1$

(b) $y_{k+1} - y_k + 1 = 0$ $y_0 = 2$

*(c) $y_{k+1} = y_k$ $y_0 = 2$

(d) $y_{k+1} + 2y_k = 1$ $y_0 = 0$

*(e) $y_{k+1} + y_k = 1$ $y_0 = 2$

(f) $3y_{k+1} + y_k = 3$ $y_0 = 2$

*(g) $4y_{k+1} - y_k = 2$ $y_0 = 3$

(h) $3y_{k+1} - y_k = 1$ $y_0 = 1$

*(i) $y_{k+1} - 2y_k = 1$ $y_0 = 2$
(j) $y_{k+1} - 3y_k - 2 = 0$ $y_0 = -2$
*(k) $y_{k+1} = 5y_k + 8$ $y_0 = -2$

7. (a) Write a BASIC program that reads values of M, C and y_0. The program should check that $M \neq 0$ and stop if it does. Compute $y_1, y_2, \ldots, y_{20}$ from (3.31) and print the results.
(b) Run the program for the difference equations in Exercise 6.
(c) Run the program for the difference equation in Exercise 5. Then change the initial condition to be slightly greater than ⅔ and run the program again. What do you observe?

3.4   TYPES OF SOLUTIONS

It often is helpful to be able to classify solutions of equations into different types. For example, regardless of the choices for a, b, and c, the solutions of the general quadratic equation

$$ax^2 + bx + c = 0 \qquad (3.37)$$

fall into one of three types: two solutions, one solution, or no solution. Which of these three types occurs depends on what is called the discriminant of the equation. The discriminant is

$$D = b^2 - 4ac$$

If $D > 0$, there are two solutions. If $D = 0$, there is one solution, and if $D < 0$, there are no solutions. To see that this is so, we need only recall that the solutions of (3.37) are given by

$$x = \frac{-b \pm \sqrt{b^2 - 4ac}}{2a}$$

Since our concern is not quadratic equations, we will not pursue a discussion of them any further. The important point is that we can classify the solutions of such equations, and *any* quadratic equation has a solution that is of one and only one of the three types. We know that there can never be three solutions to a quadratic equation, and that fact often is useful.

We can classify solutions of linear, first-order difference equations into ten types, and we now proceed to do so. Having done so we will then know that *any* such difference equation must have a solution that is one of these ten types. That alone is quite useful information. But there is more to be gained from our classification. By merely examining a difference equation, we will be able to decide which type of solution it has. Thus we will be able to say, for example, that

$$y_{k+1} = 3y_k + 4, \qquad y_0 = 0 \qquad (3.38)$$

grows rapidly and without bound.[3] Moreover, we can make such a statement *without* solving the equation. Such information about the behavior of a solution may be the fundamental point of interest to us—more so than the actual solution. For example, if (3.38) represented a nation's economy, we would know immediately that it was an expanding economy, without finding the solution at all.

We now proceed to examine methodically all possibilities for the difference equation (3.25); that is,

$$y_{k+1} = My_k + C \tag{3.25}$$

We divide the possible values of M into six cases.

I	$M > 1$
II	$M = 1$
III	$0 < M < 1$
IV	$-1 < M < 0$
V	$M = -1$
VI	$M < -1$

The only possibility not covered by these six cases is $M = 0$, but that is not allowed in any case if (3.25) is to be a difference equation. Each of these six cases will be subdivided into three subcases. Except for $M = 1$, the three subcases will be:

(a) $\quad y_0 = \dfrac{C}{1 - M}$

(b) $\quad y_0 > \dfrac{C}{1 - M}$

(c) $\quad y_0 < \dfrac{C}{1 - M}$

Again, this covers all possibilities for y_0. The decision to compare y_0 with $C/(1 - M)$ results from examining (3.31) and noting the term in parentheses. With six cases, each with three subcases, we have eighteen possibilities. However, there are only ten possible behaviors for the solutions, so there obviously is some overlap.

We will use identical strategies in each of the six cases, I through VI. That strategy is:

1. Consider the subcase where the solution as given in (3.31) does not involve k at all. This is accomplished by choosing y_0 so that the coefficient of M^k is zero (subcase a). Since no k appears in the solution, the solution does not change as k changes. Hence the solution is constant.

---

[3] By "without bound" we mean the values of y_k get steadily larger and larger and, in fact, eventually become as large as we like.

2. Next consider the subcase where the coefficient of M^k is positive (subcase b). Again we achieve this by a judicious choice of y_0. We then examine how M^k itself behaves as k gets successively larger. Does it steadily increase? Does it get larger then smaller then larger again? And so on. From these observations we deduce what y_k does as k increases.

3. Finally, we consider the subcase where the coefficient of M^k is negative (subcase c). Once again we examine what M^k does, as we did in step 2, and use this to deduce what y_k does.

The case where $M = 1$ (Case II) is somewhat different in that instead of choosing y_0 to obtain the three cases, we choose C.

CASE I: $M > 1$

The solution is, from (3.31),

$$y_k = M^k \left(y_0 - \frac{C}{1 - M} \right) + \frac{C}{1 - M} \tag{3.39}$$

(a) If $y_0 = C/(1 - M)$, then the term in parentheses is zero and

$$y_k = \frac{C}{1 - M}$$

The solution is a constant.

A graph of this solution is shown in Figure 3.1a. Notice that the graph has no lines or curves, merely points. The graph of the solution to a difference equation always has that quality (i.e., it consists of a set of points with no lines or curves). The reason for this is that there is no meaning attached to $y_{1/2}$ or $y_{1.23}$. We can only plot values of y for $k = 0$, $k = 1$, $k = 2$, and so on. This being so, the graph must be a set of points.

Unfortunately, graphs that are only points with no lines or curves are sometimes difficult to read or interpret. Looking at Figure 3.1a, for example, does that graph represent a set of points on a horizontal line as shown in Figure 3.1b or does it represent the type of bouncy behavior shown in Figure 3.1c? Of course, we know from our previous discussion that Figure 3.1b is the correct choice, but how is someone who has not had the benefit of this entire discussion to know? One way is to connect the points with line segments so that the general behavior pattern is obvious at a glance. For example, we could use Figure 3.1b instead of Figure 3.1a. Indeed, we will do so, but we must keep in mind that the line segments are not really a part of the graph. They are there simply to help us recognize the general behavior. The actual graph consists only of the points that are marked by heavy dots.

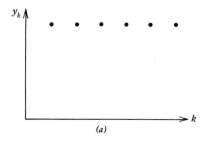

(a)

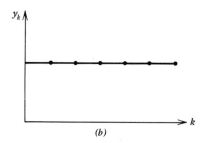

(b)

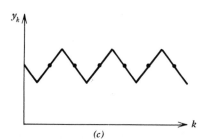

(c)

FIGURE 3.1 Graph of a set of points with two possible curves.

The graph in Figure 3.1*b* is redrawn in Figure 3.2*a* so that it appears in the same diagram with the other subcases, which are described below.

(b) If $y_0 > C/(1 - M)$, then the term in parentheses in (3.39) is positive. Now $M > 1$ so M^k gets indefinitely large as k gets large. M^k is multiplied by a positive number so the solution grows exponentially[4] without bound, as shown in Figure 3.2*b*. For one final time we call the reader's attention to the fact that only the points (heavy dots) are part of the graph. The line segments are for our convenience and to improve the visual quality of the graph.

(c) If $y_0 < C/(1 - M)$, then the term in parentheses in (3.39) is negative, so (3.39) can be written

$$y_k = -SM^k + \frac{C}{1 - M}$$

---

[4]"Exponentially" means that the rate of growth, k, appears in an exponent.

where $S > 0$. Again, M^k gets indefinitely large as k gets large, so $-SM^k$ gets very small algebraically (large in magnitude but negative). For example, if $M = 2$, then $-SM^k$ takes on the values $-2S$, $-4S$, $-8S$, $-16S$, and so on. The solution decreases exponentially without bound as shown in Figure 3.2c.

CASE II: $M = 1$

From (3.31) the solution is

$$y_k = y_0 + kC \tag{3.40}$$

We consider three subcases: $C = 0$, $C > 0$, and $C < 0$.
(a) If $C = 0$,

$$y_k = y_0$$

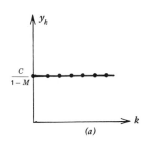

(a)

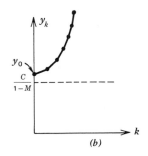

(b)

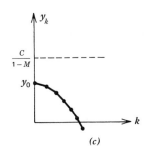

(c)   **FIGURE 3.2** Behavior of solution when $M > 1$.

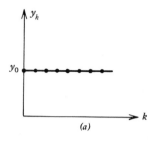

(a)

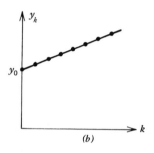

(b)

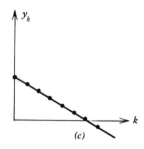

(c)   **FIGURE 3.3**   Behavior of solution when $M = 1$.

and the solution is a constant (Figure 3.3*a*).

(b) If $C > 0$, then (3.40) increases linearly without bound (see footnote 2) as k increases, as shown in Figure 3.3*b*.

(c) If $C < 0$, then (3.40) can be written

$$y_k = y_0 - Sk$$

where $S > 0$. This is a line sloping down to the right. The solution decreases linearly without bound, as shown in Figure 3.3*c*.

CASE III: $0 < M < 1$

The solution is again (3.39).

(a) If $y_0 = C/(1 - M)$, the term in parentheses in (3.39) is zero, so the solution is a constant (see Figure 3.4*a*).

(b) If $y_0 > C/(1 - M)$, the term in parentheses is positive, so

$$y_k = SM^k + \frac{C}{1 - M}$$

where $S > 0$. Now $0 < M < 1$, so $0 < M^k < 1$. Moreover, as k gets large, M^k becomes very close to zero. Thus the first term, SM^k, is always positive but becomes very small. The solution then is always greater than $C/(1 - M)$ but gets very close to $C/(1 - M)$. The solution decreases exponentially to a bound of $C/(1 - M)$, as indicated in Figure 3.4b.

(c) If $y_0 < C/(1 - M)$, the solution can be expressed as

$$y_{k+1} = -SM^k + \frac{C}{1 - M}$$

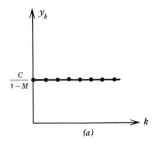

(a)

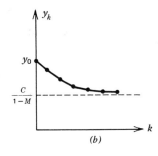

(b)

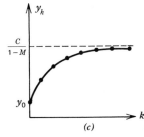

(c) **FIGURE 3.4** Behavior of solution for $0 < M < 1$.

where $S > 0$. Again, SM^k is positive for all values of k. However, SM^k is preceded by a minus sign, so the solution is always less than $C/(1 - M)$. But as k gets large, SM^k gets very small (recall that $0 < M < 1$), so the solution gets very close to $C/(1 - M)$. The solution increases [it is less than $C/(1 - M)$], but gets close to $C/(1 - M)$ exponentially to the bound $C/(1 - M)$, as shown in Figure 3.4c.

CASE IV: $-1 < M < 0$

The solution is (3.39).

(a) If $y_0 = C/(1 - M)$, the term in parentheses is zero and the solution is a constant (Figure 3.5a).

(b) If $y_0 > C/(1 - M)$,

$$y_{k+1} = SM^k + \frac{C}{1 - M}$$

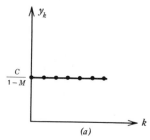

(a)

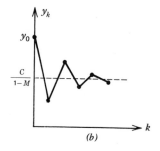

(b)

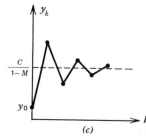

(c)

FIGURE 3.5 Behavior of solution for $-1 < M < 0$.

and $S > 0$. With each change of k by 1, the sign of SM^k changes. For example, $SM^0 > 0$, but $SM^1 < 0$. Then $SM^2 > 0$, but $SM^3 < 0$. Therefore the solution is alternately larger and smaller than $C/(1 - M)$. However, as k gets large, the magnitude of SM^k gets very small. The solution oscillates[5] on either side of $C/(1 - M)$ with decreasing amplitude,[6] as shown in Figure 3.5b.

(c) If $y_0 < C/(1 - M)$,

$$y_k = -SM^k + \frac{C}{1 - M}$$

Once more the solution oscillates about $C/(1 - M)$ with decreasing amplitude. The only difference between this subcase and subcase b above is that here the oscillations start below $C/(1 - M)$, while in subcase b, the oscillations started above $C/(1 - M)$. The graph is shown in Figure 3.5c.

CASE V: $M = -1$

The solution, from (3.31), is

$$y_k = (-1)^k \left(y_0 - \frac{C}{2} \right) + \frac{C}{2}$$

(a) If $y_0 = C/2$, the solution is a constant, as shown in Figure 3.6a.

(b) If $y_0 > C/2$ then, for $k = 1$,

$$y_1 = -y_0 + C$$

for $k = 2$,

$$y_2 = y_0$$

for $k = 3$,

$$y_3 = -y_0 + C$$

and so on. Thus the solution oscillates with constant amplitude. The oscillations are about the value $C/2$ (i.e., the distance of the peaks above $C/2$ is the same as the distance of the troughs below $C/2$). A graph of this subcase is shown in Figure 3.6b.

(c) If $y_0 < C/2$, the solution is similar to that in (b) above. That is, the solution oscillates with constant amplitude. The only difference between (b) and (c) is that in the former a peak occurs for $k = 0$, and here in (c) a trough occurs at $k = 0$. Compare the graphs in Figure 3.6b and 3.6c.

[5]Bounces up and down.
[6]Size of the bounces.

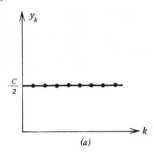

(a)

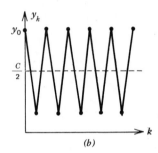

(b)

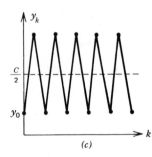

(c)

FIGURE 3.6 Behavior of solution for $M = -1$.

CASE VI: $M < -1$

The solution is again

$$y_k = M^k \left(y_0 - \frac{C}{1 - M} \right) + \frac{C}{1 - M} \tag{3.39}$$

(a) If $y_0 = C/(1 - M)$, then the term in parentheses is zero, and the solution is a constant (Figure 3.7a).

(b) If $y_0 > C/(1 - M)$, then the coefficient of M^k in (3.39) is positive. The value of M^k changes sign with each increase of 1 in k. Therefore the solution oscillates. Moreover,

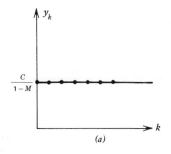

(a)

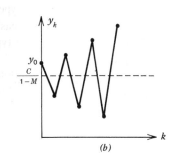

(b)

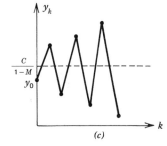

(c)

FIGURE 3.7 Behavior of solution for $M < -1$.

the magnitude of M^k gets larger and larger as k gets larger. The solution oscillates with increasing amplitude, as shown in Figure 3.7b.

(c) If $y_0 < C/(1 - M)$, then the coefficient of M^k in (3.39) is negative. The solution once again oscillates with increasing amplitude, as it did in (b); see Figure 3.7c.

With these six cases we have covered all possible values of M and, therefore, all possible linear, first-order difference equations with constant coefficients. We have found that there are only ten possible types of solutions. They are:

1. Constant.
2. Linearly increasing without bound.

3. Linearly decreasing without bound.
4. Exponentially increasing without bound.
5. Exponentially decreasing without bound.
6. Exponentially increasing to a bound.
7. Exponentially decreasing to a bound.
8. Oscillating with constant amplitude.
9. Oscillating with increasing amplitude.
10. Oscillating with decreasing amplitude.

It is not possible for the solution of a linear, first-order difference equation to behave as the graphs in either Figure 3.8a or 3.8b show, since those graphs do not fit one of these ten possibilities. The first graph, Figure 3.8a, starts increasing exponentially without bound and then switches to exponentially increasing to a bound (combination of types 4 and 6 above). The second, Figure 3.8b, begins by oscillating with decreasing amplitude and changes to oscillating with increasing amplitude (combination of types 10 and 9 above.) Such combinations cannot occur, as our discussion has shown.

We can use the results of this section to tell us the general behavior of the solution of a difference equation *without* solving the equation, if desired or necessary. Consider, for example,

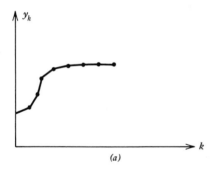

(a)

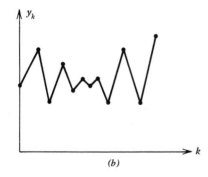

(b)

FIGURE 3.8 Two graphs showing behaviors that cannot result from first-order, linear difference equations.

$$2y_{k+1} + 4y_k - 6 = 0$$

with $y_0 = 2$. We first put this in the form of (3.25):

$$y_{k+1} = -2y_k + 3$$

so $M = -2$ and $C = 3$. We will need to compute

$$\frac{C}{1 - M} = \frac{3}{1 - (-2)} = 1$$

Now $M < -1$, so this is Case VI, and $y_0 > C/(1 - M)$, which is subcase b. Therefore the solution oscillates with increasing amplitude.

Our general plan to determine the type of solution of a difference equation is:

1. Put the equation in the form of (3.25).
2. Determine M and C.
3. Using M, determine which of the six cases covers this equation.
4. If $M = 1$, a comparison of C and y_0 determines the subcase of Case II.
5. If $M \neq 1$, compute $C/(1 - M)$ and compare this with y_0 to find the proper subcase.

EXERCISES FOR SECTION 3.4

1. *Without* solving the following difference equations, describe their behavior using one of the ten types of behavior catalogued in this section.

   *(a) $y_{k+1} = y_k + 1$ $y_0 = 0$
    (b) $y_{k+1} = y_k + 1$ $y_0 = -6$
   *(c) $2R_{k+1} = R_k + 6$ $R_0 = 0$
    (d) $2R_{k+1} = R_k + 6$ $R_0 = 6$
   *(e) $t_{k+1} + t_k = 2$ $t_0 = 1$
    (f) $t_{k+1} + t_k = 2$ $t_0 = 2$
   *(g) $2y_{k+1} - 4y_k = 0$ $y_0 = -2$
    (h) $2y_{k+1} - 4y_k = 0$ $y_0 = 0$
    (i) $3y_{k+1} + 6y_k = 0$ $y_0 = 2$
    (j) $3y_{k+1} + 6y_k = 0$ $y_0 = 0$
    (k) $6y_{k+1} + 3y_k = 9$ $y_0 = 2$
    (l) $6y_{k+1} + 3y_k = 9$ $y_0 = 1$
    (m) $6y_{k+1} + 3y_k = 9$ $y_0 = 0$

*2. Plot the solution for the population growth problem in Chapter 1 from the results shown immediately following (1.39). The difference equation (1.38) is a first-order equation. Explain why the behavior is not one of the ten catalogued in this section.

3. For the following difference equations choose a value of y_0 so that the solution is a constant.
   *(a) $y_{k+1} - 2y_k = 3$

(b) $y_{k+1} + 2y_k = 3$

*(c) $y_{k+1} = 3y_k - 2$

(d) $y_{k+1} = 3y_k + 2$

*(e) $y_{k+1} - 2y_k = 0$

(f) $y_{k+1} - 3y_k = 0$

(g) $2y_{k+1} = 4y_k + 6$

(h) $2y_{k+1} = 4y_k - 6$

(i) $3y_{k+1} + 7y_k = 20$

(j) $3y_{k+1} - 7y_k = 20$

3.5 EQUILIBRIUM VALUES AND STABILITY

We close this chapter with a discussion of two extremely important points about difference equations. We recall first that the solution of a linear, first-order difference equation will generally depend on not only the difference equation (i.e., the values of M and C), but on the initial condition as well (the value of y_0). In this section we will not concern ourselves with the entire solution, but only with the solution for very large values of k. That is, we are only interested here in what happens to the values y_k of the sequence when k is larger than some large value. This is only a portion of the solution of the difference equation. We call this portion the *long-run solution*.

Our concern is with the cases in which the long-run solution is constant. Actually, we will be content if the long-run solution is *nearly* a constant.

As a start, let us see what the long-run solution would be if it were exactly a constant. That is, we will *assume* that

$$y_{k+1} = y_k$$

for k sufficiently large and see where the assumption leads us. Introducing a bit of new notation, let us write

$$y = y_{k+1} = y_k \tag{3.41}$$

Thus y, without a subscript, will represent the long-run solution. Using (3.41) to replace both y_{k+1} and y_k in (3.25) produces

$$y = My + C \tag{3.42}$$

Two cases arise: $M = 1$ and $M \neq 1$. If $M \neq 1$,

$$y = \frac{C}{1 - M} \tag{3.43}$$

On the other hand, If $M = 1$, (3.42) is

$$y = y + C$$

This last equation has no solution unless $C = 0$. If $C = 0$, however, then there are an infinite number of solutions (i.e., *any* value of y will do).

In the previous section the possibility of a constant solution arose in each of the six possible cases for the value of M. When $M \neq 1$, a constant solution arose if

$$y_0 = \frac{C}{1 - M}$$

If $M = 1$, a constant solution arose only if $C = 0$. Comparing these results with the above results for long-run solutions, we see that they are identical. This agreement, of course, should not surprise us, because the long-run solutions we obtained in this section resulted from assuming a solution that was a constant.

A value such as y is called an *equilibrium value* of the difference equation. For $M \neq 1$ the equilibrium value is $C/(1 - M)$, whereas for $M = 1$, an equilibrium value exists only if $C = 0$ and, in that event, any value is an equilibrium value. Note that the equilibrium value depends only on the difference equation and not on the initial condition. If the solution of a first order difference equation is equal to the equilibrium value for any one value of k, then the solution remains equal to y for all succeeding values of k.

We now raise the central question of this section. Suppose that at some point the solution of a difference equation deviates from the equilibrium value. Will the solution return to the equilibrium value?

To clarify the question and to indicate why the question is of such interest to us, we recast the question in terms of a specific example. Suppose we are studying the growth of a population, as we did in Case Study 3. Suppose, also, that the population has reached a point where for all intents and purposes it is not changing. We say that the population is in equilibrium with its surroundings and that the value of the population is the population's equilibrium value. Now suppose that there is a disaster, perhaps a flood, and 10% of the population is suddenly killed. Will the population return to its original equilibrium? Will it oscillate? Will it become extinct? Will the population find a *new* equilibrium? Our intent in the remainder of this section is to provide the tools needed to answer questions like these.

Consider first $M \neq 1$. We will start counting time from this point of deviation. That is, we let $k = 0$ at a point where the solution is not equal to y. This says that

$$y_0 \neq \frac{C}{1 - M}$$

Then, from the discussion of the types of behavior in the previous section, we know that if either

$$0 < M < 1$$

or

$$-1 < M < 0$$

the solution eventually becomes very close to $C/(1 - M)$, which is the equilibrium value (see Figures 3.4 and 3.5). In all other cases, the solution gets further and further away from $C/(1 - M)$ as k increases.

Thus the long-run solution will be equal to the equilibrium value if the absolute value of M is less than 1, *regardless of the choice of* y_0. In such cases we call the equilibrium value a *stable equilibrium value*. The adjective "stable" arises from the fact that the solution will stabilize itself even if it is temporarily perturbed from its course. The difference equation is called a *stable difference equation*.

Even if the absolute value of M is not less than 1, the long-run solution may be equal to the equilibrium value. However, when the absolute value of M is not less than 1, the only way this can happen is if y_0 is chosen properly. (We are excluding $M = 1$ for the moment.) For the proper choice of y_0 $\left[\text{i.e.,} \ C/(1 - M)\right]$, the solution is a constant. In this case the equilibrium value is called an *unstable equilibrium value*. The qualifier "unstable" is used because any deviation, however slight, will prevent the solution from returning to its equilibrium value.

For $M = 1$, if $C \neq 0$, then the solution can *never* be a constant. If $C = 0$, then the solution is *always* a constant. In this latter case the equilibrium is unstable because if the solution is perturbed, it remains at its perturbed value and does not return to its original value.

In summary, a linear, first-order difference equation is stable if and only if

$$\left| M \right| < 1$$

In all other cases, the difference equation is unstable.

EXERCISES FOR SECTION 3.5

*1. Which of the thirteen difference equations in Exercise 1, Section 3.4 are stable? For those that are stable, find the equilibrium value.

2. Which of the following difference equations are stable? For those that are stable, find the equilibrium value.
    (a) $2x_{k+1} - x_k = 6$ (b) $x_{k+1} - 2x_k = 6$
    (c) $R_k + 2R_{k+1} = 4$ (d) $S_k = 3S_{k+1} + 3$

*3. Which of the eleven difference equations in Exercise 6, Section 3.3 are stable? For those that are stable, find the equilibrium value.

4. What is the equilibrium value of the learning model in (1.7)? Is it stable?

*5. What is the equilibrium value of the model of the national economy as given in (1.28)?

6. What are the equilibrium values of the population model in (1.38)?

4

PROBLEMS IN
BUSINESS SCIENCE

W e now turn to some further applications of difference equations. In Chapter 1 we discussed a learning model, a model of the national economy, and some population growth models. In this chapter we take up the computation of the interest on a savings account or a loan. We also will discuss amortization of a loan (e.g., the calculation of mortgage payments). Finally, we will take up the calculation of the ''true'' interest of an add-on or discounted loan.

4.1 INTEREST CALCULATIONS

The interest on a loan is the product of the principal, the interest rate, and the time of the loan. For example, a $1000 loan at 6% per year for one year accrues an interest of

$$I = \$1000 \times .06 \times 1 = \$60$$

Note that the interest rate is in decimal form (i.e., the percentage rate divided by 100). In general, the interest is given by

$$I = P \times R \times T \tag{4.1}$$

where P is the principal, R is the interest rate, and T is the length of the loan in time. It is essential that R and T have the same units. Thus, if the interest rate R is "per year," the time T must be measured in years as well. If R is "per month," then T must be in months. Moreover, as we noted earlier, R must be the percentage rate divided by 100 (i.e., R must be in decimal form).

Suppose we deposit $1000 in a savings account that accrues interest at 6% per year. Interest is computed quarterly (every three months). What is the balance in the account after ten years? If we think of each quarter as a period of time in the above example, we will need to compute the balance at the end of 40 periods.

Instead of looking for an answer to the specific question posed in the last paragraph, we will look at the more general problem of computing the balance in a savings account after k time periods, where k may be any nonnegative integer. Of course, if we can compute this more general balance, we can solve the specific example above by letting $k = 40$.

Based on our development of difference equations, we start by letting B_k be the balance at the end of the kth time period. A period may be three months, one year, one day, or any other fixed period of time. The balance at the end of the $(k + 1)$st period is the previous balance plus the interest. More precisely,

$$B_{k+1} = B_k + I_{k+1} \tag{4.2}$$

where I_{k+1} is the interest accumulated *during* the $(k + 1)$st period. This is very similar to a difference equation. If we had an explicit formula giving I_{k+1} in terms of B_k and/or B_{k+1}, then (4.2) would, indeed, be a difference equation for B_k. In that event we could attempt to solve this difference equation for B_k using the techniques of Chapter 3. Our next step, therefore, is to look for ways of computing I_{k+1}.

Since (4.1) is a general rule for computing interest, I_{k+1} must follow this rule. The values to be used in (4.1) for the rate, R, and the time, T, usually are straightforward. T, for example, must be the length of the time period. However, the principal, P, allows us some degree of flexibility. We could let P be the original amount of the deposit (or loan), in which case we would compute what is called simple interest. Equally well, we could let P be the current balance, in which case we would be computing compound interest. In the next two sections we take up these two cases and clarify what we mean by simple interest and compound interest. It is well to note even at this point, however, that the only difference between simple and compound interest is in the way in which P is chosen in (4.1).

EXERCISES FOR SECTION 4.1

(Note. Solutions are given for those exercises marked with an asterisk.)

*1. Using (4.1), compute the interest on $2000 for two years if the interest rate per year is
   (a) 5%

(b) 6%

(c) 7%

2. Using (4.1), compute the interest on $2500 for three years if the interest rate per year is
   (a) 5%
   (b) 6%
   (c) 7%

3. Using (4.1), compute the interest on $5000 for two years if the interest rate per year is
   (a) 4½%
   (b) 5½%
   (c) 6½%

*4. Suppose that I_{k+1} in (4.2) is given by

$$I_{k+1} = \frac{Ami}{1200}$$

where A, m, and i are positive constants. Solve the resulting difference equation if $B_0 = A$.

5. Suppose that I_{k+1} in (4.2) is given by

$$I_{k+1} = \frac{mi}{1200}B_k$$

where m and i are positive constants. Solve the resulting difference equation if $B_0 = A$ and A also is positive.

4.2 SIMPLE INTEREST

When computing simple interest, the interest in any period is based on the original deposit (or loan). Thus, if $1000 is deposited (or borrowed) at 6% per year and if the interest period is three months, the interest is

$$P \times R \times T = \$1000 \times .06 \times \frac{3}{12} = \$15$$

Notice that this is the interest in *every* period. The fact that the interest is the same in every interest period characterizes simple interest.

Suppose then that we make a deposit[1] of $A at i% per year with periods of m months duration. The interest in any period is

$$P \times P \times T = A \times \frac{i}{100} \times \frac{m}{12}$$

---

[1]Here and in what follows the word "deposit" can equally well be read "loan." Deposit implies we provide the initial money, and we collect the interest. Loan, on the other hand, implies someone else provides the money to us, and we must pay the interest. The computations are the same; only the roles of lender and borrower are interchanged.

In this last equation, since i is a percentage, we divide i by 100 so that it is in decimal form. Along these same lines, we divide m, the number of months, by 12, so that the time is given in years.

Using this for I_{k+1} in (4.2),

$$B_{k+1} = B_k + \frac{Ami}{1200} \qquad (4.3)$$

This is a linear, first-order difference equation with constant coefficients. The coefficient of B_k is $+1$. In terms of (3.25), $M = 1$. Moreover, the constant, $Ami/1200$, is positive. Hence the solution is linearly increasing without bound (see Figure 3.3b).

To solve (4.3), we need an initial condition, B_0. By definition, B_0 is the balance after zero periods, and this is the original amount of the deposit of A, so

$$B_0 = A \qquad (4.4)$$

Using (4.3) and (4.4) in (3.31),

$$B_k = A + k\frac{Ami}{1200}$$

or

$$B_k = A \left(1 + \frac{mi}{1200}k \right) \qquad (4.5)$$

Given any number of periods, we can compute the balance. For example, suppose a period is three months (i.e., $m = 3$). Suppose also that the interest rate is $5\frac{1}{2}\%$, so that $i = 5.5$. (It would be incorrect to let $i = .055$.) Suppose finally that the deposit is \$2000, or $A = 2000$. After ten years, a total of 40 periods has transpired, so

$$B_{40} = 2000 \left(1 + \frac{3 \times 5.5}{1200} \times 40 \right) = 3100$$

Of course, we could have calculated this directly from (4.3) and (4.4) by successively computing B_1, B_2, . . . , B_{39}, B_{40}. But (4.5) permits us to compute B_{40} without calculating the other 39 balances. This facility to jump over cases that may not be of interest to us makes the solution (4.5) more valuable and useful than the difference equation (4.3) itself.

EXERCISES FOR SECTION 4.2

1. Compute the simple interest after 5, 10, 15, and 20 years for the following savings deposits.
   *(a) \$1000 at 5% per year.
   (b) \$2000 at 5% per year.

*(c) $1000 at 10% per year.
 (d) $2000 at 10% per year.
*(e) $1000 at 15% per year.
 (f) $2000 at 15% per year.

*2. Suppose $1000 is deposited in a bank where simple interest is given. If after five years the balance is $1300, what is the annual interest rate?

 3. Suppose $1000 is deposited in a bank where simple interest is given. If after 18 years the balance is $1560, what is the annual interest rate?

*4. Suppose a deposit is made in a bank where simple interest is given at an annual rate of 8%. Four years later the balance is $1384. What was the amount of the original deposit?

 5. Suppose a deposit is made in a bank where simple interest is given at an annual rate of 7%. Five years later the balance is $675. What was the amount of the original deposit?

*6. A deposit of $1000 is made in a bank where simple interest is given at an annual rate of 8%. Some years later the balance is $1240. How many years was the money held in the bank?

 7. A deposit of $700 is made in a bank where simple interest is given at an annual rate of 7%. Some years later the balance is $896. How many years was the money held in the bank?

*8. Using simple interest at an annual rate of 5%, how many years elapse before the amount of the investment is at least doubled?

 9. Using simple interest at an annual rate of 8%, how many years elapse before the amount of the investment is at least doubled?

4.3 COMPOUND INTEREST

In compound interest the amount of interest in any period is based on the balance at the start of that period. (Recall that simple interest used the original deposit value to calculate the interest; i.e., simple interest used the balance at the start of the *first* period.)

For example, returning to a $1000 loan at 6% annual interest for one year, suppose the interest is compounded every six months. We first calculate the interest for one period of six months. The principal is the balance at the start of the first period (i.e., the original amount of the loan, $1000). The interest is

$$\$1000 \times .06 \times 1/2 = \$30$$

The ½ represents the time, which is six months or one-half year. The balance at the end of six months is $1000 + $30, or $1030. The interest during the next six months uses this balance as the principal. (If we were using simple interest, the principal for the calculation of the interest in the second six months would still be $1000.) Therefore the interest during the second six months of the year is

$$\$1030 \times .06 \times 1/2 = \$30.90$$

The new balance then is $1030 + $30.90, or $1060.90. Notice that this is larger (by 90¢) than the balance using simple interest for the year.

For compound interest the principal, P, to be used for the $(k + 1)$st period is the balance at the start of the $(k + 1)$st period. But the start of period $(k + 1)$ coincides with the end of period k. Thus the principal for use in computing I_{k+1} is B_k, or

$$I_{k+1} = P \times R \times T = B_k \times \frac{i}{100} \times \frac{m}{12}$$

where, again, the interest rate is $i\%$ per year, and each period is m months in length. Using this in (4.2),

$$B_{k+1} = \left(1 + \frac{mi}{1200}\right) B_k \qquad (4.6)$$

This, too, is a linear difference equation with constant coefficients, as was (4.3). The coefficient of B_k is clearly greater than 1. The constant term is zero, so if the original loan, B_0, is positive, it follows that the solution is increasing exponentially without bound (see Figure 3.2b). Therefore we should expect that the balance using compound interest will grow more rapidly than the balance using simple interest, since compound interest grows exponentially while simple interest grows linearly. Notice that we are able to make this comparison *without* solving the difference equations.

The solution of the difference equation (4.6) is

$$B_k = \left(1 + \frac{mi}{1200}\right)^k A \qquad (4.7)$$

where, again, $B_0 = A$, the original amount of the loan.

Let us look at the specific example that we examined at the close of the previous section where we were discussing simple interest. This was a deposit of $2000 at 5½% per year with an interest period of three months. This means that interest is compounded quarterly. From (4.7) the balance after 10 years ($k = 40$) is

$$B_{40} = \left(1 + \frac{3 \times 5.5}{1200}\right)^{40} \times 2000 = 3453.56$$

Notice that this balance exceeds the balance using simple interest by $353.56. This is an increase of more than 10% over the simple interest balance.

EXERCISES FOR SECTION 4.3

1. Compute the compound interest after 5, 10, 15, and 20 years for the following savings deposits.
   *(a) $1000 at 5% per year compounded annually.
   (b) $2000 at 5% per year compounded annually.
   *(c) $1000 at 10% per year compounded annually.

(d) $2000 at 10% per year compounded annually.

*(e) $1000 at 5% per year compounded semiannually.

(f) $2000 at 5% per year compounded semiannually.

*2. Suppose $1000 is deposited in a bank where interest is compounded annually. If after two years the balance is $1102.50, what is the annual interest rate?

3. Suppose $2000 is deposited in a bank where interest is compounded annually. If after two years the balance is $2242, what is the annual interest rate?

*4. Suppose a deposit is made in a bank where interest is compounded annually at 8% per year. Four years later the balance is $1496.54. What was the amount of the original deposit?

5. Suppose a deposit is made in a bank where interest is compounded annually at 7% per year. Five years later the balance is $701.28. What was the amount of the original deposit?

*6. A deposit of $1000 is made in a bank where interest is compounded annually at a rate of 8% per year. Some years later the balance is $1259.71. How many years was the money held in the bank? (Compare with the result of Exercise 6, Section 4.2.)

7. A deposit of $700 is made in a bank where interest is compounded annually at a rate of 7% per year. Some years later the balance is $917.56. How many years was the money held in the bank? (Compare with the result of Exercise 7, Section 4.2.)

*8. Using interest compounded annually at 5% per year, how many years elapse before the amount of the investment is at least doubled? (Compare with the result of Exercise 8, Section 4.2.)

9. Using interest compounded annually at 8% per year, how many years elapse before the amount of the investment is at least doubled? (Compare with the result of Exercise 9, Section 4.2.)

4.4 COMPARISON OF SIMPLE AND COMPOUND INTEREST

We already have made a comparison of simple and compound interest for one specific example. In this section we will use a computer to extend our comparison. We also will look at the case where the rate (or speed) of compounding increases. This latter case will enable us to investigate the claim made by some banking institutions that "interest is compounded instantaneously."

Consider first a numerical comparison of simple and compound interest. We will compute the balance at the end of each year for 10 years and will use an interest period of three months. Thus $m = 3$ in both (4.5) and (4.7). We allow the amount of the deposit, A, and the annual interest rate, i, to be selected by the user of the computer program.

At the end of one year, four periods will have elapsed, so $k = 4$. At the end of the second year, $k = 8$. Thus we wish to evaluate both (4.5) and (4.7) for $k = 4, 8, 12, \ldots, 36, 40$ in order to compute the balances are 1, 2, $\ldots$, 9, 10 years. A flowchart to describe the computations is shown in Figure 4.1.

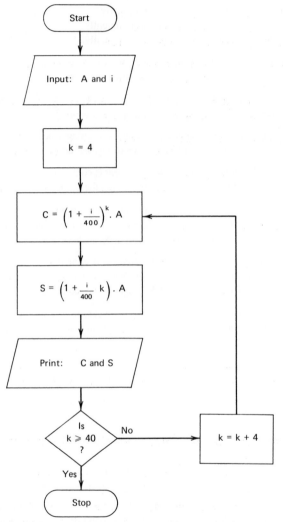

FIGURE 4.1 Flowchart for simple and compound interest.

A computer program which follows this flowchart is shown in Figure 4.2. The reader might wish to insert additional PRINT statements in the program to make the output more readable. For example,

```
15 PRINT "YEAR", "CØMPØUND", "SIMPLE"
```

will produce useful headings for the three columns of numbers.

Suppose we run the program in Figure 4.2 for the example that we already discussed in the previous two sections (i.e., $A = \$2000$, $i = 5.5\%$, $m = 3$ months). The results

```
10   INPUT A,I
20   FØR K=4 TØ 40 STEP 4
30   LET C=A*(1+I/400)↑K
40   LET S=A*(1+I*K/400)
50   PRINT K/4,C,S
60   NEXT K
70   END
```

FIGURE 4.2 BASIC program to compute simple and compound interest.

of doing so are shown in Figure 4.3. Notice that the final row agrees with our hand computations. This agreement is some reassurance (but not proof) that the computer program is correct.

We now return to the problem of increasing the rate of compounding so that the period becomes shorter and shorter. Many lending institutions compound daily, and some claim to compound instantaneously.[2]

```
?2000,5.5
  1          2112.29        2110
  2          2230.89        2220
  3          2356.14        2330
  4          2488.43        2440
  5          2628.14        2550
  6          2775.7         2660
  7          2931.54        2770
  8          3096.13        2880
  9          3269.96        2990.
 10          3453.56        3100
```

FIGURE 4.3 Simple and compound interest as computed in Figure 4.2.

We will concentrate our attention on the balance after *one year* using different rates of compounding and compare these different results. If there is one period per year then there are 12 months in the period. If there are two periods per year, there are six months in each period. For various numbers of periods in a year, the appropriate number of months per period is tabulated in the following table.

Number of Compounding Periods in One Year	Number of Months in Each Period
1	12
2	6
4	3
12	1
24	½

Notice that the product of the two corresponding entries in each row of this table is always 12. Thus, if we compound the interest n times per year and there are m months in each period,

$$mn = 12$$

[2] Ask yourself if such a claim could be justifiable. What does "instantaneously" mean in this context? Is someone or some thing actually registering the interest every moment?

For daily compounding there are 365 periods per year, so the length of each period is 12/365 months. When we say "compounding instantaneously," we mean that m becomes very, very small. Therefore n becomes large. Nevertheless, the product mn remains equal to 12 because we are concerned with the balance at the end of 12 months.

If we compound n times per year, then B_n will be the balance at the end of one year. From (4.7), with $m = 12/n$,

$$B_n = \left(1 + \frac{i}{100n}\right)^n A$$

If we double the value of A, we double B_n. Thus, if we wish to study the behavior of B_n, we might as well let $A = 1$. Then, if we wish the result for, say $A = \$1876$, we can multiply all of the results by 1876. With this in mind we will study

$$B_n = \left(1 + \frac{i}{100n}\right)^n \tag{4.8}$$

as n becomes larger and larger. That is, we will study the "long-run" behavior of (4.8), which means we will study instantaneously compounding.

```
10    INPUT I
20    FØR N=10 TØ 100 STEP 10
30    LET B=(1+I/(100*N))↑N
40    PRINT N,B
50    NEXT N
60    END
```

FIGURE 4.4 BASIC program for instantaneous compounding.

A BASIC program, which takes as input a value for i and computes and prints B_n for $n = 10, 20, \ldots , 100$, is shown in Figure 4.4. The result of running this program with an interest rate of 5.5% is shown in Figure 4.5. Notice in Figure 4.5 that the balance seems to have stabilized at something less than 1.06. If the basic annual interest rate is 5½%, no matter how fast we compound the interest, we do not seem to be able to achieve a 6% rate of interest. Thus we can say that from an investor's point of view, 6% compounded yearly is preferable to 5½% compounded instantaneously.

```
?5.5
  10              1.05638
  20              1.05644
  30              1.05649
  40              1.0565
  50              1.05651
  60              1.05652
  70              1.05653
  80              1.05653
  90              1.05652
 100              1.05652
```

FIGURE 4.5 Run of program in Figure 4.4.

An interesting result is obtained if we try $i = 100$, as seen in Figure 4.6. The balance again seems to be stabilizing, but we must be careful not to draw hasty conclusions

```
?100
   10                2•59374
   20                2•6533
   30                2•67431
   40                2•68507
   50                2•69159
   60                2•69597
   70                2•69913
   80                2•70148
   90                2•70331
  100                2•70481
```
FIGURE 4.6 Instantaneous compounding for 100% interest.

from such computer results. Suppose we allow n to become still larger. In particular, suppose we replace statement 20 in Figure 4.4 with

```
20 FØR N = 100 TØ 1000 STEP 100
```

If we run this modified BASIC program again with $i = 100$, the results are those shown in Figure 4.7. From this last figure we can see that we were correct in assuming that the results were stabilizing. However, while Figure 4.6 seemed to imply that the "long-run" solution was approaching 2.70, it now appears that 2.717 might be more appropriate. Actually, the correct result to four figures is 2.718. In any case, what we have obtained is a reasonably good approximation to the number "e," the base of the natural or Naperian logarithms. There are, as you might imagine, more efficient and more accurate ways to compute e. The value of e to six digits is 2.71828. Other than π, the number e is perhaps the most well-known and most often encountered special number in mathematics.

```
?100
  100                2•70481
  200                2•7116
  300                2•71379
  400                2•71492
  500                2•7157
  600                2•7163
  700                2•71648
  800                2•71671
  900                2•71657
 1000                2•71667
```
FIGURE 4.7 Instantaneous compounding for 100% interest with modification to program in Figure 4.4.

EXERCISES FOR SECTION 4.4

*1. Suppose a savings account is opened with a deposit of $\$A$. Interest at $i\%$ per year is computed at the end of each period of m months. At the end of each period, a deposit of $\$D$ is made. If interest is compounded each period, write a difference equation similar to (4.6) relating B_{k+1}, the balance at the end of the $(k + 1)$st period, to B_k, the balance at the end of the kth period.

2. Solve the difference equation derived in Exercise 1.

*3. (a) Write a BASIC program to compute the balance on a savings account at the end of each of the first 10 years, assuming an interest period of 3 months for the situation described in Exercise 1.

(b) Run the program in part a for an initial deposit of $2000 at 5.5% interest per year and with an additional $50 deposit each quarter. Compare your results with those in Figure 4.3, where only the initial $2000 deposit was made.

(c) What is the total amount deposited in part b?

(d) If only an initial deposit of X is made and no quarterly deposits are made, what value of X produces the same final balance after 10 years as the balance achieved in part b?

4. (a) Run the BASIC program in Figure 4.2 for an initial deposit of $2000 and an interest rate of 6.5% per year. Compare your results with those given in Figure 4.3 for an interest rate of 5.5%.

(b) Run the program in Exercise 3a for an initial deposit of $2000 at 6.5% per year and with an additional deposit of $50 each quarter. Compare your results with those in Exercise 3b.

(c) What is the total amount deposited in part b?

(d) If only an initial deposit of X is made and no quarterly deposits are made, what value of X produces the same final balance after 10 years as the balance achieved in part b?

5. Do all the parts of Exercise 4, but change the interest rate to 4.5% per year.

*6. (a) Write a BASIC program to compute the balance on a savings account at the end of each of the first 10 years, assuming an interest period of one day (daily compounding). Assume also an initial deposit of A and an interest rate of i% per year. Both A and i should appear in an INPUT statement.

(b) Run the program in part a for a deposit of $2000 and an annual interest rate of 5.5%. Compare the results with those for quarterly compounding given in Figure 4.3.

7. Run the program in Exercise 6a for a deposit of $2000 and an annual interest rate of 6.5%. Compare your results with those in Exercise 6b.

8. (a) Write a BASIC program to compute the balance at the end of every two years for the first 20 years, assuming an interest period of one day (daily compounding). Assume also an initial deposit of A and an interest rate of i% per year. Both A and i should appear in an INPUT statement. (*Hint.* See Exercise 6a.)

(b) Run the program for $A = \$2000$ and $i = 5.5\%$.

*9. Using the program in Figure 4.2 for quarterly compounding, determine the interest rate required in order to double the original investment after ten years (i.e., $B_{40} = 2A$).

10. Same as Exercise 9 except the investment is to triple in 10 years (i.e., $B_{40} = 3A$).

4.5 AMORTIZATION OF LOANS

When a loan has been repaid, the loan is said to be *amortized*. One of the most common methods of amortizing a loan is to repay the loan in equal installments. A mortgage loan on a house is an example of such an amortization scheme. Mortgage loans are generally repaid by a monthly installment and each payment (except perhaps the final payment) is in the same amount as every other payment. Each payment (or installment) includes the interest that has accrued since the most recent payment. Moreover, if the loan is to be amortized, then each payment should repay part of the principal of the loan. In this section we will study the amortization of loans through equal installments paid at equal intervals.

In a typical situation we are given (1) the amount of the loan, (2) the annual interest rate, (3) the length or duration of the loan in months or years, and (4) the time between payments. For example we might wish to amortize a loan of (1) \$22,000 at (2) 6% per year for (3) 25 years to be repaid (4) monthly. The problem is to calculate the amount of each payment so that the loan will be completely amortized in the specified time.

Before we tackle this problem, we pause to consider the way in which the interest will be computed (i.e., the compounding period). Usually interest is compounded with the same period as the period with which payments are made. Thus, if the installment payments are made monthly, then the interest is compounded monthly. We will assume that this is always the case (i.e., *the payment periods and the interest periods are identical*). While this is not always true, it is the situation in most cases and, in particular, it is true for home mortgage loans.

Consider then the balance B_{k+1} at the end of the $(k + 1)$st period. At this time a payment in the amount of \$$X$ is made where X is, at least for the moment, unknown. Thus the balance is reduced by an amount X. However, interest has accrued during the period, and this increases the balance. If the interest is at $i\%$ per year and the period is m months in length, then the interest accruing during the $(k + 1)$st period is (as in the earlier sections of this chapter)

$$P \times R \times T = B_k \times \frac{i}{100} \times \frac{m}{12} = \frac{mi}{1200} B_k$$

Notice that the value used for the principal in this computation is B_k, the balance at the start of the $(k + 1)$st period. The new balance is the old balance plus the interest less the payment, or

$$B_{k+1} = B_k + \frac{mi}{1200} B_k - X$$

which can be rewritten

$$B_{k+1} = \left(1 + \frac{mi}{1200}\right) B_k - X \qquad (4.9)$$

for $k = 0, 1, 2, \ldots$. The reader should compare this difference equation with (4.6) for compound interest. The only change is the addition of the term $-X$.

Suppose that the loan is for \$$A$. Then

$$B_0 = A \qquad (4.10)$$

is the initial condition that allows us to find the solution to (4.9). We could, of course, find a solution of (4.9) and (4.10) using the techniques of Chapter 3, and we will do so. However, before carrying out the solution, we will examine the amortization process more carefully.

We have started with a loan of $\$A$ at $i\%$ per year. Each payment period, hence each compounding period, is m months in length. For mortgages m is usually one month. A payment of $\$X$ is made at the end of every period. We now suppose that we wish to amortize the loan after n payments. That is, the balance at the end of the nth period should be zero.

$$B_n = 0 \tag{4.11}$$

We will use (4.11) to compute the value of X, the amount of each payment. Note that the duration of the loan is $m \times n$ months.

To make the discussion a little more concrete, suppose that a mortgage is taken out on a home in the amount of $\$22,000$. Then $A = 22000$. If the lending institution charges 6% per year, then $i = 6$. We will assume that the home owner makes a payment each month, so $m = 1$. If the mortgage is for 25 years, then $25 \times 12 = 300$ payments will be made, so $n = 300$. The length of the loan is $m \times n = 1 \times 300 = 300$ months.

We now proceed to calculate the value of X, which produces a balance of zero after n payments. First, we write down the solution of the difference equation (4.9) with the initial condition (4.10). The solution is

$$B_k = \left(A - \frac{1200X}{mi}\right)\left(1 + \frac{mi}{1200}\right)^k + \frac{1200X}{mi} \qquad k = 0, 1, 2, \ldots \tag{4.12}$$

Now, if the loan is to be amortized after n payments, it must be that B_n, the balance after the nth payment, is zero, as indicated in (4.11). Therefore, from (4.12) with $k = n$,

$$0 = \left(A - \frac{1200X}{mi}\right)\left(1 + \frac{mi}{1200}\right)^n + \frac{1200X}{mi}$$

The values of A, m, i, and n are all specified. The only unknown quantity in this last equation is X. Therefore we solve the equation for X and obtain

$$X = \frac{miA}{1200} \frac{\left(1 + \dfrac{mi}{1200}\right)^n}{\left[\left(1 + \dfrac{mi}{1200}\right)^n - 1\right]} \tag{4.13}$$

and from this we can calculate X. Notice that the arithmetic is not exactly trivial, since we must raise a quantity, $1 + (mi/1200)$, to the nth power, where n may be very large. For a 25-year mortgage, for example, $n = 300$. Of course, we could calculate the result using logarithms, but we will use a computer instead. A BASIC program to perform this calculation is shown in Figure 4.8.

If this program is run with $A = \$22,000$, $i = 6\%$, $m = 1$, and $n = 300$, the result should be $X = \$141.7463$. This is the monthly payment on a 25-year mortgage at 6% for $\$22,000$.

```
10    PRINT "TYPE AMØUNT ØF LØAN"
20    INPUT A
30    PRINT "TYPE ANNUAL INTEREST IN PERCENT"
40    INPUT I
50    PRINT "TYPE NØ. ØF MØNTHS PER PAYMENT"
60    INPUT M
70    PRINT "TYPE NØ. ØF PAYMENTS"
80    INPUT N
90    LET S=M*I/1200
100   LET Y=(1+S)↑N
110   LET X=(A*S*Y)/(Y-1)
120   PRINT X
999   END
```

FIGURE 4.8 BASIC program to compute mortgage payments.

Of course, no one can pay $141.7463 because it is not an even number of pennies. In general, what happens in such a case is that the mortgagee pays $141.75 or perhaps $142 each month, and there is some adjustment made in the final month to take care of the slight overpayment. The procedure is to make the payment slightly larger so that it is a precise number of cents or dollars. The "round off" function described in Chapter 2 can be used for this purpose.

EXERCISES FOR SECTION 4.5

*1. Consider a conventional loan to be amortized in equal installments every two months over a period of three years. The interest rate is 5% per year.
 (a) Compute the amount of the payment for loans of $1000, $2000, and $4000.
 (b) What conclusions can you draw from the results of part a about the change of the payment, X, as the loan value, A, increases?

2. Consider a conventional loan to be amortized in equal installments every three months over a period of four years. The rate of interest is 5% per year.
 (a) Compute the amount of the payment for loans of $500, $1000, and $2000.
 (b) What conclusions can you draw from the results of part a about the change of the payment, X, as the loan value, A, increases?

*3. Consider a conventional loan of $1000 to be amortized in equal installments every two months over a period of three years.
 (a) Compute the amount of each payment for interest rates of 5%, 10%, and 15%.
 (b) What conclusions can you draw from the results of part a about the change in the payment, X, as the interest rate, i, increases?

4. Consider a conventional loan of $500 to be amortized in equal installments every three months over four years.
 (a) Compute the amount of each payment for interest rates of 6%, 9%, and 12%.
 (b) What conclusions can you draw from the results of part a about the change in the payment, X, as the interest rate, i, increases?

*5. Consider the conventional loan described in Exercise 1.
 (a) Compute the total interest paid on loans of $1000, $2000, and $4000.
 (b) What conclusions can you draw about the way in which the total interest payments change as the amount of the loan increases?

6. Consider the loan described in Exercise 2.

   (a) Compute the total interest paid on loans of $500, $1000, and $2000.

   (b) What conclusions can you draw about the way in which the total interest payments change as the amount of the loan increases?

*7. Suppose in the amortization of a loan that the interest is compounded twice each payment period. That is, if payments are made every two months, the interest is compounded monthly. Consider a loan of A at $i\%$ per year to be repaid in n equal installments in periods of m months.

   (a) Would you expect these payments to be larger or smaller than the payment given by (4.13)? Explain.

   (b) Find a formula for the amount of each payment.

8. Same as Exercise 7 except that interest is compounded three times in each repayment period.

*9. Suppose that upon retirement a person wishes to receive X every m months for a total of n payments (a time of mn months). How much must be deposited in a savings account if the bank will guarantee $i\%$ per year on the money? $\left[Hint. \text{ Use (4.13).} \right]$ Such a repayment scheme is often called an *annuity plan*, and the amount S is called the *annuity*.

10. Suppose a person wishes to receive $250 each month for 10 years. The bank guarantees 5% per year on investments. How much must be invested? (See Exercise 9.)

*11. The *straight-line method of depreciation* results in equal depreciation in each year. Suppose a piece of equipment is purchased for A. It is to be depreciated over n years. The salvage value at the end of n years is S. The straight-line method results in $(A - S)/n$ dollars depreciation each year.

   (a) Let B_k be the undepreciated value at the end of k years, so that $B_0 = A$ and $B_n = S$. Write a difference equation relating B_{k+1} to B_k.

   (b) Solve the difference equation.

   (c) Suppose the equipment costs $6000, is depreciated over five years, and has a salvage value of $1000. Compute the depreciation in each year.

*12. The *sum-of-the-years'-digits* method of depreciation results in less depreciation in each succeeding year. The total depreciation is $A - S$ where A is the purchase price and S is the salvage value after n years. In the first year n/X of the total depreciation is depreciated where $X = 1 + 2 + \ldots + n = n(n + 1)/2$. In the second year $(n - 1)/X$ of the total is depreciated. In general, in the kth year, $(n - k + 1)/X$ of the total is depreciated.

   (a) Write a difference equation relating the balance B_{k+1} at the end of the $(k + 1)$st year to the balance B_k at the end of the kth year. (*Note.* The developments in Chapter 3 *cannot* be used to solve this difference equation.)

   (b) Suppose the equipment to be depreciated is purchased for $6000, is depreciated over five years, and has a salvage value of $1000. Compute the depreciation in each year.

*13. The *double-declining balance* method of depreciation results in $2/n$ times the remaining undepreciated balance being depreciated in each year where n is the number of years of life of the equipment. If the purchase price is A and the salvage value is S, then $(2/n) \cdot (A - S)$ is depreciated during the first year. The remaining undepreciated balance after one year is $(A - S) - (2/n)(A - S)$, so

$$\frac{2}{n}\left[(A - S) - \frac{2}{n}(A - S)\right]$$

is depreciated in the second year.

(a) Write a difference equation relating the undepreciated balance B_{k+1} at the end of the $(k + 1)$st year to the undepreciated balance B_k at the end of the kth year.

(b) Solve the difference equation.

(c) Suppose a piece of equipment is purchased for $6000, is to be depreciated over five years, and has a salvage value of $1000. Compute the depreciation schedule in each year.

14. Suppose a piece of equipment is purchased for $10,000. It is to be depreciated over six years and has a salvage value of $1000. Compute the depreciation in each year using:

(a) The straight-line method (see Exercise 11).

(b) The sum-of-the-years'-digits method (see Exercise 12).

(c) The double-declining balance method (see Exercise 13).

4.6 CASE STUDY 4: TRUE INTEREST ON A LOAN

Loans are not always amortized in the way described in the previous section. One typical situation that arises in practice is that the total interest is calculated at the time the loan is made, and this interest is added onto the loan. The borrower signs a pledge to repay the loan plus the interest in equal installments. As a specific example, suppose $1000 is borrowed for one year at a 6% interest rate. The interest is computed to be

$$I = P \times R \times T = \$1000 \times .06 \times 1 = \$60$$

Thus $60 is added onto the amount of the loan, $1000, and the borrower signs a note for $1060. Upon signing the note the borrower receives $1000. He must repay $1060.

Suppose that the borrower and the lender agree that the borrower will repay the loan in two installments of $530 each at six-month intervals. More precisely, the borrower receives $1000 at the start of the loan. Six months later he repays $530. Yet another six months later (one year after the start of the loan), the borrower repays another $530. At this point the loan is amortized.

We now pose the following question: What is the "true" rate of interest that the borrower is paying? The prevalence of such loans has given rise to so-called "truth-in-lending" laws that require the lender to quote the true interest rate regardless of the rule used to determine the payment amount. However, the laws seldom define true interest and hence allow for wide variations in interpretation.

In this case study we will be concerned with the type of loan described above where the interest is added onto the loan before the note is signed. We will define true interest to be consistent with our earlier discussions of amortization, and we will compute the true interest based on our definition.

Our concern is with loans where the following items are specified.

1. The amount received by the borrower.
2. The number of payments (all of identical amounts) needed to amortize the loan.
3. The amount of each payment.
4. The number of months in each payment period.

In the numerical example (1) the amount received was $1000, (2) the number of payments was two, (3) each payment was $530, and (4) there were six months in each payment period. We now recall the discussion of the previous section, and ask "Is there an interest rate that, if applied to a $1000 loan repaid in two installments in six-month intervals, would produce a payment of $530?" The answer is yes, and we will compute the appropriate interest rate shortly. Having done so, we will call the result the "true" interest rate.

Recall that in our earlier discussions the compounding period and the repayment period were identical. Indeed, the correspondence of these periods will be taken as our definition of the true interest rate.[3] With this as background we now proceed to calculate the true interest for the numerical example above.

We will look at the two payment periods separately. Let i represent the true interest which, of course, we do not know at the moment. At the end of the first six months, the interest accrued is

$$P \times R \times T = \$1000 \times (i/100) \times (1/2) = 5i$$

A payment of $530 is made, so the amount repaid on the principal is 530 less the interest or

$$530 - 5i \tag{4.14}$$

The unpaid balance at the end of this six-month period is

$$1000 - (530 - 5i) = 470 + 5i$$

During the second six months, interest should be calculated based on this new balance of $470 + 5i$. Therefore the interest due at the end of the second six months is

$$P \times R \times T = (470 + 5i) \times (i/100) \times (1/2)$$

The amount repaid on the principal is 530 less this last interest or

$$530 - \frac{i}{200}(470 + 5i) \tag{4.15}$$

The total repaid on the principal in both payments is the sum of (4.14) and (4.15) or

[3] As we have noted, truth-in-lending laws seldom define true interest. This definition seems appropriate, since home mortgages have traditionally used it.

$$1060 - 5i - \frac{i}{200}(470 + 5i) \qquad (4.16)$$

But, since the loan is now amortized, this last figure must equal $1000. Setting (4.16) equal to 1000 and rearranging we get

$$i^2 + 294i - 2400 = 0 \qquad (4.17)$$

This is a quadratic equation, and we could use the well-known quadratic formula to compute the values of i, which are solutions of (4.17). However, if we were to consider the case of five repayment periods, we could not find a formula for the true interest. Our main concern is with many more than five payments, so we will develop a method for finding a solution of an equation that will work not only for (4.17), but also for more general cases. This more general method will not produce a formula but, instead, will require the use of a computer and a BASIC program. We will first develop the method using (4.17).

Suppose we simply evaluate the left-hand side of (4.17) for $i = 7$. For convenience we will call the value of the left-hand side of (4.17) f—for formula! If we let $i = 7$, then $f = -193$. On the other hand, if we use $i = 8$, we arrive at $f = +16$. With this information we will attempt to sketch a graph of the formula. We can plot two points on the graph, as indicated in Figure 4.9a.

We now make an important observation. The graph of the formula f is a "smooth" curve (i.e., we could draw it without lifting our pencil from the paper). This being the case, the graph must cross the abscissa (i-axis) somewhere between $i = 7$ and $i = 8$.

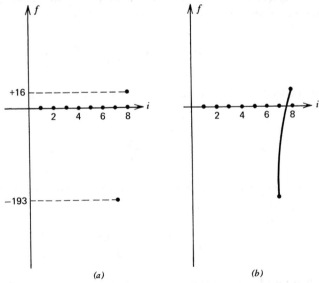

FIGURE 4.9 Graph of two points from (4.17) and one possible curve connecting them.

One possibility is shown in Figure 4.9*b*. In this case the graph crosses the abscissa precisely once between $i = 7$ and $i = 8$. In general, however, if we find a point on the graph below the abscissa and another point above the abscissa, then the graph must cross the abscissa at least once between the two points.

At the point where the graph crosses the *i*-axis, the value of *f* is zero. That is, for the value of *i* where the graph crosses the axis, (4.17) is satisfied. Hence that value of *i* is the true interest rate. From the fact that the graph crosses the abscissa between $i = 7$ and $i = 8$, we can conclude that the true interest rate is between 7 and 8.

Summarizing our discussion, if we find two values for *i*, one producing a negative value for the left-hand side of (4.17) and one producing a positive value for that left-hand side, then the true interest rate is somewhere between these two values.

With this as background we try to "guess" at a value of *i* between 7 and 8 that will be a zero of *f*. Since the function's value is closer to zero at $i = 8$ than it is at $i = 7$, it is reasonable to guess at some value near 8. We try $i = 7.9$. Then, from (4.17), $f = -14.99$. Reasoning as before, there must be a zero for some value of *i* between 7.9 and 8. We could continue to narrow down the interval that contains the zero of *f*. Since we will soon have a computer program to do that, however, we postpone any further refinements until that time. Suffice it to note that the true interest is close to 8%, while the quoted interest was 6%.

We turn now to a more general discussion of add-on loans and the true interest on such loans. We consider a loan of $\$A$ at a quoted interest rate of $q\%$ per year. The loan is to be paid back in *n* payments every *m* months. The total duration of the loan is *mn* months.

The interest is calculated as follows. Since the loan is for *mn* months, it is for $mn/12$ years. Since *q* is the annual interest in percent, the interest is

$$P \times R \times T = A \times \frac{q}{100} \times \frac{mn}{12}$$

This amount is added onto the original amount, *A*, thus the name *add-on loan*. The note is for

$$A \left(1 + \frac{mnq}{1200}\right)$$

Since there are to be *n* equal payments, each payment is the amount of the note just stated divided by *n*, or

$$X = \frac{A}{n} \left(1 + \frac{mnq}{1200}\right) \tag{4.18}$$

Now recall from (4.12) that the remaining balance on an amortized loan after *k* payment periods each of *m* months length is

$$B_k = \left(A - \frac{1200X}{mi}\right) \left(1 + \frac{mi}{1200}\right)^k + \frac{1200X}{mi}$$

provided the interest compounding periods and the repayment periods coincide. After n periods, this balance should be zero because the loan is to be completely amortized by that time. Thus

$$\left(A - \frac{1200X}{mi}\right) \left(1 + \frac{mi}{1200}\right)^n + \left(\frac{1200X}{mi}\right) = 0 \qquad (4.19)$$

In the previous section we used this last equation to calculate the payment X from the values of A, m, i, and n. With add-on loans the situation is slightly different, although (4.19) is still valid. For add-on loans we know the values of A, the amount of the loan; m, the number of months in each payment period; and n, the number of payments. In addition we know the payment, X, or at least we can calculate it from (4.18). Notice that (4.18) requires that the values of A, m, and n be given as well as the quoted interest rate, q. Thus, for add-on loans, we know all of the quantities in (4.19) except for i, the true interest rate. Therefore we will use (4.19) to compute the true interest rate, i.

It should be clear that (4.19) is a generalization of the problem of calculating the value of i that satisfies (4.17). In fact, if we use $A = 1000$, $m = 6$, $n = 2$, and $X = 530$, then (4.19) becomes (4.17).

However, (4.19) is not particularly easy to solve for i, especially in many cases of interest where n is rather large. When $n = 2$, (4.19) is a quadratic equation. When $n = 3$, it is cubic equation that involves i^3. We will use the technique of "guessing" values of i and evaluating the left-hand side of (4.19) until we arrive at a value of i that produces zero. A good first guess is something larger than q, the quoted interest rate. We will then calculate the left-hand side of (4.19). If this left-hand side is zero, we have made a very good guess. In fact, we have guessed the true interest rate exactly. More likely, the left-hand side of (4.19), when evaluated for our guess at the value of i, will be positive or negative instead of zero.

In this latter case we make a second "guess" at the true interest rate and again evaluate the left-hand side of (4.19) using this new guess. We continue to make guesses until we find two guesses that make the left-hand side of (4.19) take on opposite signs. That is, we seek two values of i such that for one of these values the left-hand side of (4.19) is negative and for the other value of i the left-hand side of (4.19) is positive. We then know that the true interest rate lies between these two values of i. The reader unsure of this last statement should reread the discussion of the solution of (4.17) and study Figure 4.9.

We will use a computer to help us with the evaluation of the left-hand side of (4.19).

A flowchart describing the evaluation process is shown in Figure 4.10. In that flowchart the amount of the payment, X, is calculated from (4.18) and rounded off to the nearest penny. The loop at the bottom of the flowchart allows us to guess at values of the true interest rate and evaluates (and prints) the left side of (4.19). If, in the

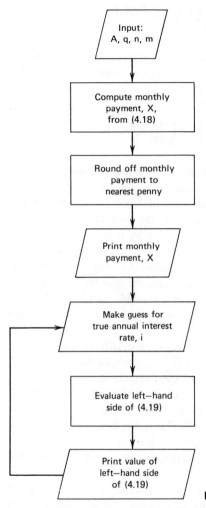

FIGURE 4.10 Flowchart for determining true interest.

process of using this flowchart, we find two values of i that are very close together (e.g., 7.89 and 7.91) and that produce left-hand sides of (4.19) that have opposite signs, then we can stop the calculations. In this example we would know that the value of i to the nearest one tenth of 1% is 7.9.

However, the flowchart, as shown in Figure 4.10, does not allow for stopping the guessing game when we are satisfied with our last guess. Hence we will have to make some changes when we write a BASIC program to compute true interest. The proper time to make a decision on whether or not to stop is immediately after printing the left side of (4.19) (i.e., bottom of the flowchart). If we are satisfied with this last printed result, we should stop. The flowchart, however, asks for a new guess. It is at this point (when the flowchart asks for a new guess) that we will stop the process.

```
10    REM  **  INPUT: A,Q,N AND M  **
20    PRINT "TYPE AMOUNT OF LOAN"
30    INPUT A
40    PRINT "TYPE QUOTED ANNUAL INTEREST IN PERCENT"
50    INPUT Q
60    PRINT "TYPE NO. OF PAYMENTS"
70    INPUT N
80    PRINT "TYPE NO. OF MONTHS IN EACH PAYMENT"
90    INPUT M
100   REM  **  COMPUTE MONTHLY PAYMENT, X, FROM (4.18)  **
110   LET X=A*(1+M*N*Q/1200)/N
120   REM  **  ROUNDOFF MONTHLY PAYMENT TO NEAREST PENNY  **
130   LET X=INT(100*X+.5)/100
140   REM  **  PRINT MONTHLY PAYMENT  **
150   PRINT "EACH PAYMENT IS";X
160   PRINT
170   REM  **  MAKE GUESS FOR TRUE ANNUAL INTEREST RATE, I  **
180   PRINT "TYPE GUESS FOR TRUE ANNUAL INTEREST RATE IN PERCENT"
190   INPUT I
200   REM  **  IF GUESS IS ZERO, STOP  **
210   IF I=0 THEN 260
220   REM  **  EVALUATE LEFT-HAND SIDE OF (4.19)  **
230   LET L=(A-1200*X/(M*I))*(1+M*I/1200)+N+1200*X/(M*I)
240   PRINT "LEFT-SIDE OF (4.19) =";L
250   GOTO 160
260   END
```

FIGURE 4.11 BASIC program to compute true interest on an add-on loan—Case Study 4.

Zero will never be a reasonable guess (unless of course the quoted interest rate, q, is zero). Therefore we arbitrarily decide that typing a zero for a guess is a signal that the computations should cease. If the guess is zero, we stop. If the guess is not zero, we proceed to calculate yet another left-hand side of (4.19) using this new guess for the value of i.

A computer program that follows Figure 4.10 and in addition stops if a guess of zero is made is shown in Figure 4.11.

```
TYPE AMOUNT OF LOAN
?1000
TYPE QUOTED ANNUAL INTEREST IN PERCENT
?6
TYPE NO. OF PAYMENTS
?12
TYPE NO. OF MONTHS IN EACH PAYMENT
?1
EACH PAYMENT IS 88.33

TYPE GUESS FOR TRUE ANNUAL INTEREST RATE IN PERCENT
?10
LEFT-SIDE OF (4.19) =-5.21484

TYPE GUESS FOR TRUE ANNUAL INTEREST RATE IN PERCENT
?11
LEFT-SIDE OF (4.19) = .650391
```

FIGURE 4.12 Case Study 4—First portion of computer results for true interest on a loan of $1000 at 6% for one year with monthly payments

The user of the program should choose the successive guesses based on the values printed for LEFT-SIDE OF (4.19). If the original quoted interest, q, is guessed as a value for i, then the left side of (4.19) will be negative. On the other hand, if a value of 100 is chosen as a guess, then the left side of (4.19) will be positive. Therefore q and 100 provide starting guesses if the user has trouble finding guesses that produce two values of the left side of (4.19) with opposite signs.

We will use this program to compute the true interest on a loan of $1000 at a quoted interest rate of 6% per year. The loan is to be repaid in one year in equal monthly installments. The input and the first portion of the results are shown in Figure 4.12. Each payment is computed to be $88.33 to the nearest penny. The first guess at the true interest is 10% and the second guess is 11%. Since the former produced a negative value for the left side of (4.19), while the latter produced a positive value, the true interest rate is between 10 and 11%. To obtain a more accurate value for the true interest rate, we continue guessing values between 10 and 11. From Figure 4.12 we can see that the value of the left side of (4.19) is much smaller in magnitude for $i = 11\%$ than it is for $i = 10\%$ (.65 versus 5.2). Therefore it is reasonable to guess that the true interest rate is closer to 11% than to 10%. Hence we guess $i = 10.8\%$ (see Figure 4.13). This produces a negative result, so the true interest rate is between 10.8 and 11%. Our next guess is 10.9% (Figure 4.13), which yields a positive result. Moreover, the value of the left side of (4.19) is smaller for 10.9% than it is for either 10.8 or 11%. Therefore the true annual interest rate to the nearest one tenth of 1% is 10.9%. If this is sufficiently accurate for our purposes we can end the calculations by guessing a value of zero.

Suppose for the sake of argument that we need the true interest rate to the nearest one hundredth of 1%. From the first two results in Figure 4.13 we can see that the true interest rate is between 10.8 and 10.9% and is likely to be closer to 10.9%. Therefore we guess at 10.88 and then 10.89. Notice that the value of the left side of (4.19) is

```
TYPE GUESS FOR TRUE ANNUAL INTEREST RATE IN PERCENT
?10.8
LEFT-SIDE OF (4.19) =-.535156

TYPE GUESS FOR TRUE ANNUAL INTEREST RATE IN PERCENT
?10.9
LEFT-SIDE OF (4.19) = 6.83594E-02

TYPE GUESS FOR TRUE ANNUAL INTEREST RATE IN PERCENT
?10.88
LEFT-SIDE OF (4.19) =-5.07812E-02

TYPE GUESS FOR TRUE ANNUAL INTEREST RATE IN PERCENT
?10.89
LEFT-SIDE OF (4.19) = 1.95312E-02

TYPE GUESS FOR TRUE ANNUAL INTEREST RATE IN PERCENT
?0
```

FIGURE 4.13 Case Study 4—final computer results for true interest on a loan of $1000 at 6% for one year with monthly payments.

smaller for 10.89% than it is for either 10.88 or 10.90%. We conclude from these results that to the nearest one hundredth of 1% the true interest rate is 10.89%. To stop the computations, we type a zero in response to a request for a new guess.

Notice that in this case the true annual interest rate is almost 11% for a yearly loan at a quoted rate of 6% repaid monthly. Previously we found that a yearly loan at 6% repaid semiannually produced a true interest rate of about 8%. Therefore it appears that it is better to keep the number of payments small. In fact, if there is one payment, then the true interest is equal to the quoted interest. Can you see why? Look back at (4.18) and (4.19) for the case $n = 1$.

EXERCISES FOR SECTION 4.6

1. (a) Using (4.18), show that for an add-on loan when $n = 1$ (one payment), the payment X is the amount of the loan plus the simple interest for m months.
   (b) Using (4.18) and (4.19), show that when $n = 1$ (one payment) the quoted interest rate, q, is the true interest rate, i.

2. Modify the flowchart in Figure 4.10 so that it agrees with the BASIC program in Figure 4.11.

*3. Use the BASIC program in Figure 4.11 for an add-on loan of $1000 for two years at a quoted annual interest rate of 6%. Compute the true interest rate if the loan is repaid in
   (a) 24 payments.
   (b) 12 payments
   (c) Based on the results of parts a and b, is it better for the repayments to be more or less frequent?

4. Use the BASIC program in Figure 4.11 for an add-on loan of $3600 for three years at a quoted annual interest rate of 7%. Compute the true interest rate if the loan is repaid in
   (a) 18 payments.
   (b) 12 payments.
   (c) Based on the results of parts a and b, is it better for the repayments to be more or less frequent?

*5. Use the BASIC program in Figure 4.11 for an add-on loan for two years at a quoted annual interest rate of 6% to be repaid in 24 payments. Compute the true interest rate if the loan is for
   (a) $1000.
   (b) $2000.
   (c) Based on the results of parts a and b, does the true interest rate increase or decrease with the amount of the loan?

6. Use the BASIC program in Figure 4.11 for an add-on loan for three years at a quoted annual interest rate of 7% to be repaid in 18 payments. Compute the true interest rate if the loan is for
   (a) $3600.
   (b) $7200.
   (c) Based on the results of parts a and b, does the true interest rate increase or decrease with the amount of the loan?

*7. The Small Loan Company has made a loan of $2000 to Mr. A at 7% (true) annual interest to be repaid monthly over two years. Mr. A's payments are $89.55 per month. After five payments have been made, the Small Loan Company approaches the Large Loan Company and asks the latter company to buy the note. If the Large Loan Company buys the latter note they will receive $89.55 per month from Mr. A for 19 months, since that is Mr. A's sole commitment. The manager of the Large Loan Company decides his company will buy the note if they can realize 9% return on their investment. How much should the Large Loan Company offer to pay the Small Loan Company for the note? (*Hint*. Use (4.19) and note that X, m, n, and i are known.)

8. Same as Exercise 7 except the Small Loan Company asks the Large Loan Company to buy Mr. A's note after the latter has made 12 payments.

9. Same as Exercise 7 except that the manager of the Large Loan Company wishes to realize a 10% return on the investment.

*10. Mr. B wishes to acquire a $20,000 mortgage on a house. He can afford to pay $120 a month. The present annual interest rate is 6%. How long (in months) a mortage does he need?

11. Same as Exercise 10 except that the present annual interest rate is 7%.

12. Same as Exercise 11 except that Mr. B can afford $150 a month.

*13. You borrow $4000 to pay for your final year at college. Of that amount the lending institution gives you $2000 on the first of July and another $2000 on the first of January to pay your tuition, dormitory fees, and so forth. You agree to repay the loan with 12 monthly payments of $355.35 each starting August 1. What is the true annual interest rate you are paying?

*14. You borrow $16,000 to pay for your four years at college. Of that amount the lending institution gives you $2000 on the first of July each of four years and another $2000 on the first of January in each of four years. You agree to repay the loan with 72 monthly payments of $263.55 starting August 1 of your first year. What is the true annual interest rate?

15. Same as Exercise 14 except you agree to repay the loan with 60 monthly payments of $300.46.

16. Same as Exercise 14 except that you receive $4000 on the first of July each of four years.

5

PROBLEMS IN BIOLOGY AND MEDICINE: Studies of Nonlinear Difference Equations

5.1 A SINGLE-SPECIES POPULATION

In Case Study 3 we discussed the growth of a single species of life in a closed environment. We observed two different types of behavior of the population. The first type of behavior was a population that became larger and larger without bound. It arose from the difference equation (1.33), which we reproduce here as (5.1).

$$N_{k+1} = (1 + A)N_k \tag{5.1}$$

Recall that N_k is the number of individuals alive at the end of the kth time period and that A is positive. The behavior of the solution of this difference equation is graphically exhibited in Figure 5.1. We note that the graph in reality is a collection of points. In Figure 5.1 we have taken the liberty of joining the points by line segments and then smoothing out the curve.

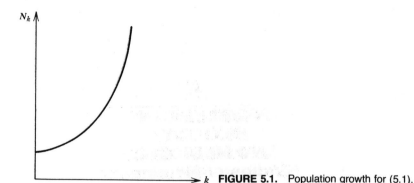

k **FIGURE 5.1.** Population growth for (5.1).

The second type of population behavior noted in Case Study 3 was one that grew rapidly at first but then tapered off to a more or less constant value. This behavior is described in Figure 5.2 Algebraically, this type of population behavior was described by the difference equation (1.38), which is repeated here as (5.2).

$$N_{k+1} = (1 + A - BN_k)N_k \tag{5.2}$$

N_k has the same interpretation as it did in (5.1), and now both A and B are positive.

Actually, the behaviors as exhibited in Figure 5.1 and 5.2 were the result of numerical experiments. All we can really say with regard to the solution of (5.2) is that for $A = .4$ and $B = .0002$, the behavior follows the sketch in Figure 5.2. We also know this to be the case for $A = .232912$ and $B = 0.000671071$ from the United States census study in Chapter 2. The question is: Will the solution to (5.1) always look like Figure 5.1, and will the solution to (5.2) always look like Figure 5.2?

The first part of this question is easily answered, since (5.1) is a linear, first-order difference equation. Its solution, from (3.31), is

$$N_k = (1 + A)^k N_0 \tag{5.3}$$

We may verify that for $A > 0$ the solution behaves as shown in Figure 5.1 by reviewing the discussion in Section 3.4.

The second part of the above question regarding (5.2) is not so easily answered. Since that difference equation is nonlinear, we cannot write down a solution similar to (5.3). Hence we turn to the computer to assist us in our analysis. We will use the computer to determine the behavior of the solution for various values of the parameters A and B. Of course, even if we find that the behavior mimics that of Figure 5.2 for all of our choices for A and B, we will not be able to answer the question posed above. However, if we can find one pair of values for A and B that produces a different type of behavior, then the answer is: "No, the solution does not always follow an S-shaped curve as sketched in Figure 5.2." Indeed, we will find values of A and B that produce a behavior different from that in Figure 5.2, so the answer to our question is "no."

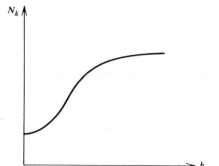

FIGURE 5.2 Population growth for (5.2).

But there is more to be gained from computer experiments than a simple "no" to our question. Clearly there are some values for A and B that do produce S-shaped curves. Our experiments will lead us to guesses for the range of choices that will produce such curves. We then will turn to a mathematical analysis that will verify that our guesses were indeed correct.

In numerical experiments it is usually wise to keep all but one of the parameters fixed and to vary that one parameter. If we vary more than one parameter, we will not know which of the parameters is causing any changes that may occur. Therefore we at first will keep B fixed at .0001 and try different values for A.

```
100    PRINT "TYPE VALUE FØR A"
200    INPUT A
300    PRINT "TYPE VALUE FØR B"
400    INPUT B
500    PRINT "TYPE INITIAL PØPULATIØN"
600    INPUT N
700    PRINT "TYPE NØ. ØF PREDICTIØNS"
800    INPUT M
900    PRINT
1000   PRINT "PERIØD","PØPULATIØN"
1100   FØR I=0 TØ M
1200   PRINT I,N
1300   LET N=(1+A-B*N)*N
1400   NEXT I
1500   END
```

FIGURE 5.3 BASIC program for population growth, (5.2).

A computer program to carry out the experiments is shown in Figure 5.3. The result of running this program with $A = .5$ and $B = .0001$ is shown in Figure 5.4. A sketch of these results would produce again an S-shaped curve such as the one shown in Figure 5.2. Next we try some larger values for A. In particular we try $A = 1.5$ and $A = 4$. The results are shown in Figure 5.5. For $A = 1.5$ (Figure 5.5a) the population goes above 15,000; then below; then above; eventually, at period 17, it settles down to precisely 15,000, where it remains. The behavior is shown in Figure 5.6a. On the other

```
TYPE VALUE FØR A
?0.5
TYPE VALUE FØR B
?0.0001
TYPE INITIAL PØPULATIØN
?1000
TYPE NØ. ØF PREDICTIØNS
?30

PERIØD              PØPULATIØN
  0                 1000
  1                 1400
  2                 1904.
  3                 2493.48
  4                 3118.47
  5                 3705.22
  6                 4184.97
  7                 4526.06
  8                 4740.57
  9                 4863.55
 10                 4929.91
 11                 4964.47
 12                 4982.11
 13                 4991.02
 14                 4995.5
 15                 4997.75
 16                 4998.87
 17                 4999.44
 18                 4999.72
 19                 4999.86
 20                 4999.93
 21                 4999.96
 22                 4999.98
 23                 4999.99
 24                 5000.
 25                 5000.
 26                 5000.
 27                 5000.
 28                 5000.
 29                 5000.
 30                 5000.
```

FIGURE 5.4   Result of running BASIC program in Figure 5.3 for small value of A, 0.5.

hand, for $A = 4$, the population rapidly rises to over 60,000 and then plummets to negative values.[1] Figure 5.6*b* describes this behavior.

Having determined that the solution does not always behave as shown in Figure 5.2, we ask some additional questions.[2]

Are there other types of behavior besides those in Figure 5.2 and 5.6?

[1] Negative values have no interpretation in our problem. However, the mathematics—as we have seen—does not prevent them from appearing. This is another case where we must use our judgment in interpreting mathematical results. One interpretation here is to set any negative values to zero automatically.

[2] In mathematics and computing the answer to one question often only serves to raise additional questions that are more penetrating than the original question and shed more light on the general problem than did the original question. Indeed, the most significant result of a mathematical analysis may be the ability to pose a more meaningful question.

```
TYPE VALUE FØR A
?1.5
TYPE VALUE FØR B
?0.0001
TYPE INITIAL PØPULATIØN
?1000
TYPE NØ. ØF PREDICTIØNS
?25
```

PERIØD	PØPULATIØN
0	1000
1	2400
2	5424.
3	10618.
4	15270.8
5	14857.3
6	15069.3
7	14964.9
8	15017.4
9	14991.2
10	15004.4
11	14997.8
12	15001.1
13	14999.5
14	15000.3
15	14999.9
16	15000.1
17	15000.
18	15000.
19	15000.
20	15000.
21	15000.
22	15000
23	15000
24	15000
25	15000

FIGURE 5.5a Result of running BASIC program in Figure 5.3 for moderate value of A, 1.5.

```
TYPE VALUE FØR A
?4.0
TYPE VALUE FØR B
?0.0001
TYPE INITIAL PØPULATIØN
?1000
TYPE NØ. ØF PREDICTIØNS
?8
```

PERIØD	PØPULATIØN
0	1000
1	4900
2	22099
3	61658.4
4	-71884.
5	-876151.
6	-8.11448E+07
7	-6.58853E+11
8	-4.34087E+19

FIGURE 5.5b Result of running BASIC program in Figure 5.3 for large value of A, 4.0.

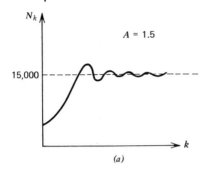

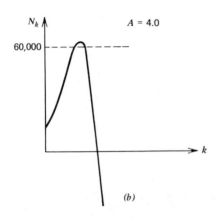

FIGURE 5.6 Graphical representation of results shown in Figure 5.5(a) and (b). $A = 1.5$, $A = 4.0$.

Is there any way to determine the behavior of the solution other than by computing numerical results?

We will not answer these questions directly. Instead we will try to phrase a more limited question or questions for which we will obtain answers. To assist us in this, we first study the results that we have achieved already.

Notice that for $B = .0001$ and $A = .5$ or $A = 1.5$ the long-run solution settles down. In the first case, $A = .5$, the population settles down to 5000, while in the latter it becomes 15,000. But what happens for other values of A? Keeping B at .0001, Exercises 4 to 8 compute the population's behavior for $A = 1.0, 1.25, 2.0, 2.5$, and 3. Solutions are given for three of those exercises.

We now can say something about the different types of behavior which are possible. The following possibilities exist.

1. The solution may grow rapidly and then slow its growth and approach a bound (e.g., $A = 0.5$, $B = .0001$).
2. The solution may grow rapidly and then approach a bound, oscillating about the bound with oscillations of decreasing amplitude (e.g., $A = 1.5$, $B = .0001$).

3. The solution may grow rapidly and then oscillate with more or less constant amplitude [e.g., $A = 2.0$, $B = .0001$ (Exercises 6 and 17)].
4. The solution may oscillate with increasing amplitude until the population becomes extinct (e.g., $A = 4.0$, $B = .0001$).
5. The solution may oscillate erratically [e.g., $A = 3$, $B = .0001$ (Exercise 8)].

Without attempting any further cataloging of the behavior of solutions, we note that behaviors 1 and 2 are of particular interest since, in both of cases, the population eventually settles down to some value. One or the other of these two types of behavior resulted for the following values of A: 0.5, 1.0, 1.25, and 1.5. For our other choices of A (i.e., 2.5, 3.0, or 4.0) the population did not settle down. For $A = 2.0$ the behavior appears to be somewhat like that in Figure 5.6a but, in fact, the population never settles down (see Exercises 17 and 18).

Even though our results were achieved for only one value of B, we can make some tentative guesses based on these behaviors. It appears from these experiments that for $A < 2$, the population settles down reasonably quickly. For $A \geq 2$, on the other hand, the population does not settle down and sometimes becomes extinct.

To test these guesses, we should try this same range of values for A but for a different choice of B. Exercises 9 to 16 ask you to do just that for $B = .0002$. These results give us greater faith in our guess that 2 is the critical value of A, but they do not *prove* that populations will settle down if $A < 2$. We could, of course, repeat the experiments for other values of B. The more numerical results we obtain, the more confidence we should have in our guesses, *provided* we do not find any values of A and B that contradict the guesses.

Even assuming that the population does settle down, there still remains the question of what value the population eventually reaches. In Section 3.5 we discussed such long-run solutions of linear difference equations. There we referred to a long-run solution that was nearly constant as the equilibrium value of the difference equation. We also discussed the conditions under which the equilibrium value was stable. We now turn to a discussion of equilibrium and stability for (5.2). We will use the results of our computer experiments as a guide.

EXERCISES FOR SECTION 5.1

(Note. Solutions are given for those exercises marked with an asterisk.)

 *1. *Without* solving (5.1) discuss why the solution should grow exponentially without bound if $A > 0$.

 2. *Without* solving (5.1) find values of A for which the solution is
   (a) Exponentially decreasing to zero.
   (b) Oscillating with decreasing amplitude to zero.
   (c) Oscillating with constant amplitude.
   (d) Oscillating with increasing amplitude.
 (Note. c and d do not produce reasonable populations—some values are negative—*but* mathematically the behaviors can and do exist.)

3. (a) Change the program in Figure 5.3 so that if the population ever becomes negative it is set to zero.
   (b) Test the program by running it for $A = 4.0$, $B = .0001$, and $N_0 = 1000$. The results should agree with Figure 5.5b through N_3 and should be zero thereafter.

*4. (a) Run the program in Figure 5.3 for $A = 1.0$ and $B = .0001$. Start with $N_0 = 1000$.
   (b) Sketch the behavior of the solution and compare your results with Figures 5.2 and 5.6.

5. Same as Exercise 4 except $A = 1.25$.

*6. Same as Exercise 4 except $A = 2.0$.

7. Same as Exercise 4 except $A = 2.5$.

*8. Same as Exercise 4 except $A = 3.0$.

9. (a) Run the program in Figure 5.3 for $A = 0.5$ and $B = .0002$. Start with $N_0 = 1000$.
   (b) Sketch the behavior of the solution and compare your results with Figures 5.2 and 5.6.

*10. Same as Exercise 9 except $A = 1.0$.

11. Same as Exercise 9 except $A = 1.25$.

*12. Same as Exercise 9 except $A = 1.5$.

13. Same as Exercise 9 except $A = 2.0$.

*14. Same as Exercise 9 except $A = 2.5$.

15. Same as Exercise 9 except $A = 3.0$.

*16. Same as Exercise 9 except $A = 4.0$.

*17. (a) Run the computer program in Figure 5.3 for $A = 2.0$, $B = .0001$, and $N_0 = 20,010$.
   (b) What would you conclude about the behavior of the population when $A = 2$?

18. Same as Exercise 17 except $B = .0001$ and $N_0 = 19,990$.

19. Suppose that instead of (5.2) the following difference equation describes the growth of a population.

$$N_{k+1} = (1 - A + BN_k - CN_k^2)N_k$$

Suppose $A = 0.2$, $B = 0.0022$, and $C = 0.000002$.
(a) Find N_1 and N_2 when $N_0 = 50$.
(b) Find N_1 and N_2 when $N_0 = 100$.
(c) Find N_1 and N_2 when $N_0 = 500$.
(d) Find N_1 and N_2 when $N_0 = 1000$.
(e) Find N_1 and N_2 when $N_0 = 1500$.

5.2 EQUILIBRIUM POPULATIONS

If the population reaches equilibrium then the population is not changing or at least is changing very little. In fact, negligible changes in the population may be taken as the definition of equilibrium. Hence, if there is an equilibrium population, say N, it must be that

$$N = N_{k+1} = N_k \tag{5.4}$$

for k very large.[3] Using (5.4) in (5.2)

$$N = (1 + A - BN)N \tag{5.5}$$

One possible solution of this algebraic equation is

$$N = 0$$

Thus, if the population is zero, it will remain at zero. However, if $N \neq 0$, then we can divide both sides of (5.5) by N and arrive at

$$N = A/B \tag{5.6}$$

For (5.2), therefore, there are two equilibrium populations 0 and A/B. Notice that if for any value of k, $N_k = A/B$, thereafter the population remains at A/B. Similarly, if $N_k = 0$ for any k, then all succeeding populations are zero.

Let us now reexamine some of our numerical results and compare them with the equilibrium populations. For $A = 0.5$ and $B = .0001$ (5.6) produces

$$N = .5/.0001 = 5000$$

From Figure 5.4 we see that in this case the population does approach 5000. For $A = 1.5$ and $B = .0001$,

$$N = 15,000$$

and Figure 5.5a shows that the population approaches this value of N, the equilibrium population.

Suppose now we return to $A = 0.5$ and $B = .0001$ so that the equilibrium population is 5000. If $N_0 = 1000$ the population reaches its equilibrium value (see Figure 5.4). Suppose, however, that we start with a different initial population, one that exceeds the equilibrium value (e.g., $N_0 = 15,000$). The results of running the program of Figure 5.3 for these values of A, B, and N_0 are shown in Figure 5.7a. Notice that the population immediately becomes extinct. But $A < 2$! Does this mean that our guess that the population settles down (reaches the equilibrium value A/B) if $A < 2$ is wrong?

Before jumping to any conclusions, we should note that this erratic behavior occurred when $N_0 > N$. Our previous numerical examples all used $N_0 < N$. Could it be that the population settles down only when the initial population is less than the equilib-

---

[3]Equation 5.4 is only approximately true, since the population may be changing slightly. However, we are assuming that any changes are negligible compared with the population N itself.

```
TYPE VALUE FØR A
?0.5
TYPE VALUE FØR B
?0.0001
TYPE INITIAL PØPULATIØN
?15000
TYPE NØ. ØF PREDICTIØNS
?10

PERIØD          PØPULATIØN
   0              15000
   1                0
   2                0
   3                0
   4                0
   5                0
   6                0
   7                0
   8                0
   9                0
  10                0
```

FIGURE 5.7a Result of running BASIC program shown in Figure 5.3. Starting population exceeds equilibrium.

```
TYPE VALUE FØR A
?8.5
TYPE VALUE FØR B
?0.0001
TYPE INITIAL PØPULATIØN
?7500
TYPE NØ. ØF PREDICTIØNS
?25

PERIØD          PØPULATIØN
   0              7500
   1              5625
   2              5273.44
   3              5129.24
   4              5062.95
   5              5031.08
   6              5015.44
   7              5007.7
   8              5003.84
   9              5001.92
  10              5000.96
  11              5000.48
  12              5000.24
  13              5000.12
  14              5000.06
  15              5000.03
  16              5000.01
  17              5000.01
  18              5000.
  19              5000.
  20              5000.
  21              5000
  22              5000
  23              5000
  24              5000
  25              5000
```

FIGURE 5.7b Result of running BASIC program shown in Figure 5.3. Starting population exceeds equilibrium.

rium population *and* $A < 2$? Let's try another value of N_0 that exceeds N. In particular we try $N_0 = 7500$. The results in Figure 5.7*b* clearly indicate that the population does settle down to the equilibrium value, A/B. So the population may reach equilibrium if it initially is greater than N. Apparently whether or not equilibrium is reached depends on N_0 in a more complicated way. For $N_0 = 15,000$ the population became extinct but, for one half that value ($N_0 = 7500$), the population reached equilibrium.

Instead of continuing to guess first this and then that, we turn to some mathematical analysis in search of an answer. Do not think that our computer experiments have been a waste of time, however. They have given us some valuable clues and, moreover, they have given us some specific cases that we can use to test the results of our analysis in the next section.

EXERCISES FOR SECTION 5.2

*1. Compute the equilibrium population from (5.6) for $A = 1.0$ and $B = .0001$. Compare this result with Exercise 4 of Section 5.1.

2. Compute equilibrium population from (5.6) for $A = 1.25$ and $B = .0001$. Compare this result with Exercise 5 of Section 5.1.

3. Compute the equilibrium population from (5.6) for $A = 1.5$ and $B = .0002$. Compare this result with Exercise 12 of Section 5.1.

4. Compute the equilibrium population from (5.6) for $A = 1.25$ and $B = .0002$. Compare this result with Exercise 11 of Section 5.1.

*5. (a) Run the program in Figure 5.3 for $A = 1.0$, $B = .0001$, and $N_0 = 25,000$.
(b) Does the population reach equilibrium?
(c) Experiment to find the largest value of N_0 so that the population does reach equilibrium?

6. Same as Exercise 5 except $A = 1.5$, $B = .0001$, and $N_0 = 40,000$.

*7. Same as Exercise 5 except $A = .5$, $B = .0002$, and $N_0 = 7500$.

8. Same as Exercise 5 except $A = 1.0$, $B = .0002$, and $N_0 = 15,000$.

9. Suppose instead of (5.2) the following difference equation describes the growth of a population.

$$N_{k+1} = (1 - A + BN_k - CN_k^2)N_k$$

Find the equilibrium population for
*(a) $A = 0.2$, $B = 0.0022$, and $C = 0.000002$.
(b) $A = 0.2$, $B = 0.0042$, and $C = 0.000004$.
(c) $A = 0.2$, $B = 0.0024$, and $C = 0.000004$.

10. Find the equilibrium values for the following difference equations.
*(a) $4y_{k+1} = y_k^2 + 3$
(b) $2y_{k+1} = y_k^2 + y_k$
(c) $2y_{k+1} = y_k^2 - 2y_k - 12$

5.3 STABILITY

The problem we have been discussing in the preceding section is the problem of *stability* (i.e., if the population does not start at its equilibrium value, under what conditions will it return to equilibrium?). We encountered this problem for linear, first-order difference equations in Section 3.5. There we found that for the difference equation

$$y_{k+1} = My_k + C$$

with $M \neq 1$ that the equilibrium value was

$$y = C/(1 - M)$$

and that this equilibrium value was stable if

$$|M| < 1$$

In Section 5.2 we found that

$$N = A/B \qquad (5.6)$$

was an equilibrium value of

$$N_{k+1} = (1 + A - BN_k)N_k \qquad (5.2)$$

if $A > 0$ and $B > 0$. It remains to be seen when (5.6) is stable (i.e., under what conditions N_k will become approximately equal to N for large values of k).

To investigate the stability of N we let

$$N_k = N + n_k \qquad (5.7)$$

where n_k is small. That is, if $N = 5000$ (as it does when $A = 0.5$ and $B = .0001$), and $N_0 = 7500$,

$$n_0 = N_0 - N = 7500 - 5000 = 2500$$

Similarly, if $N = 6000$ and $N_4 = 5500$,

$$n_4 = N_4 - N = -500$$

We have said that n_k is to be small. But small compared to what? Is -500 small? Is 7500? Later we will say precisely what we mean by small but, for now, we simply think of n_k as small compared with N. In any case, using (5.7) in (5.2),

$$N + n_{k+1} = \left[1 + A - B(N_k + n_k)\right](N + n_k)$$

Multiplying out the terms and recalling that $N = A/B$, we get

$$n_{k+1} = (1 - A)n_k - Bn_k^2 \tag{5.8}$$

On the surface, this is hardly much of an improvement over (5.2). But now we use the fact that n_k is small. In fact, we suppose that n_k is so small that we can neglect the term in n_k^2 without seriously affecting the solution of (5.8). Neglecting this term reduces (5.8) to the linear equation

$$n_{k+1} = (1 - A)n_k \tag{5.9}$$

and this is a considerable improvement, since we can analyze linear equations in considerable depth. As we noted at the start of this section, (5.9) has an equilibrium value of $n = 0$ and is stable if

$$\left|1 - A\right| < 1$$

or

$$0 < A < 2 \tag{5.10}$$

But if n_k is zero for large values of k, then from (5.7) $N_k = N$ for these same values of k. In other words, the solution of our original difference equation, (5.2), has a long-run solution of A/B if $0 < A < 2$.

To some extent this confirms the results of our numerical experiments. But recall that for $A = 0.5$, $B = .0001$, and $N_0 = 15000$ the solution was anything but stable (see Figure 5.7a). Could it be that the answer lies in clarifying what we mean by n_k being small? The answer to this last question is "yes." We need to decide how small n_k must be in order for the term Bn_k^2 in (5.8) to be negligible compared to the other terms in that equation. Indeed, the only place where the smallness of n_k was used was in neglecting the term Bn_k^2.

By neglecting Bn_k^2 we were really saying that that term is smaller, much smaller, than the other terms in the difference equation. Of course, -10 is small (algebraically) compared to $+2$, but this is not the type smallness we mean. Instead, we mean that the size or absolute value of Bn_k^2 is small. In other words,

$$\left|Bn_k^2\right| << \left|(1 - A)n_k\right|$$

where the double inequality, $<<$, is read "is much smaller than." But $B > 0$ and if $n_k \neq 0$, this last relationship implies

$$|n_k| << \frac{|1 - A|}{B} \tag{5.11}$$

We are now in a position to say precisely what we mean by n_k being small. We mean that the absolute value of n_k is much smaller than $|1 - A|/B$ [i.e., we mean (5.11)].

Our conclusion is that the solution of (5.2) is stable provided

$$0 < A < 2 \tag{5.10}$$

and

$$|n_k| << \frac{|1 - A|}{B} \tag{5.11}$$

Let us examine a few numerical examples to try to gain a better insight into the meaning of (5.11) and the meaning of "is much smaller than." First, note that for $A = 0.5$ and $B = .0001$ that (5.10) is satisfied and (5.11) becomes

$$|n_k| << 5000 \tag{5.12}$$

Consider $N_0 = 1000$. Then $n_0 = N_0 - N = -4000$ so $|n_0| = 4000$. Is this sufficiently smaller than 5000? Evidently it is, since Figure 5.4 shows that the solution is stable.

Next consider $N_0 = 7500$ so that $|n_0| = 2500$. This, too, is sufficiently less than 5000, as Figure 5.7b shows. We next return to an unstable case: $A = 0.5, B = .0001$, and $N_0 = 15000$ (see Figure 5.7a). Here

$$n_0 = N_0 - N = 15000 - 5000 = 10000$$

and now n_0 certainly violates (5.12). This resolves our dilemma about why this case was unstable even though $0 < A < 2$. Stability depends on the value of N_0 as well as on the value of A.

But we still have not completely resolved the stability problem, as the following example shows. We continue with $A = 0.5$ and $B = .0001$. We choose a value of n_0 that violates (5.12) (i.e., $n_0 = 7500$). Thus $N_0 = 12500$, and we might expect an unstable behavior. The numerical results shown in Figure 5.8 clearly show a stable behavior.

Have we contradicted our statement about stability? Not at all. The statement said that stability was guaranteed *if* both (5.10) and (5.11) were satisfied. The statement did not say that everything else was unstable.

Mathematicians would say that our statement gave *sufficient conditions* for stability. That is, conditions (5.10) and (5.11) were sufficient to assure stability. The conditions

```
TYPE VALUE FØR A
?0.5
TYPE VALUE FØR B
?0.0001
TYPE INITIAL PØPULATIØN
?12500
TYPE NØ. ØF PREDICTIØNS
?25

PERIØD          PØPULATIØN
  0              12500
  1               3125
  2               3710.94
  3               4189.3
  4               4528.93
  5               4742.27
  6               4864.49
  7               4930.41
  8               4964.72
  9               4982.24
 10               4991.09
 11               4995.54
 12               4997.77
 13               4998.88
 14               4999.44
 15               4999.72
 16               4999.86
 17               4999.93
 18               4999.96
 19               4999.98
 20               4999.99
 21               5000.
 22               5000.
 23               5000.
 24               5000.
 25               5000.
```

FIGURE 5.8 Example that violates (5.12) but is stable.

are not *necessary conditions* (i.e., they are not necessary for stability as the example $A = 0.5$, $B = .0001$, and $N_0 = 12500$ shows).

It turns out that in addition to (5.10) it is sufficient if

$$\frac{A}{B} < n_0 < \frac{1}{B} \tag{5.13}$$

or, equivalently,

$$0 < N_0 < \frac{A + 1}{B} \tag{5.14}$$

While these results can be obtained using an algebraic analysis, the argument is tedious and not particularly instructive, so we will ignore it here.

In summary, the solution of (5.2) is stable and the population approaches the value A/B provided

$$0 < A < 2$$

and

$$0 < N_0 < \frac{A + 1}{B}$$

We can interpret the last of these as saying that the value of B cannot be too large.

We close this section with two observations. First, while the stability of a linear, first-order difference equation depends only on the equation, the stability of a nonlinear equation may depend on the equation *and* the initial condition.

Second, our method of stability analysis of nonlinear equations is quite general. The procedure is to:

1. Compute an equilibrium value N by setting $N_{k+1} = N_k = N$.
2. Let $N_k = N + n_k$ where n_k is small.
3. Use the replacement described in step 2 to eliminate N_k, N_{k+1}, and so on, and produce a difference equation in n_k.
4. Discard all nonlinear terms in n_k. That is, neglect terms in n_k^2, n_k^3. and so on.
5. Examine the linear equation in n_k that results from step 4 and determine the stability conditions.
6. Look closely at the discarded nonlinear terms to decide how small n_k must be.
7. Realize that the conditions for stability are sufficient to guarantee stability, but that stability may occur even in other instances.

EXERCISES FOR SECTION 5.3

*1. Run the program in Figure 5.3 for $A = 0.5$, $B = .0001$, and the following values for N_0.
   (a) $N_0 = 1$
   (b) $N_0 = 10$
   (c) $N_0 = 100$
   (d) What can you conclude about the smallest value of N_0 for which the solution is stable?

2. Same as Exercise 1 except $A = 1.5$ and $B = .0001$.

3. Same as Exercise 1 except $A = 0.5$ and $B = .0002$.

*4. Run the computer program in Figure 5.3 for $A = 0.5$, $B = .0001$, and the following values for N_0.
   (a) 14,500
   (b) 14,900
   (c) 15,100
   (d) 15,500
   (e) What can you conclude about an upper bound for N_0 (and n_0) if the solution is to be stable? Select some additional values of N_0 if you are in doubt.

5. Same as Exercise 4 except $A = 1.5$, $B = .0001$, and the following values for N_0.
   (a) 24,500
   (b) 24,900
   (c) 25,100

(d) 25,500

(e) Same as Exercise 4e.

*6. Run the program in Figure 5.3 for $A = 2.5$, $B = .0001$, and $N_0 = 10,000$. Explain the behavior of the solution.

7. Same as Exercise 6 except $A = 3.0$, $B = .0002$, and $N_0 = 5000$.

*8. For arbitrary values of A and B in (5.2), what choice of N_0 will produce equilibrium in one time period (i.e., $N_1 = A/B$). (*Hint.* In (5.8) let $n_1 = 0$.)

9. Using an analysis similar to that in Section 5.3, determine under what conditions the equilibrium value $N = 0$ is stable.

*10. Consider the difference equation (5.2). Suppose

$$0 < N_k < A/B$$

(a) Show that if $A > 0$,

$$N_{k+1} > N_k$$

(b) Show that if $0 < A < 3$,

$$N_{k+1} < \frac{A}{B} + \frac{1}{B}$$

(c) What can you conclude about the stability of (5.2) from these results?

*11. Consider the difference equation

$$4y_{k+1} = y_k{}^2 + 3$$

(a) Find the equilibrium values.

(b) Which of the equilibrium values are stable?

12. Consider the difference equation

$$2y_{k+1} = y_k{}^2 + y_k$$

(a) Find the equilibrium values.

(b) Which of the equilibrium values are stable?

13. Consider the difference equation

$$2y_{k+1} = y_k{}^2 - 2y_k - 12$$

(a) Find the equilibrium values.

(b) Which of the equilibrium values are stable?

5.4 COMPUTING THE PARAMETERS A AND B

So far we have assumed that the values of A and B in (5.2) were available to us. We turn now to the question of how in any specific case we might compute these values from data that a biologist or ecologist might be able to supply.

One possibility is that we may be given the equilibrium population, N. For example, a biologist might be able to predict the maximum population that can be sustained in a certain forest preserve. This population is N. But this only determines the ratio A/B and not the values of A and B themselves. We need one additional piece of datum. Suppose we can obtain the growth rate that the species has in the absence of any retarding effects. This is the value of A, and we call it the *unrestricted growth rate*. Given N and A, we can easily calculate B from

$$B = A/N \tag{5.15}$$

However, in some cases it may not be reasonable to ask for estimates of the equilibrium population and the unrestricted growth rate. In such cases we might be able to obtain the population or census figures in three consecutive time periods. In particular suppose we could find N_0, N_1, and N_2. We will indicate how these figures can be used to obtain A and B.

First, we write out (5.2) for $k = 0$ and $k = 1$.

$$N_1 = (1 + A - BN_0)N_0 \tag{5.16}$$

$$N_2 = (1 + A - BN_1)N_1 \tag{5.17}$$

The only unknown quantities in these two equations are A and B. Since we have two equations and two unknowns, we should be able to find the values of the unknowns.

After some rearranging, (5.16) and (5.17) become

$$N_0A - N_0^2B = N_1 - N_0 \tag{5.18}$$

$$N_1A - N_1^2B = N_2 - N_1 \tag{5.19}$$

We multiply (5.18) by N_1^2 and obtain

$$N_1^2N_0A - N_1^2N_0^2B = N_1^2(N_1 - N_0) \tag{5.20}$$

Next we multiply (5.19) by N_0^2 to get

$$N_0^2N_1A - N_0^2N_1^2B = N_0^2(N_2 - N_1) \tag{5.21}$$

Notice that the coefficient of B in each of these last two equations is the same. Therefore, if we subtract one of the equations from the other, we will obtain an equation that does not contain B. We can then solve this resulting equation for A. To this end we subtract (5.21) from (5.20) and, with some rearranging, obtain

$$N_0 N_1 (N_1 - N_0) A = N_1^3 - N_0^2 N_2 - N_0 N_1 (N_1 - N_0)$$

Dividing both sides by $N_0 N_1 (N_1 - N_0)$:

$$A = -1 + \frac{N_1^3 - N_0^2 N_2}{N_0 N_1 (N_1 - N_0)} \qquad (5.22)$$

Our strategy was to multiply each of the two equations (5.18) and (5.19) by something that would make the coefficients of B identical. We then subtracted the two equations to get an equation that involved A but not B. Finally, we solved that equation for A. To find the value of B, we proceed in an entirely analogous way. We multiply each of (5.18) and (5.19) by quantities that make the coefficients of A identical. Thus we multiply (5.18) by N_1, and we multiply (5.19) by N_0.

$$N_1 N_0 A - N_1 N_0^2 B = N_1 (N_1 - N_0)$$
$$N_0 N_1 A - N_0 N_1^2 B = N_0 (N_2 - N_1)$$

Subtracting the second of these from the first,

$$(-N_1 N_0^2 + N_0 N_1^2) B = N_1 (N_1 - N_0) - N_0 (N_2 - N_1)$$

Thus,

$$B = \frac{N_1^2 - N_0 N_2}{N_0 N_1 (N_1 - N_0)} \qquad (5.23)$$

Equations 5.22 and 5.23 can be used to compute A and B from the values of N_0, N_1, and N_2. Throughout our development we have assumed that

$$N_0 \neq 0$$
$$N_1 \neq 0$$
$$N_0 \neq N_1 \qquad (5.24)$$

EXERCISES FOR SECTION 5.4

*1. (a) Suppose three succeeding populations are 10, 20, and 30. Compute values of A and B from (5.22) and (5.23).
   (b) Use the values of A and B to predict the next two succeeding populations.
   (c) What is the equilibrium population?

2. (a) Suppose three succeeding populations are 10, 20, and 25. Compute values of A and B from (5.22) and (5.23).
   (b) Use the values of A and B to predict the next two succeeding populations.
   (c) What is the equilibrium population?

3. (a) Write a computer program that:
   (1) Reads three population values N_0, N_1, and N_2.
   (2) Computes A and B from (5.22) and (5.23).
   (3) Computes and prints the equilibrium population.
   (4) Uses (5.2) to predict the succeeding populations N_3, N_4, ...
   (5) Prints the values of N_k together with k for $k = 0, 1, 2, . . . ,10$.
   (b) Test the program by using $N_0 = 1000$, $N_1 = 1400$, and $N_2 = 1904$. Compare your results with Figure 5.4.

*4. (a) Use the program in Exercise 3 and the United States census figures for the 48 contiguous states for 1890, 1900, and 1910 (i.e., $N_0 = 62.948$, $N_1 = 75.995$, and $N_2 = 91.972$ in millions of people). Predict the population through 1990.
   (b) Discuss your results. Do you think (5.2) is a good representation in this case? What value, if any, do you think the population will settle at?

5. Same as Exercise 4 except use the populations in 1950, 1960, and 1970 and predict through 2050 (i.e., use $N_0 = 150.697$, $N_1 = 178.464$, and $N_2 = 199.208$).

6. Use the values of $y_0, y_1,$ and y_2 to find a formula for M and C in

$$y_{k+1} = My_k + C$$

(*Note. You may assume that $y_1 \neq y_0$.*)

7. Use the values of $y_0, y_1,$ and y_2 to find a formula for a and b in

$$y_{k+1} = ay_k^2 + by_k$$

5.5 A SIMPLE MODEL OF EPIDEMICS

We now turn our attention to the spread of epidemics. We will see that the difference equations and the behavior of the solutions bear a striking resemblance to the population equations that we encountered earlier in this chapter.

First, however, we raise some questions that health authorities might ask about epidemics.

When a new infection (e.g., a new strain of an influenza virus) is introduced into a population, will the infection spread? At what rate?

How many cases of a new disease must enter a population to produce an epidemic?

If an epidemic develops how serious will it become? That is, in any given week, how many new victims will the disease claim?

When will the epidemic be at its worst?

How quickly must infected persons be detected and quarantined to prevent an epidemic from developing?

Is there an equilibrium value for the number of persons who contract the disease?

Is the equilibrium, if it exists, stable?

Are there cycles to epidemics? In other words, does the number of diseased persons rise and fall periodically?

Try to think of other questions that might be of interest to medical or health authorities.

In the remainder of this chapter and in the exercises we will answer these questions at least for some special types of epidemics. The present section is concerned with a population whose size never changes (i.e., no people leave and no new people enter). We investigate what happens when a new virus to which everyone in the population is susceptible appears. An example of when this can occur is when students at a boarding school return from the Christmas holidays and one or two students bring with them cases of a new flu virus.

Consider a population of fixed size and all of whose members are susceptible to a certain disease.[4] Some, but not all, of the population become infected with the disease.

We will assume that if an individual is susceptible and comes in contact with an infected individual, then the first individual becomes infected. Each person who is infected at the start of a period will be responsible for some number of new infections during the period. How many new infectives this individual is responsible for depends on the number of susceptible persons with whom he comes in contact. This, in turn, depends on the number of susceptibles in the population as a whole. If we let S_k be the number of susceptible (but not infected) persons at the end of the kth period, then during period $k + 1$ each infective produces CS_k new infectives, where C is the fraction of the population that comes in contact with any one individual. If there are I_k infected persons at the end of period k, then during period $k + 1$ the total number of newly infected persons is $CS_k I_k$. Thus

$$I_{k+1} - I_k = CS_k I_k \qquad (5.25)$$

where $C > 0$ and is called the *contact rate*. This parameter C depends on things such as the population density, sanitary conditions, and the like. For example, in overcrowded populations, C will be relatively large, because any one individual comes in contact with many other persons.

Because we have assumed that the entire population is susceptible to the disease, each individual is either infected or susceptible; that is,

$$S_k + I_k = N \qquad (5.26)$$

---

[4]In the exercises we will discuss cases where some of the population is immune and cases where the number of people in the population changes.

where N is the total number of individuals in the population. The value of N is fixed. Using (5.26) to eliminate S_k in (5.25), we arrive at

$$I_{k+1} = (1 + CN - CI_k)I_k \tag{5.27}$$

This last difference equation is quite similar in structure to (5.2), and we will make use of our knowledge of (5.2) to analyze epidemics. Before plunging into our analysis, however, we recapitulate the assumptions that led us to (5.27).

First, we assumed that there was a fixed number of individuals, N, in the population. Thus no new individuals entered the population nor did any leave. Second, we assumed that every individual was susceptible to the disease. Third, we assumed that in any given period of time the number of newly infected persons depended on the state of things at the beginning of the time period. Finally, we assumed that the number of newly infected persons depended in a positive way on the number of infectives and on the number of susceptibles.

Suppose that we replace CN by A and replace C by B. Then (5.27) becomes

$$I_{k+1} = (1 + A - BI_k)I_k$$

which is (5.2), except that here I_k appears in place of N_k. We can, therefore, use all of our analysis regarding the equilibrium and stability of (5.2) to draw some conclusions about our epidemic model. For example, recall that from (5.6) that one equilibrium population for (5.2) was

$$N = A/B$$

In (5.27) CN plays the role of A and C plays the role of B, so

$$I = CN/C = N$$

Our epidemic model has two equilibrium solutions, 0 or N (i.e., either no infectives or an entire population infected).

We already have noted that the contact rate, C, is positive. But the value of C cannot be chosen indiscriminately. Suppose the population consists of 100 individuals ($N = 100$) and there are ten infectives ($I_0 = 10$). If $C = 1$, then (5.27) produces $I_1 = 910$. But the total population is only 100, so there cannot be 910 infected persons. If $C = \frac{1}{2}$, then $I_1 = 230$, which is equally unacceptable. Values of C larger than 1 result in the same behavior—more infectives than individuals. On the other hand, if $C = 1/100$, then $I_1 = 19$. It appears, therefore, that there is some upper limit to the value of C. If C exceeds this upper limit our model produces the undesirable result of more infectives than there are individuals. To determine this critical value of C, we return to our discussions of (5.2) and its comparison with (5.27).

When we discussed the stability of the equilibrium population A/B, we arrived at the difference equation

$$n_{k+1} = (1 - A)n_k \tag{5.9}$$

where n_k represented the difference between the actual population N_k, and the equilibrium population, A/B. The solution of this linear difference equation is

$$n_k = (1 - A)^k n_0 \tag{5.28}$$

So if

$$\left| 1 - A \right| < 1$$

the solution approaches zero in the long run and N_k approaches A/B (i.e., the population is stable). Replacing A by CN and replacing n_k by i_k, (5.28) becomes

$$i_k = (1 - CN)^k i_0 \tag{5.29}$$

The values of i_k approach zero if

$$\left| 1 - CN \right| < 1$$

and, under these conditions, I_k approaches N. But suppose $1 - CN$ is negative; then successive values of i_k alternate in sign: positive, negative, positive, and so on. Thus I_k, which is $N + i_k$, is first larger than N, then smaller than N, and so on. But we cannot allow I_k to become larger than N. It follows that $1 - CN$ cannot be negative. Suppose then that

$$0 < 1 - CN < 1 \tag{5.30}$$

Then all values of i_k in (5.29) have the same sign. If $i_0 < 0$, all $i_k < 0$. Moreover, each succeeding value of i_k is smaller than its predecessor. In this case I_k never exceeds N.

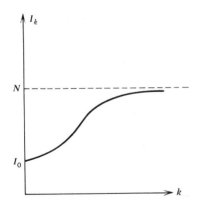

FIGURE 5.9 Typical curve for simple epidemic model.

```
100    PRINT "TYPE NØ. ØF INDIVIDUALS"
200    INPUT N
300    PRINT "TYPE CØNTACT RATE"
400    INPUT C
500    PRINT "TYPE INITIAL NØ. ØF INFECTIVES"
600    INPUT I0
700    PRINT "TYPE NØ. ØF PERIØDS TØ BE PREDICTED"
800    INPUT M
900    PRINT
1000   PRINT "PERIØD","INFECTIVES","NEW INFECTIVES"
1100   FØR J=1 TØ M
1200   LET I1=(1+C*N-C*I0)*I0
1300   PRINT J,I1,I1-I0
1400   LET I0=I1
1500   NEXT J
1600   END
```

FIGURE 5.10   BASIC program for simple epidemic model.

We can conclude that (5.30) is sufficient to ensure stability and a population that never exceeds N. The inequalities (5.30) may be rewritten

$$0 < C < \frac{1}{N} \tag{5.31}$$

```
TYPE NØ. ØF INDIVIDUALS
?1000
TYPE CØNTACT RATE
?0.0005
TYPE INITIAL NØ. ØF INFECTIVES
?10
TYPE NØ. ØF PERIØDS TØ BE PREDICTED
?20
```

PERIØD	INFECTIVES	NEW INFECTIVES
1	14.95	4.95
2	22.3132	7.36325
3	33.2209	10.9077
4	49.2796	16.0587
5	72.7051	23.4256
6	106.415	33.7095
7	153.96	47.5453
8	219.088	65.1282
9	304.632	85.5443
10	410.548	105.916
11	531.547	120.999
12	656.05	124.502
13	768.874	112.824
14	857.728	88.8535
15	918.743	61.0155
16	956.07	37.3271
17	977.07	21.0001
18	988.272	11.2019
19	994.067	5.79517
20	997.016	2.94885

FIGURE 5.11   Run of BASIC program in Figure 5.10.

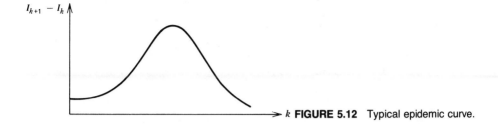

$I_{k+1} - I_k$

k **FIGURE 5.12** Typical epidemic curve.

This is a restriction that must be placed on the value of the contact rate if our model is to make sense. In any actual example the value of C may be larger than $1/N$. If this is the case, our model is not appropriate.

If we use (5.27) with the restriction (5.31) we always will get an S-shaped curve such as the one shown in Figure 5.9. This being the case, is there anything of interest which we might examine?

One piece of datum that is of interest to health authorities is the number of new infectives in any one period. This number, $I_{k+1} - I_k$, is a measure of the intensity of the epidemic. If the infectives require medication or hospitalization, then $I_{k+1} - I_k$ is the number of new patients in period $k + 1$. To examine these numbers of new infectives, we turn to the computer program in Figure 5.10.

```
TYPE NØ. ØF INDIVIDUALS
?1000
TYPE CØNTACT RATE
?0.0005
TYPE INITIAL NØ. ØF INFECTIVES
?20
TYPE NØ. ØF PERIØDS TØ BE PREDICTED
?20
```

PERIØD	INFECTIVES	NEW INFECTIVES
1	29.8	9.8
2	44.256	14.456
3	65.4047	21.1487
4	95.9681	30.5634
5	139.347	43.3791
6	199.312	59.9648
7	279.105	79.7934
8	379.708	100.603
9	497.473	117.765
10	622.47	124.997
11	739.971	117.501
12	836.178	96.207
13	904.67	68.4924
14	947.791	43.1212
15	972.533	24.7416
16	985.889	13.3564
17	992.845	6.95593
18	996.397	3.55188
19	998.192	1.79504
20	999.095	.902466

FIGURE 5.13 Second run of program in Figure 5.10.

The results of running this program for $N = 1000$ and $C = .0005$ are shown in Figure 5.11. Notice that the number of infectives does follow a curve similar to Figure 5.9 and is approaching the value of N as expected. The number of new infectives, on the other hand, rises to a peak of 124.5 in period 12 and then drops off. This behavior is shown in Figure 5.12. This curve often is referred to as the *epidemic curve*. At the peak of the epidemic curve, the epidemic is at its most intense.

If we keep all the data except I_0 fixed and rerun the program, we will obtain quite similar results. For $I_0 = 20$ the results are given in Figure 5.13. Notice that the peak occurs earlier, period 10, than in the first case, but is approximately the same height.

Our results therefore raise the following questions:

Will the epidemic curve always take on the shape shown in Figure 5.12?

If so, how does the height and location of the peak depend on the number of individuals, the contact rate, and the initial number of infectives?

We will investigate these questions and obtain partial answers to them in Case Study 5. We close this section by noting that the model presented by (5.27) and (5.31) can be enlarged to allow for the detection and removal of infectives, for addition of new individuals into the population, and so forth. Some of these extensions are discussed in Exercises 4 and 12 at the close of this section.

EXERCISES FOR SECTION 5.5

*1. If $N = 100$ and $I_0 = 10$, what value of C produces a completely infected population in one time period (i.e., $I_1 = 100$)?

2. If $N = 100$ and $I_0 = 99$, what value of C produces a completely infected population in one time period (i.e., $I_1 = 100$)?

3. Alter the program in Figure 5.10 so that when N and C have been typed, the program checks to verify that $C < 1/N$. If this is not true print an error message and ask for the two values to be retyped.

*4. Suppose that during any given time period a certain fraction, H, of those individuals infected at the period's start are detected and removed from the general population. Start with (5.25) and (5.26). Let R_k be the number of removals in the kth time period. Write a difference equation for the change in the number of removals during the $(k + 1)$st period. Alter (5.25) and (5.26) to take into account removals.

*5. Write a computer program that takes as input C, the contact rate; H, the removal rate (see Exercise 4); N, the initial population; I_0, the initial number of infectives; and uses the three difference equations to compute the number of infectives, susceptibles, and removals in succeeding periods. Assume the number of removals is initially zero.

6. Run the program in Exercise 5 with $C = .0005$, $H = .3$, and $N = 1000$ with (a) $I_0 = 10$, and (b) $I_0 = 20$.

7. Change the program in Exercise 5 so that only every tenth period is printed.

*8. Run the program in Exercise 7 for the data given in Exercise 6.

*9. Change the program in Exercise 5 so that it prints only the maximum number of infectives and the period in which the maximum occurs.

10. Run the program in Exercise 9 for $C = .0005$, $H = .3$, and $N = 1000$ with (a) $I_0 = 10$, (b) $I_0 = 20$, (c) $I_0 = 40$.

11. Show, for the equations in Exercise 4, that no epidemic develops if $S_0 \leq H/C$.

*12. Suppose, in addition to the detection and removal of some portion of the infectives as described in Exercise 4, that in each period a fixed number, M, of new individuals enter the population. You may think of these new additions as births or immigrants. Assume that all new individuals are susceptible. Write three difference equations that describe the number of infectives, susceptibles, and removals in period $k + 1$ in terms of the numbers of these types of individuals in period k.

13. Show that if the number of infectives reaches M/H and the number of susceptibles reaches H/C, the model in Exercise 12 is in equilibrium.

5.6 CASE STUDY 5: THE HEIGHT OF AN EPIDEMIC

Epidemics that obey the mathematical relationships given in (5.27) and (5.31) will follow an S-shaped curve such as the one shown in Figure 5.9. Moreover, the number of new infectives in any period will produce an epidemic curve similar to Figure 5.12. In this case study we ask: is there any way we can predict the intensity of the epidemic? In other words, can we derive a formula for the height of the epidemic curve?

To answer this question, we construct another computer program. First, however, we construct a flowchart. We will build up the flowchart in stages from a crude one to a detailed one. The translation of the last of these to a BASIC program will be straightforward and virtually mechanical. The first flowchart is shown in Figure 5.14

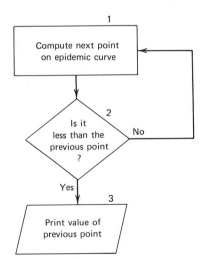

FIGURE 5.14 Crude flowchart outlining strategy for finding peak of epidemic curve.

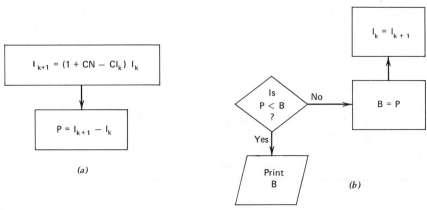

FIGURE 5.15. Expansions of boxes in flowchart of Figure 5.14.

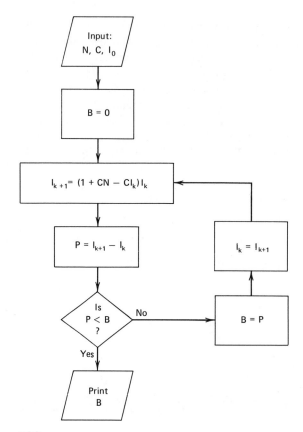

FIGURE 5.16 Final flowchart for computing peak of epidemic curve.

and describes our overall strategy. As we proceed, we compute successive points on the epidemic curve (i.e., we compute $I_{k+1} - I_k$) (Box 1). Since the curve rises to a peak and then drops, as soon as the point computed drops below its predecessor we can stop computing (Box 2). When we stop, we print the value of the predecessor, which must be the highest point on the epidemic curve (Box 3). We now proceed to expand each of these three boxes.

Figure 5.15a is an expansion of Box 1. Notice that P represents the current height of the epidemic curve. Figure 5.15b is an expansion of Boxes 2 and 3. In this latter figure B represents the height of the previous point on the epidemic curve. It is assumed that at the start B is zero, since the epidemic curve starts at zero. We will attend to that point in a moment. First we note that if the present point, P, exceeds its predecessor, B, then we make P the predecessor (i.e., let $B = P$), so that B will still be the predecessor when we return to compute a new point. Similarly, in this event, we replace I_k by I_{k+1} so that when we return to the first box in Figure 5.15a, we will use the latest value for the number of infectives in the computation.

We now piece together Figure 5.15a and 5.15b and, at the same time, attend to starting B at zero and to providing values for N, C, and I_0 (See Figure 5.16).

A BASIC program is easily constructed from this flowchart and is shown in Figure 5.17. The only point to note is that I_k in the flowchart is represented by IO in the BASIC program, and I_{k+1} is represented by I1.

```
100   REM  **  INPUT N, C, IO  **
200   PRINT "TYPE NØ. ØF INDIVIDUALS"
300   INPUT N
400   PRINT "TYPE CØNTACT RATE"
500   INPUT C
600   PRINT "TYPE INITIAL NØ. ØF INFECTIVES"
700   INPUT IO
800   PRINT
900   LET B=0
1000  REM  **  CØMPUTE NEW NØ. ØF INFECTIVES  **
1100  LET I1=(1+C*N-C*IO)*IO
1200  REM  **  CØMPUTE NEW PØINT ØN EPIDEMIC CURVE  **
1300  LET P=I1-IO
1400  REM  **  IF EPIDEMIC CURVE TURNS DØWNWARD, STØP AND
1500  REM       PRINT PREVIØUS PØINT                        **
1600  IF P<B THEN 2200
1700  REM  **  USE CURRENT PØINT ØN CURVE AS PREDECESSØR  **
1800  LET B=P
1900  LET IO=I1
2000  REM  **  RETURN TØ CØMPUTE NEXT PØINT ØN EPIDEMIC CURVE  **
2100  GØTØ 1100
2200  PRINT "PEAK ØF EPIDEMIC CURVE IS";B
2300  END
```

FIGURE 5.17 BASIC program for finding height of epidemic curve.

How will we use this program to obtain a formula for B, the maximum number of infectives in any period? There are only three things on which B could depend: I_0, N, and C. Our strategy is to keep two of these three fixed and to change the third one. We

```
TYPE NØ. ØF INDIVIDUALS
?1000
TYPE CØNTACT RATE
?0.0005
TYPE INITIAL NØ. ØF INFECTIVES
?10

PEAK ØF EPIDEMIC CURVE IS 124.502

DØNE
RUN
FIG517

TYPE NØ. ØF INDIVIDUALS
?1000
TYPE CØNTACT RATE
?0.0005
TYPE INITIAL NØ. ØF INFECTIVES
?20

PEAK ØF EPIDEMIC CURVE IS 124.997

DØNE
RUN
FIG517

TYPE NØ. ØF INDIVIDUALS
?1000
TYPE CØNTACT RATE
?0.0005
TYPE INITIAL NØ. ØF INFECTIVES
?40

PEAK ØF EPIDEMIC CURVE IS 124.424
```

FIGURE 5.18 Height of epidemic curve as initial number of infectives changes.

will then observe what happens to B. For example, if we double C, we also double B, then we might suspect that B is something multiplied by C. Similarly, if when we double N we quadruple B, then B is not something multiplied by N, but perhaps B depends on N^2. It is important, however, that we only change one quantity at a time. Moreover, we should make our changes systematically so that we can draw some conclusions later. For example, we should double or halve the quantity that is being varied.

In Figure 5.18 we see the results of keeping C and N fixed but changing I_0 from 10 to 20 to 40. That is, we doubled I_0 twice. The value of B remains essentially constant in all three cases. Hence we conclude that B does not depend on I_0 at all. Of course, we have not proved that this is so, but we take it to be quite likely that changes in I_0 do not affect B.

We next fix both I_0 and N but allow C, the contact rate, to vary. Letting $N = 1000$ and $I_0 = 10$ and letting C be successively .0005, .00025, and .000125, we obtain Figure 5.19.

We have halved C twice. Each time we halved C, we also halved B. Therefore B seems to be something multiplied by C. The something may involve N, since N is

```
TYPE NØ. ØF INDIVIDUALS
?1000
TYPE CØNTACT RATE
?0.0005
TYPE INITIAL NØ. ØF INFECTIVES
?10

PEAK ØF EPIDEMIC CURVE IS 124.502

DØNE
RUN
FIG517

TYPE NØ. ØF INDIVIDUALS
?1000
TYPE CØNTACT RATE
?0.00025
TYPE INITIAL NØ. ØF INFECTIVES
?10

PEAK ØF EPIDEMIC CURVE IS 62.4988

DØNE
RUN
FIG517

TYPE NØ. ØF INDIVIDUALS
?1000
TYPE CØNTACT RATE
?0.000125
TYPE INITIAL NØ. ØF INFECTIVES
?10

PEAK ØF EPIDEMIC CURVE IS 31.2343
```

FIGURE 5.19 Height of epidemic curve as contact rate changes.

constant throughout the three cases. We can express the results displayed in Figure 5.19 by writing

$$B = 250,000 \times C \tag{5.32}$$

For example, when $C = .0005$ (the first case in Figure 5.19), the value produced by (5.32) is 125. This is quite close to the computer result.

If we change our choice of N, the constant 250,000 might change. To guard against this we replace (5.32) by

$$B = S \times C \tag{5.33}$$

where S itself may depend upon the value of N. For $N = 1000$ it must be that $S = 250,000$.

Finally, we keep C and I_0 fixed at .0005 and 10, respectively, and successively double N twice. The results are shown in Figure 5.20. Notice that doubling N results in multiplying B by 4. Thus B certainly is not proportional to N. To see what the relationship between B and N might be, we digress to look at a simpler problem.

```
TYPE NØ. ØF INDIVIDUALS
?1000
TYPE CØNTACT RATE
?0.0005
TYPE INITIAL NØ. ØF INFECTIVES
?10

PEAK ØF EPIDEMIC CURVE IS 124.502

DØNE
RUN
FIG517

TYPE NØ. ØF INDIVIDUALS
?2000
TYPE CØNTACT RATE
?0.0005
TYPE INITIAL NØ. ØF INFECTIVES
?10

PEAK ØF EPIDEMIC CURVE IS 498.601

DØNE
RUN
FIG517

TYPE NØ. ØF INDIVIDUALS
?4000
TYPE CØNTACT RATE
?0.0005
TYPE INITIAL NØ. ØF INFECTIVES
?10

PEAK ØF EPIDEMIC CURVE IS 1999.9
```

FIGURE 5.20 Height of epidemic curve as total number of individuals changes.

Suppose that either by using a computer or by some other means we develop the following table of numbers.

X	F
1	3
2	12
4	48

How does F depend on X? Can we find a formula relating F and X? Notice that doubling X results in multiplication of F by 4. We rewrite the F column as follows.

X	F
1	3×1
2	3×4
4	3×16

and note that the terms that multiply 3 are the squares of the corresponding values of X. Thus we might guess that F was 3 times X^2.

$$F = 3X^2$$

We now return to the results in Figure 5.20 and express them in tabular form.

N	B
1000	125
2000	500
4000	2000

We have altered the values of B slightly so that they are each precisely four times their predecessor. We rewrite this as

N	B
1000	$(.000125) \times (1000)^2$
2000	$(.000125) \times (2000)^2$
4000	$(.000125) \times (4000)^2$

On the basis of this observation, we guess that

$$B = .000125 \times N^2 \qquad (5.34)$$

Once again a change in the value of C may change the value of the constant $.000125$, so we replace (5.34) with

$$B = R \times N^2 \qquad (5.35)$$

Let us now look at (5.33) and (5.35). These two formulas tell us that B, the height of the epidemic, depends on the values of both C and N. They also tell us the nature of this dependence. We can summarize and combine both (5.33) and (5.35) in

$$B = ACN^2 \qquad (5.36)$$

where A depends on neither C nor N. Since B does not depend on the value of I_0 and there are no other quantities that could affect B, it must be that A is a constant (i.e., a number). To determine the value of the number A, we look at any one specific case. For example, from the first case in Figure 5.18, the value of B is 125 (rounded off), while $N = 1000$ and $C = .0005$. Using these in (5.36),

$$125 = A(.0005)(1000)^2$$

so

$$A = 1/4$$

Thus

$$B = CN^2/4 \tag{5.37}$$

The formula we have been seeking is given by (5.37).

We now can predict the maximum intensity of an epidemic given values of C and N simply by using (5.37). We emphasize that we have not proved that (5.37) is valid, but our numerical results have led us to believe that it will produce the approximate epidemic height.

We close by recalling for the reader that (5.37), indeed all of the last two sections, are based on some specific assumptions about the way in which epidemics behave: a fixed population all of whose members are susceptible, the newly infected in a time period depends only on the state of the individuals at the period's beginning, and the number of new infectives is proportional to the number of presently infected and to the number of susceptibles. We cannot conclude that (5.37) is valid if any of these assumptions is violated. Moreover, and even more significant, if (5.37) is not true for some epidemic, then at least one of the above assumptions must be false.

EXERCISES FOR SECTION 5.6

1. (a) Run the program in Figure 5.17 for $N = 2000, C = .00025$, and $I_0 = 10, I_0 = 20$, and $I_0 = 40$.
   (b) Is the maximum of the epidemic curve dependent on I_0?

*2. (a) Run the program in Figure 5.17 for $N = 1000, I_0 = 20$, and $C = .0005, C = .00025$, and $C = .000125$.
   (b) Does the relationship (5.33) hold?

3. (a) Run the program in Figure 5.17 for $C = .00025, I_0 = 20$, and $N = 500, N = 1000$, and $N = 2000$.
   (b) Does the relationship (5.35) hold?

*4. Given the following table, guess at a formula relating x and F.

x	F
2	8
4	64
8	512

5. Given the following table, guess at a formula relating t and Q.

t	Q
1	−5
2	−20
4	−80

6. It is not always possible to guess at a relationship between two quantities by simply doubling one of them. Consider the following table.

w	f
1/4	3/16
1/2	1/4
1	0
2	−2

In such cases it often is helpful to try values of w that are equally spaced. If we enlarge the table to include

w	f
0	0
1/4	3/16
1/2	1/4
3/4	3/16
1	0

a pattern is discernible. Find a relationship between w and f.

7. Given the following table, guess at a formula relating t and d.

t	d
0	0
1	−16
2	−64
3	−144
4	−256

8. Given the following table, guess at a formula relating x and y.

x	y
−9	2
−6	5/2
−3	4
3	−2
6	−1/2
9	0

9. Use (5.31) and (5.37) to show that the maximum number of new infectives in any one period cannot exceed 25% of the total population.

6
ELEMENTS OF PROBABILITY

6.1. INTRODUCTION

We begin our study of probability by posing a problem that is typical of the ones encountered in the applications of this branch of mathematics.

An employee of a company has been injured on the job while operating a piece of mechanical equipment. The employee is suing his employer on the grounds that the machine that he was operating was malfunctioning at the time of his injury in a way that made it dangerous to a human operator. The employee claims, therefore, that he sustained his injuries through the negligence of the company. The machine produces light bulbs. From observations both before and after the accident it has been verified that the part of the machine that caused the injury functions properly 95% of the time. When the machine functions properly, on the average, 9 out of 10 light bulbs are without defects. On the other hand, when the critical part of the machine is malfunctioning, only 5 out of 10 bulbs are acceptable.

When the accident occurred, a quality control inspector was examining the output of the machine. He reported that he tested three bulbs and found two to be defective.

The question that arises is: Was the machine malfunctioning in a dangerous way at the time of the injury? Or, equivalently, Is the employee's claim valid or not?

It should be obvious that such questions cannot be answered definitively. There is some chance that the part of the machine that caused the injury was in good working order, and there is some chance that it was not. A more appropriate question, therefore, is: Which is more probable, that the machine was malfunctioning or that the machine was in good working order?

Probability is concerned with answering questions such as this last one. Throughout this chapter and several that follow we will be concerned with occurrences and events where the outcome is in doubt and with estimating the degree or the extent of the doubt.

While our main interest is with problems such as the legal one given above, we will use simpler problems to introduce and illustrate the ideas and concepts of probability. Hence we will turn to events such as tossing a coin, rolling a die, or drawing marbles from an urn. For example, we will look at such simple questions such as:

An urn contains three marbles: one red, one white, and one black. We draw a marble from the urn, observe its color and replace it. We draw another marble from the urn and observe its color. What is the likelihood that both marbles will be red?

We also will consider more complicated examples such as:

These are two urns. The first contains five marbles: one red, two black, and two white. The second urn contains four marbles: one red, two black, and one white. We draw a marble from one urn. The marble drawn is white. What is the likelihood that the urn from which the marble was drawn is the first urn (i.e., the one with five marbles)?

Keep in mind that these urn models are used only to introduce and illuminate ideas that we will use later on other more relevant problems such as:

A medical patient exhibits certain symptoms that are common to several diseases. What is the likelihood that he has any one of the diseases, and which disease does he most likely have?

6.2 PROBABILITY AND ODDS

Probability is concerned with the study of random or nondeterministic experiments. For example, if we toss a coin, it is uncertain whether it will land with the head facing up or with the tail facing up. Tossing the coin is a nondeterministic experiment.

Suppose we toss a coin a number of times and observe the number of times that a head appears. If N is the number of tosses and S the number of times a head appears, then we call S/N the *relative frequency* of a head. Now suppose we perform this experiment of tossing the coin N times over and over again and continue to observe the relative frequency of a head. If this relative frequency is stable (i.e., does not change much from experiment to experiment) we say that the probability of a head appearing is equal to the relative frequency.

This is an intuitive description of probability and is used here to give the reader an idea of what lies behind the definition of probability. Of course, a series of experiments such as those described in the previous paragraph are not likely to produce identical relative frequencies. For example, if $N = 5000$, typical values of S (the number of heads) are 2479, 2561, 2399, 2510, 2556, and 2484. These are "close" to 2500, but none of them actually are equal to 2500. The relative frequencies are "close" to ½ but are not actually equal to ½. Nevertheless, we will say that the probability of a head appearing is ½. In the following paragraphs we will give a more rigorous definition of probability based on these intuitive concepts.

Let us consider another experiment that is nondeterministic and that has more than one possible outcome. In the simple coin toss there are two possible outcomes: heads or tails. We denote these two outcomes by the two symbols H and T. In the roll of a six-sided die there are six possible outcomes: the upper face shows a 1, the upper face shows a 2, the upper face shows a 3, the upper face shows a 4, the upper face shows a 5, and the upper face shows a 6. We denote these by 1, 2, 3, 4, 5, and 6, respectively. In general there are m possible outcomes to a nondeterministic experiment, and they are denoted by $e_1, e_2, \ldots, e_m$. For the toss of a coin $m = 2$, and for the roll of a die $m = 6$.

We will find it convenient to represent these outcomes as points in space. We will refer to them as *simple* or *elementary events* and, representing them as points, will refer to them as *sample points*. The sample points taken together make up what we will call the *sample space*. The sample space must include sample points corresponding to all possible outcomes of the experiment. Moreover, each possible outcome must be included only once in the sample space.

For example, for the toss of one coin, there are two sample points, and the sample space is made up of these two points. We represent this sample space schematically in Figure 6.1a. Similarly, for the roll of a die, there are six sample points, and the sample space is shown in Figure 6.1b. The particular physical arrangement of the points in the sample space is immaterial. For example, Figure 6.1c suffices equally well for the roll of a die. The points are placed in some physical location for definiteness and, in some cases, for convenience. In general, we will designate a sample space as shown in

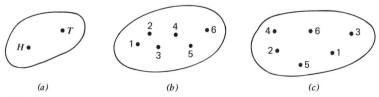

(a)   *(b)*   *(c)*

FIGURE 6.1 Sample spaces for toss of a coin and roll of a die.

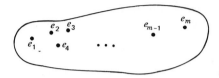

FIGURE 6.2 General sample space.

Figure 6.2. We will sometimes not even indicate the individual points at all but merely draw the boundary of the area enclosing the sample points, as shown in Figure 6.3. Of course, if we only draw the boundary, we must then be certain that we understand which sample points are enclosed in any given area.

Suppose for the moment that each of the simple or elementary events is equally likely. In the two examples (the toss of a coin and the roll of a die) it appears that each simple event is equally likely. That is, in the toss of one coin, it is equally likely that we will get a head as it is we will get a tail. In the roll of a die, each of the six possibilities is equally likely to occur.

If each of the m events is equally likely, then each event should be expected to occur once out of each m experiments. That is, the relative frequency of each event should be $1/m$. Therefore we will define the probability of each event to be $1/m$.

In the case of the toss of one coin, the probability of a head is ½, as is the probability of a tail. For the roll of a die there are six outcomes ($m = 6$), and each of the events has a probability of 1/6. That is, we should expect each of these events to occur one sixth of the time.

Suppose now that we wish to consider events that are not one of the simple or elementary events. In the roll of a die, for example, suppose we are interested in the event "the number on the top face is odd and not equal to 1." Two of the simple events will produce this new event. We say that there are two outcomes that are *favorable* to the outcome of this event. The event itself we represent by E and denote by

$$E = \{3, 5\}$$

Now this event E should occur on the average two out of six times. Thus the probability of E should be 2/6.

If there are k different simple events that are favorable to an event E, the probability of E is

$$P(E) = k/m \tag{6.1}$$

where $P(E)$ is read "probability of E."

As one further example, consider the event "the number on the top face is divisible by 3." Once again there are two events favorable to the outcome of this event. They are 3 and 6. Therefore the probability of this event is 2/6.

If we choose an event such that all of the m outcomes are favorable to the event, then $k = m$ and, from (6.1),

FIGURE 6.3 General sample space with points not shown.

$$P(E) = m/m = 1$$

The event E is called *certain,* since it is inevitable that it occur.

On the other hand, if an event is chosen so that none of the m outcomes is favorable, then $k = 0$ and, from (6.1),

$$P(E) = 0/m = 0$$

Such an event is called *impossible.*

To illustrate certain and impossible events, consider the die rolling experiment once more. If we choose an event "the number on the top face is odd or it is greater than or equal to 2," this event is certain and has probability one. On the other hand, if we consider the event "the number is odd and greater than 5," this event is impossible. Its probability is zero.

All of the above assumed that each simple event was equally likely to occur. We need not be bound by such a restriction. Suppose in the die rolling experiment we consider only three simple events: the number is even, the number is less than 2, the number is an odd prime number.[1] These three simple events cover all possible outcomes for the roll of one die. The first of these simple events is three times as likely to occur as the second, and the third is twice as likely as the second. With a little reflection we can see that the probabilities that should be assigned to these events are 1/2, 1/6, and 1/3, respectively. As another example, consider an unfair coin that produces a head three times for each tail. Then the probability of a head is ¾ and the probability of a tail is ¼.

In general, for any event (simple or otherwise), we can assign to the event a probability that is between zero and one; that is,

$$0 \leqslant P(E) \leqslant 1 \tag{6.2}$$

Moreover, if $e_1, e_2, \ldots, e_m$ are the simple events that cover all possible outcomes of the experiment once and only once,

$$P(e_1) + P(e_2) + \ldots + P(e_m) = 1 \tag{6.3}$$

Any set of numbers $P(e_1), P(e_2), \ldots, P(e_m)$ that satisfy (6.2) and (6.3) represent probabilities. Of course, in any particular case, we must choose the set of numbers in a way that has meaning for the problem we have under consideration.

---

[1]Prime numbers are integers greater than one whose only divisors are one and the number itself. The first five prime numbers are 2, 3, 5, 7, and 11. The only even prime number is 2.

If all events are equally likely,

$$P(e_1) = P(e_2) = \ldots = P(e_m)$$

Using this in (6.3)

$$m \cdot P(e_1) \neq 1$$

so

$$P(e_1) = 1/m$$

which agrees with our previously stated result for the probability of events that are equally likely.

The restriction that the simple events cover all possible outcomes once and only once is essential to (6.3). To see that this is so, consider once more the roll of a die. Suppose as simple events we choose: $e_1 = \{$the number is even$\}$ and $e_2 = \{$the number is an odd prime number$\}$. Then $P(e_1) = \frac{1}{2}$ and $P(e_2) = \frac{1}{3}$, so

$$P(e_1) + P(e_2) = 5/6 \neq 1$$

While this may appear to be in violation of (6.3), it is not, since one possible outcome, the number 1, has not been covered by e_1 and e_2. Similarly, suppose in addition to e_1 and e_2, as defined above, we add $e_3 = \{$the number is odd$\}$. Then $P(e_3) = \frac{1}{2}$, so

$$P(e_1) + P(e_2) + P(e_3) = 4/3 \neq 1$$

Once again this contradicts (6.3), but now two outcomes, 3 and 5, are covered twice—once by e_2 and once by e_3.

Given an event E, we often will find it convenient to consider the event "E does not occur." For example, in the die rolling experiment, we already considered the event "the number on the top face is divisible by 3" and designated it by

$$E = \{3, 6\}$$

In this case the event "E does not occur" is the event "the number is not divisible by 3." We designate it by $\sim E$ and, in this case,

$$\sim E = \{1, 2, 4, 5\}$$

Now it is clear that for any event E either the event occurs or it does not occur. Therefore,

$$P(E) + P(\sim E) = 1 \tag{6.4}$$

In the above example $P(E) = 2/6$ and $P(\sim E) = 4/6$, so (6.4) is valid.

The symbols $\sim E$ are read "not E" or "E complement." If we know $P(E)$ we can calculate $P(\sim E)$ from (6.4). Conversely, given $P(\sim E)$, we can compute $P(E)$.

We close this section with a discussion of "odds." One often hears that the odds that a certain team will win the championship game are 3 to 1 or 6 to 5. When the odds of an event are 3 to 1, it simply means that the ratio of (1) the probability of the event occurring to (2) the probability of the event not occurring are 3 to 1. Thus, given an event E, the odds for E occurring are

$$O(E) = P(E)/P(\sim E) \tag{6.5}$$

Again, by way of example, consider the die rolling experiment. The event E will be chosen to be "the number is not divisible by 3." Thus $P(E) = 4/6$, as we previously noted. Similarly, $P(\sim E) = 2/6$. The odds that the number is not divisible by 3 are

$$O(E) = \frac{4/6}{2/6} = \frac{2}{1}$$

Thus the odds of a number not divisible by three occurring are 2 to 1. If you were to make a "fair" bet with someone that the event E will occur (i.e., bet that the number will not be divisible by 3), you should bet twice as much money as your opponent does. If someone is willing to make such a bet and asks that you bet less than twice the amount he bets, you should quickly accept the bet—the laws of probability are on your side.

Combining (6.4) and (6.5), we find that

$$O(E) = \frac{P(E)}{1 - P(E)} \tag{6.6}$$

If $O(E)$ is a/b we say "the odds that event E will occur are a to b." Thus, if $O(E) = 5/6$, the odds that E will occur are 5 to 6. The odds that E will not occur are 6 to 5. When the odds are even, then $O(E) = 1$. Notice that in this last case the probability that E will occur is ½ (i.e., it is just as likely that E will not occur as it is that E will occur).

EXERCISES FOR SECTION 6.2

(*Note. Solutions are given for these exercises marked with an asterisk.*)
1. Construct a sample space in which each simple event is equally likely for the following experiments.
    *(a) A marble is drawn from an urn that contains three green marbles and two orange marbles.
    (b) A student is selected from a class that has 10 girls and 7 boys.
    *(c) An integer between 1 and 10 is picked.
    (d) A piece of fruit is selected from a bowl containing two apples, one orange, and three bananas.

2. The probabilities of some particular events are given below. Compute the odds that the events will occur.
   *(a) 1/2
   (c) 2/5
   (b) 2/3
   (d) 1/10

3. The odds on some particular events are given below. Compute the probabilities that the events will occur.
   *(a) 3 to 1
   (c) 1 to 2
   (b) 6 to 5
   (d) even

*4. The probability of drawing a king from a deck of 52 playing cards is 1/13. If someone offers to bet $25 to your $2 that you cannot draw a king on one try, should you accept the bet?

5. The probability of drawing a face card (jack, queen, king, or ace) from a deck of 52 playing cards is 4/13. If someone offers to bet $4 to your $2 that you cannot draw a face card in one try, should you accept the bet?

*6. Suppose you draw marbles from an urn containing 10 marbles that are either black or white. Each time you draw a marble it is replaced before the next drawing. You draw 100 marbles and obtain 42 white marbles. If you do this again and again and obtain 39, 37, 41, 41, and 38 white marbles, what would you guess is the number of white marbles in the urn?

7. Same as Exercise 6 except that the number of black marbles in each experiment is 61, 61, 58, 63, 57, 65, 70, and 51. What would you guess for the number of white marbles?

8. If a single six-sided die is tossed what is the probability that
   *(a) The number is not divisible by 3?
   (b) The number is a perfect square?
   (c) The number is a multiple of four?
   (d) The number is odd?

9. The probability of some particular event occurring are given below. What are the probabilities that the events will not occur?
   *(a) 1/3
   (c) 9/10
   (b) 1/2
   (d) 0

6.3 CARDINALITY OF SETS AND SAMPLE SPACES

We have seen that we can consider a sample space as a set of points. An event, simple or otherwise, is a subset of the set of all points in the sample space.

The number of points in a set (or subset) is called the *cardinal number* of the set (or subset). We denote the cardinal number of a set X by $n(X)$.

Consider two events. Can we find the cardinal number of the set representing these two events from the cardinal numbers of the individual events? For example, in the roll of a die, we consider the two events X and Y defined as follows:

$$X = \{\text{the number is divisible by 3}\} = \{3, 6\}$$
$$Y = \{\text{the number is divisible by 2}\} = \{2, 4, 6\}$$

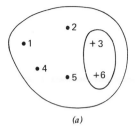 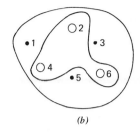

(a)   (b)

FIGURE 6.4 Sample spaces for the roll of a single die.

Then

$$n(X) = 2 \quad \text{and} \quad n(Y) = 3$$

The sample points in X are indicated by $+$ in Figure 6.4a. The sample points in Y are indicated by O in Figure 6.4b. In each case the points representing the event in question are enclosed in a region bounded by an irregular curve.

The *union* of two events is the set of outcomes that are favorable to either event. Thus, in our example of rolling a die, the union of X and Y is the event "the number is divisible by 3 or by 2 or by both." We designate the union of X and Y by $X \cup Y$ (read as "X union Y" or "X or Y"). There are four points in $X \cup Y$, and they are marked either by a $+$ or a O or both in Figures 6.4a and 6.4b. The figure is redrawn showing all such points together in Figure 6.5a. Notice that

$$n(X \cup Y) = 4$$

If we omit the points and simply draw the boundaries of the events X and Y, we can exhibit the sets X and Y and their union, as indicated in Figure 6.6. In that figure X is represented by the region with horizontal stripes and Y is represented by the region with vertical stripes. The event $X \cup Y$ is the set striped either way or both ways.

The *intersection* of two events is the set of outcomes that are favorable to both events. Thus, in our die rolling example, the intersection of X and Y is the event "the number is divisible by 3 and by 2.' We represent the intersection of X and Y by $X \cap Y$

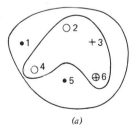 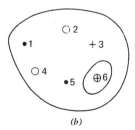

(a)   (b)

FIGURE 6.5 Union and intersection of events in the roll of a single die.

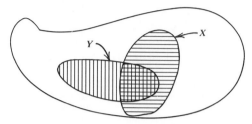

FIGURE 6.6 Union and intersection of two events.

(read "X intersect Y" or "X and Y"). In this example $X \cap Y$ is the single sample point marked by both $+$ and O, as shown in Figure 6.5b. If we again omit the points themselves and simply draw the boundaries of the events X and Y, we can refer once more to Figure 6.6. The event $X \cap Y$ is the region striped both horizontally and vertically. Diagrams of sets such as the one in Figure 6.6 are called *Venn diagrams*.

If we know the values of $n(X)$, $n(Y)$, and $n(X \cap Y)$, we can compute $n(X \cup Y)$, as we will now indicate. Consider first two events with no sample points in common, as shown in the Venn diagram in Figure 6.7. Then the number of sample points in X or Y is just the sum of the sample points in X and Y. That is,

$$n(X \cup Y) = n(X) + n(Y)$$

If the events X and Y have some simple events in common, the regions representing the two events overlap, as shown in the Venn diagram in Figure 6.8. The cardinal number of $X \cup Y$ is the number of points in the striped region (both horizontal and vertical). If we simply add the number of points in X to the number of points in Y, the points that lie in the doubly striped region will be counted twice. Therefore the number of points in $X \cup Y$ is the number in X plus the number in Y less the number in the doubly striped region. But the doubly striped region is just $X \cap Y$. Therefore,

$$n(X \cup Y) = n(X) + n(Y) - n(X \cap Y) \tag{6.7}$$

Returning to our die rolling example, recall that $n(X) = 2$ and $n(Y) = 3$. Moreover, $n(X \cap Y) = 1$. Using (6.7),

$$n(X \cup Y) = 2 + 3 - 1 = 4$$

which agrees with our previous result.

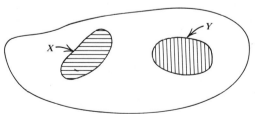

FIGURE 6.7 Venn diagram of two sets with no points in common.

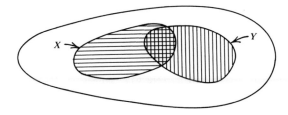

FIGURE 6.8 Venn diagram for two sets with overlapping regions.

EXERCISES FOR SECTION 6.3

*1. A deck of playing cards has four suits with 13 cards in each suit.
(a) What is the cardinal number of the set of sample points where each sample point is a different card?
(b) What is the cardinal number of the set of cards in any one suit?
(c) What is the cardinal number of the set of all red cards?

2. Suppose all the face cards (jack, queen, king, and ace) are removed from a deck of playing cards (see Exercise 1). What is the cardinal number of the set of sample points when each sample point is a different card?

3. What is the cardinal number of the following sets when a card is drawn from a complete set of playing cards (see Exercise 1)?
*(a) The number is not a face card and is divisible by 2.
(b) The number is not a face card and is not divisible by 2.
(c) The card is a queen.
(d) The card is a black king.
(e) The card is a heart.

4. The cardinal number of a set A is 21, and the cardinal number of a set B is 16. What is the cardinal number of $A \cup B$ if the cardinal number of $A \cap B$ is
*(a) 4?   (b) 12?
(c) 0?   (d) 1?

5. Suppose a set A has 12 members and its union with a set B has 15 members. What is the cardinal number of B if the cardinal number of $A \cap B$ is
*(a) 1?   (b) 0?
(c) 10?   (d) 12?

6. Why is it not possible for
*(a) $n(X) = 10$, $n(Y) = 12$, and $n(X \cup Y) = 8$?
(b) $n(C) = 6$, $n(C \cap D) = 2$, and $n(C \cup D) = 3$?
(c) $n(A) = 3$, $n(B) = 2$, and $n(A \cap B) = 4$?
(d) $n(A \cup B) = 4$ and $n(A \cap B) = 6$?

*7. In a student survey of 1000 students there were 452 enrolled in a computing course, 312 enrolled in a mathematics course, and 113 enrolled in both courses.
(a) How many students were enrolled in one or the other of the two courses?
(b) How many students were not taking either of the two courses?

(c) How many students were enrolled in only one of the courses? (*Hint.* Draw a Venn diagram. How many members are in one of the two sets, but not in their intersection?)

8. Same as Exercise 7 except that there are 500 students and 193 are enrolled in the computing course, 206 in the mathematics course, and 72 in both courses.

*9. A student survey of 1000 students indicated that there were 617 enrolled in freshman English, 592 enrolled in freshman history, and 113 enrolled in both courses. Why would you dispute these figures?

10. A student survey of 1500 freshmen students indicates that 687 were enrolled in Introduction to Sociology and 709 were enrolled in Introduction to Psychology. All freshmen are required to take at least one of these "intro" courses. Do you believe these figures? Why?

6.4 PROBABILITY AND CARDINALITY

We can make use of cardinal numbers to compute probabilities. Consider a sample space in which all events are equally likely. We will name the sample space U. Suppose the cardinal number of U is m. Then the probability of each simple event is $1/m$.

Next consider an event X that includes k of the equally likely sample points. The cardinal number of the space representing X is k. Moreoever, from (6.1),

$$P(X) = k/m$$

But $k = n(X)$ and $m = n(U)$, so

$$P(X) = \frac{n(X)}{n(U)} \tag{6.8}$$

We will use (6.8) to compute the probability of any event X. Recall that to use (6.8) all simple events must be equally likely, and we must be able to find the cardinal numbers of both the entire sample space and the event in question.

EXERCISES FOR SECTION 6.4

*1. A deck of playing cards has four suits with 13 cards in each suit. If one card is drawn from the deck, what is the probability that
   (a) The card is a spade?
   (b) The card is red?
   (c) The card is a face card (jack, queen, king, or ace)?
   (d) The card is red and not a face card?

2. Suppose all of the face cards (jack, queen, king, and ace) have been removed from a deck of playing cards (see Exercise 1). If one card is drawn from this reduced deck, what is the probability that

(a) The card is a spade?

(b) The card is red?

(c) The card is a face card?

(d) The card is red and not a face card?

3. In a deck of 52 playing cards (see Exercise 1), if one card is drawn from the deck, what is the probability that

(a) The card is not a face card and is divisible by 2?

(b) The card is not a face card and is not divisible by 2?

(c) The card is a queen?

(d) The card is a black queen?

(e) The card is a heart?

*4. A set U is a sample space that contains 100 members. A and B are events that are subsets of U. The cardinal number of A is 21, and the cardinal number of B is 16. What is the probability that either the event A or the event B or both events A and B occur if the cardinal number of $A \cap B$ is

(a) 4? (b) 12?

(c) 0? (d) 1?

5. A set U is a sample space that has 50 members. A and B are events that are subsets of U. The cardinal number of A is 12. The cardinal number of $A \cup B$ is 15. What is the probability that the event B occurs if the cardinal number of $A \cap B$ is

(a) 1? (b) 0?

(c) 10? (d) 12?

*6. Of 1000 students, 452 are enrolled in a computing course, 312 are enrolled in a mathematics course, and 113 are enrolled in both courses. If a student is selected at random, what is the probability that he or she is enrolled in

(a) The computing course?

(b) The mathematics course?

(c) Both courses?

(d) One of the other of the two courses?

(e) Neither of the two courses?

(f) Only one of the two courses? (*Hint.* See Exercise 7 of Section 6.3).

7. Same as Exercise 6 except there are 500 students and 193 are enrolled in the computing course, 206 in the mathematics course, and 72 in both courses.

*8. An urn contains two red marbles, three black marbles and an unspecified number of white marbles. How many white marbles must the urn contain in order that the probability that a black marble is drawn from the urn on one draw is 1/2?

9. Same as Exercise 8 except the probability that a black marble is drawn is to be 3/8?

6.5 COMPOUND PROBABILITIES

We now turn to the problem of computing the probabilities of multiple events. That is, given two events X and Y, we will compute the probability that either X or Y occurs.

In a like way we will find the probability that both X and Y occur. Such probabilities are called *compound probabilities*.[2]

We will try to compute these compound probabilities from the probabilities of X and Y themselves. Be forewarned, however, that it is not always possible to compute such compound probabilities in a straightforward way. When they can be computed, however, they save us a great deal of time and effort over the alternative of constructing the entire sample space. For example, the sample space for tossing eight coins would contain 256 sample points. At the close of this section, we will be able to compute the probabilities of various events in such an experiment without constructing this large sample space.

The parts of an experiment are said to be *unrelated* or *independent* if the outcome of one part of the experiment does not affect the outcome of any other part. For example, suppose we have an urn containing four marbles: one red, one green, one white, and one black. Suppose we conduct the following experiment.

1. (a) Draw a marble from the urn and observe its color.
   (b) Replace the marble in the urn.
2. Draw another marble from the urn and observe its color.

The parts of this experiment are unrelated, since the color of the marble drawn in part 2 does not depend in any way on the color of the marble drawn in part 1, and vice versa.

The sample space for this two part experiment can be represented as indicated in Figure 6.9, where RR indicates that both marbles drawn were red, RB indicates that the first marble was red and the second black, and so on. Since there are 16 sample points, each equally likely, the probability of any one of these simple events is 1/16.

If we omit step 1b above, the two parts of the experiment are *related*. Indeed, if we draw a red marble in part 1, there are only three possibilities on the second draw: green, white, or black. Similarly, if the first draw produces a black marble, the second can only produce red, white, or green. The sample space for this related experiment (drawing without replacement) may be drawn as shown in Figure 6.10. Notice that it is similar to Figure 6.9 except that the points lying along the diagonal that runs from the

FIGURE 6.9 Sample space for drawing a marble from an urn containing four different-colored marbles.

---

[2]The adjective ''compound'' derives from the use of that word as it refers to English sentences. The conjunctions *and* and *or* when connecting two sentences produce a compound sentence.

● ● ●
RG   RB   RW

●   ●   ●
GR   GB GW

●   ●   ●
BR BG   BW

●   ●   ●
WR WG WB

FIGURE 6.10 Sample space for drawing a marble from an urn without replacement.

upper left to the lower right is missing. The probability of each of the sample points in Figure 6.10 is 1/12. Compare this with 1/16 in the related experiment of drawing with replacement.

Let us now return to Figure 6.9. Let

$$R_1 = \{\text{first marble drawn is red}\}$$
$$B_2 = \{\text{second marble drawn is black}\}$$

Then

$$R_1 \cap B_2 = \{\text{first marble drawn is red } and \text{ second marble drawn is black}\}$$

Now

$$P(R_1) = 1/4 \qquad P(B_2) = 1/4.$$

From Figure 6.9, since only one of the 16 sample points is favorable to $R_1 \cap B_2$, that is,

$$n(R_1 \cap B_2) = 1$$

and all 16 sample points are equally likely,

$$P(R_1 \cap B_2) = 1/16$$

It follows that in this example

$$P(R_1 \cap B_2) = P(R_1) \cdot P(B_2)$$

In other words, the probability that both of two independent events occur is the product of the probabilities that the two events occur individually. This is true for all pairs of independent or unrelated events.

To see this, consider two independent events X and Y. Since they are independent, the probability that X occurs should be the same *regardless of whether or not Y occurs.*[3] We will calculate $P(X)$ in two different ways and equate the two results. From (6.8),

$$P(X) = \frac{n(X)}{n(U)} \tag{6.9}$$

where, as usual, $n(X)$ is the cardinal number of the set X and $n(U)$ is the cardinal number of the entire sample space.

Suppose now that the event Y has occurred. Then some of the points in the original sample space are ruled out as possibilities. Indeed, the only possibilities are those in the event Y, and this set of points Y now makes up the sample space. We now wish to calculate the probability that X occurs. The sample points favorable to X are those points in $X \cap Y$ (recall that we have assumed that Y has occurred). Thus the number of sample points favorable to X is $n(X \cap Y)$ and

$$P(X) = \frac{n(X \cap Y)}{n(Y)} \tag{6.10}$$

If X and Y are independent events, the two values of $P(X)$ in (6.9) and (6.10) are equal, i.e.,

$$\frac{n(X)}{n(U)} = \frac{n(X \cap Y)}{n(Y)}$$

Multiplying both sides of this equation by $n(Y)/n(U)$,

$$\frac{n(X)}{n(U)} \cdot \frac{n(Y)}{n(U)} = \frac{n(X \cap Y)}{n(Y)} \cdot \frac{n(Y)}{n(U)}$$

or

$$P(X) \cdot P(Y) = P(X \cap Y) \tag{6.11}$$

This verifies our earlier claim that for independent events the probability that both events occur is the product of the probabilities that each event occurs separately.

We turn to one more example before proceeding. Suppose there are two groups of people. Group I has one man and two women. Group II has two men and three women. We select one person from each group. What is the probability that both are women?

A sample space is shown in Figure 6.11. If

$$X = \{\text{woman selected from Group I}\}$$
$$Y = \{\text{woman selected from Group II}\}$$

---

[3]Indeed, this is our definition of independent events.

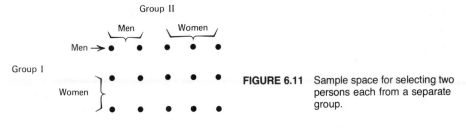

FIGURE 6.11 Sample space for selecting two persons each from a separate group.

then

$$P(X) = 2/3 \qquad P(Y) = 3/5$$

and X and Y are clearly independent events.

The sample points favorable to $X \cap Y$ are the six points in the lower right corner (bottom two rows and three right-most columns). Hence,

$$P(X \cap Y) = \frac{6}{15}$$

and (6.11) is satisfied.

Equation 6.11 is referred to as the *product law*. It is only valid if X and Y are *unrelated* or are *independent* of one another. We could define events to be independent if their probabilities satisfy the product law, (6.11), and this often is used as the definition of unrelated events.

Whether events are independent or not we found in the previous section that

$$n(X \cup Y) = n(X) + n(Y) - n(X \cap Y)$$

Dividing both sides of this equation by $n(U)$, we get

$$P(X \cup Y) = P(X) + P(Y) - P(X \cap Y) \qquad (6.12)$$

This last equation relates the probability of either or both of the events X or Y occurring to the probabilities of X, Y, and both X and Y occurring. This last equation, (6.12), we repeat, is valid for all events, related or unrelated.

If the events X and Y are independent (unrelated), combining (6.11) and (6.12),

$$P(X \cup Y) = P(X) + P(Y) - P(X) \cdot P(Y) \qquad (6.13)$$

Returning to the example of selecting one person from each of two groups, since $P(X) = 2/3$ and $P(Y) = 3/5$

$$P(X \cup Y) = 2/3 + 3/5 - 2/5 = 13/15$$

This is the probability that one or both of the two persons selected is a woman. The only sample points that do not satisfy these requirements are the two in the first two columns of the first row. All other 13 sample points do satisfy the requirement, and the probability is 13/15.

Finally, we turn to *mutually exclusive* events. Two events are said to be mutually exclusive if the occurrence of one event precludes the occurrence of the other.

Returning to the urn example with replacement in Figure 6.9, the events $X = \{$the first marble drawn is red$\}$ and $Y = \{$the first marble drawn is white$\}$ are mutually exclusive. It should be clear that we cannot choose mutually exclusive events, one of which refers to the first draw and the other of which refers to the second draw. However, in the problem where the first marble is *not* replaced (see Figure 6.10), we can choose one event concerning the first draw and another event concerning the second draw in such a way that the two events are mutually exclusive. For example, we can choose the events $X = \{$the first marble drawn is red$\}$ and $Y = \{$the second marble drawn is red$\}$. Since both of these events cannot occur (there is no replacement), they are mutually exclusive.

For two mutually exclusive events, X and Y, the event $X \cap Y$ is the impossible event (i.e., it cannot occur). Thus,

$$P(X \cap Y) = 0 \qquad (6.14)$$

From (6.12)

$$P(X \cup Y) = P(X) + P(Y) \qquad (6.15)$$

It is important to note that (6.14) is only valid when X and Y are mutually exclusive.

To recap some of the discussion of this section we note:

1. Events are independent if the outcome of one has no effect upon the outcome of the other.
2. Events are mutually exclusive if the occurrence of one precludes the occurrence of the other.

For independent events,

$$P(X \cap Y) = P(X) \cdot P(Y) \qquad (6.11)$$

For mutually exclusive events,

$$P(X \cap Y) = 0 \qquad (6.14)$$

For all events,

$$P(X \cup Y) = P(X) + P(Y) - P(X \cap Y) \qquad (6.12)$$

EXERCISES FOR SECTION 6.5

*1. A student estimates that his probability of passing his mathematics course is .5 and his probability of passing English is .6. If the two grades are independent of each other,
  (a) What is the probability that he will pass both courses?
  (b) What is the probability that he will pass at least one of the courses?

2. Same as Exercise 1 except that the probability of passing mathematics is .7 and the probability of passing English is .4.

*3. An investor has $1000 invested in each of two securities, United Biscuits and National Oxygen. He estimates that the probability that United Biscuits will increase in value during the next year is 0.6, the probability that at least one of the two will increase in value is 0.8, but the probability that both will increase in value is only 0.2.
(a) What is the probability that National Oxygen will increase in value?
(b) What are the odds that National Oxygen will increase in value?

4. Same as Exercise 3 except that the probability that United Biscuits will increase is 0.7, the probability that at least one of the two will increase is 0.9, and the probability that both will increase is 0.1.

*5. An investor is studying two securities. The changes in value of the two securities are independent of one another. The investor estimates that the probability that the first security will increase in value is 0.7 and the probability of the second security increasing in value is 0.6.
(a) What is the probability that both securities will increase in value?
(b) What is the probability that at least one of the securities will increase in value?
(c) What is the probability that neither will increase in value?

6. Same as Exercise 5 except that the probabilities of the two securities increasing in value is 0.9 and 0.5, respectively.

*7. A card is drawn at random from a deck of 52 playing cards. The card is replaced and a second card is drawn. What is the probability that
(a) The first card is a king?
(b) The second card is a king?
(c) Both cards are kings?
(d) At least one of the two cards is a king?
(e) Neither card is a king?

8. Two six-sided dice are rolled simultaneously. What is the probability that
(a) The first die shows a 6?
(b) The second die shows a 6?
(c) Both dice show a 6?
(d) At least one 6 is showing?
(e) No 6's are showing?

9. A quality control expert examines 10 light bulbs from each of two different boxes. He finds two defective bulbs in the first box and three defective bulbs in the second box. If one bulb is later selected from each box, what would you estimate to be the probability that
(a) Both bulbs are defective?
(b) Neither bulb is defective?

*10. A candidate for political office is evaluating the wisdom of taking a stand on a vital issue. He estimates that if he takes this stand the probability that he will increase his popularity with male voters is 3/4 and with women voters is 2/3. His estimate of his probability of increasing his popularity with at least one of these groups is 5/6.
(a) What is the probability that he will increase his popularity with both groups?

(b) Are the opinions of the two groups, men and women, independent of one another? Explain.

11. Same as Exercise 10 except the estimate of increasing his popularity with men is 2/3, with women 1/2, and with at least one of the two groups 5/6,

12. When is it possible for two events, A and B, to be both independent and mutually exclusive?

6.6. A PROBLEM IN INFORMATION THEORY

One of the concerns of information theory is the reliability of the transmission of information. For example, if certain pieces of information are transmitted by word-of-mouth through a succession of people, how reliable is the information received by the last recipient? (A children's game in which one person whispers a word to his neighbor who then whispers to his neighbor and so on through a number of stages exhibits the unreliability of such transmission of information.) Even in written communication each person who reads a message and then rewrites it distorts the information in some way. Orders or directives that are disseminated from a company president through his vice-presidents to division heads and other levels of management until they reach the person or persons who are to take direct action are also subject to loss of reliability.

The transmitting devices need not be humans. For example, a message could be typed by a Western Union operator and sent by telegraph. The operator may type the message perfectly but, through electronic or mechanical difficulties, the message is not received correctly. In such cases it often is possible to correct the message by using a little common sense. For example, if a telegram reads "MEETING TO BE HELD FAB 22," then it is likely that FAB should have been FEB and the meeting is scheduled for February 22. However, if the message reads "MEETING TO BE HELD JUK 22," it is difficult to decide whether the meeting is to be held in June or July.

We now pose a specific problem in determining the reliability of a transmitted message. The reader should keep in mind that while this is a simple example, it exhibits many of the factors present in more complex problems.

Suppose we wish to send a message consisting of either the numeral 0 or 1. We will do this by either sending an electrical pulse over a wire or by not sending such a pulse. If we send a pulse, the receiver understands that we mean a 1. If no pulse is sent, the receiver interprets this to mean a 0. This is a technique often used in computer operations.

Suppose further that we know that the probability that the message is received incorrectly is 1/10. That is, it is possible for the receiver to think it detects a pulse when none was sent or, correspondingly, the receiver may not detect a pulse even if a pulse is sent. In either case we are assuming that 1 out of every 10 times, the receiver makes an error. How can we increase the reliability of the reception? Or, what can we do to reduce the error rate from 1 out of 10 to, say, 1 out of 50?

One solution would be to send each message twice or even three times. If we send the message twice and receive the same message twice, then we could assume that the message was correct. If, however, we receive two different messages, we are at a loss as to what the correct message was.

Suppose we send the same message three times. If all three received messages are identical, we would feel even more confident that we had the correct message. If two out of the three messages were the same, then we might reasonably assume the two that agreed represented the correct message. Thus we would always have a way of determining what we thought was the correct message. Our method of determination is simply a majority rule. Presumably this would be more reliable than sending the message once, but is it?

Suppose we transmit a 1 and, to do so, we send a pulse over three wires simultaneously. This is equivalent to sending three successive pulses over one wire. If we wish to send a zero, we do not send a pulse over any of the three wires. At the receiving end the following eight messages may be received:

000	001	010	100
110	101	011	111

where 000 represents no pulses on any line, 001 represents no pulses on the first two lines and a pulse on the third line, and so on.

Of course, the only messages that would have been sent are either 000 or 111, but any one of the other six messages above could be received through an error in transmission. The four messages in the first row will be interpreted as a 0 (no pulse), since there are more zeroes than ones in each of those four messages. The four messages in the second row will be interpreted as a 1, since ones predominate in those messages.

There are four ways in which an error could occur in the three line system: (1) errors could occur on all three lines (then a 000 would be read as 111, and vice versa): (2) an error could occur on lines 1 and 2; (3) an error could occur on lines 1 and 3; or (4) an error could occur on lines 2 and 3. Errors on a single line or no errors on any line would not result in an error in the decoding process.

We will consider each of these four cases that lead to an error at the receiving end in the order listed above. First suppose that errors occur on each of the three lines. Now each line is independent of the other two. Therefore, from (6.11), the probability of all three lines being in error is the product of the probabilities of each line being in error. If we let e_i represent an error in line i,

$$P(e_1 \cap e_2 \cap e_3) = P(e_1) \cdot P(e_2) \cdot P(e_3)$$
$$= 0.1 \cdot 0.1 \cdot 0.1$$
$$= 0.001$$

Now consider the second case where errors occur in lines 1 and 2. If c_i represents a correct transmission on line i, we wish to calculate $P(e_1 \cap e_2 \cap c_3)$. Again, this is the product of the probabilities of each individual event, or

$$P(e_1 \cap e_2 \cap c_3) = P(e_1) \cdot P(e_2) \cdot P(c_3)$$
$$= 0.1 \cdot 0.1 \cdot 0.9$$
$$= 0.009$$

The other two cases produce the same probability of 0.009.

We now wish to calculate the probability of any one of the four events occurring. That is, we wish to compute the probability of the union of all four events. But these four events are mutually exclusive. There is no case where errors occur on all three lines and also occur on precisely two of the lines but not the third. Thus, from (6.15), the probability of any one of the four occurring is just the sum of the probabilities of the four individual events.

The probability of an erroneous reading at the receiving end is

$$0.001 + 0.009 + 0.009 + 0.009 = 0.028$$

We have increased the reliability of the communications system. If we sent only one signal, 1 out of 10 times (or 10 out of 100 times) we would receive the incorrect signal. On the other hand, in the three-line system, an incorrect signal is received only 3 out of 100 times. We have increased the reliability by a factor of three (3 errors compared to 10) at three times the cost.

EXERCISES FOR SECTION 6.6

*1. In the three-line transmission system described above, suppose the probability of an error on a single line is 0.4. What is the probability of an error in the three line system?

2. Same as Exercise 1 except the probability of an error on one line is 0.5.

3. Same as Exercise 1 except that the probability of an error on one line is 0.6.

*4. In practice, signals that are 1 are subject to error more frequently than signals that are 0. Suppose on a single line the probability of a 1 being misinterpreted as a 0 is 0.2, but that the probability of a 0 being misinterpreted as a 1 is 0.1. Using the three-line system:
   (a) What is the probability of a 0 being received as a 1?
   (b) What is the probability of a 1 being received as a 0?

5. Same as Exercise 4 except the probability of a 1 being misinterpreted is 0.02 and the probability of a 0 being misinterpreted is 0.01.

6. A special committee of the United Nations is voting on whether to send a resolution on military intervention to the General Assembly. There are three members of the committee, and they will vote by secret ballot. A political observer of the United Nations estimates that the probability that any one of the committee members will vote to send the resolution to the General Assembly is 0.4.
   (a) If a simple majority is required to send the resolution to the General Assembly, what is the probability that it will actually be sent?

(b) If the vote must be unanimous for the resolution to be sent to the General Assembly, what is the probability that it will be sent?

7. Suppose a five-line transmission system is used (i.e., to transmit a zero, there are five zeroes sent on each of five lines). If three, four, or five zeroes are received, the transmission is assumed to have been a zero. On the other hand, zero, one, or two zeroes (which is equivalent to three, four, or five 1's) is assumed to have been a one. If each of the five individual lines fails with probability 0.1, what is the probability that an erroneous message is received?

6.7 THE BLOOD DONOR PROBLEM

The blood of any one individual may or may not contain one or more of a number of antigens. Three of these antigens are called A, B, and Rh. If an individual has the two antigens A and Rh, he is referred to as A+. If he has the antigen A only, he is referred to as A−. If an individual has neither the A nor the B antigen, he is called O. Thus blood type O is merely the absence of A and B.

There are certain restrictions on the donation of blood, depending on the antigens in the blood of the donor and the donee. In particular, if one of the antigens A, B, or Rh is absent from the blood of a person, he cannot receive blood from someone whose blood contains that antigen. For example, a person with blood type A+ cannot receive blood from someone who is B+, and vice versa. The first person has the A antigen and the second does not, so the second cannot receive blood from the first. Similarly, the second person has the B antigen and the first does not, so the first cannot receive blood from the second.

The percentages of persons in the United States (1970) with the eight possible blood types are shown in Figure 6.12.

O+	37.4%	O−	6.6%
A+	35.7%	A−	6.3%
B+	8.5%	B−	1.5%
AB+	3.4%	AB−	0.6%

FIGURE 6.12 Percentages of persons in the United States with given blood types (1970).

We pose the following question. If a person has blood of type O+, what are that person's chances of finding someone to donate blood to him? A more general form of this question is: What is the probability that a person selected at random in the United States can donate blood to a person with a given blood type?

The converse question is: What is the probability that a randomly selected individual can receive blood from a person with a given blood type?

We proceed to answer the first of these questions now. The others are left as exercises for the reader.

Of all individuals some subset of them have the antigen A in their blood. We will consider the set A to be comprised of all those individuals with the A antigen. Similar-

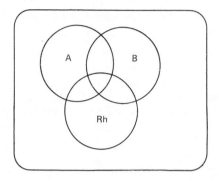

FIGURE 6.13 Venn diagram for three blood antigens: A, B, and Rh.

ly, the set B is the set of all individuals containing the antigen B. The set Rh is the set of individuals with the Rh antigen. There are, of course, some individuals whose blood contains none of the three antigens. We can represent these sets in a Venn diagram, as shown in Figure 6.13. Notice that there are regions that are common to the sets A and B, to the sets A and Rh, to the sets B and Rh, and common to all three sets.

We have redrawn the Venn diagram in Figure 6.14 and, in this figure, the regions are numbered for reference. Persons in region 1 are A−, since they have the A antigen but neither B nor Rh. Persons in region 2 are AB−, since they have both A and B antigens but not Rh. On the other hand, persons in region 8 are O−, since they have none of the three antigens. The correspondence between the regions and the eight blood types is shown in the following table.

Region	Blood Type	Region	Blood Type
1	A−	5	AB+
2	AB−	6	B+
3	B−	7	O+
4	A+	8	O−

Suppose we have a group of 1000 individuals.[4] We will think of each of these individuals as a sample point. From the figures given earlier we should expect to find 374 individuals (sample points) in region 7 (blood type O+). We should expect to find only 6 in region 2 (blood type AB−).

Let $P(X)$ be the probability that a person selected randomly can donate blood to a person with blood type X. For example, $P(A+)$ is the probability that a randomly selected person can donate blood to a person with A+ blood.

[4]The number 1000 is quite arbitrary. We could just as well have selected 10,000 or 100,000. The point is that the number of individuals in the group should be "large," so large that if one or two of any particular blood type are selected and removed from the group that the probabilities of selection are not significantly changed.

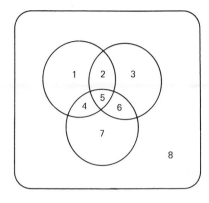

FIGURE 6.14 Relabeled Venn diagram for three blood antigens with eight blood types marked.

Now a person with A+ blood can receive blood from any person except a person with the B antigen present in his (the donor's) blood. Thus, to compute $P(A+)$, we would like to know the number of persons who do not have the B antigen. If we let $n(X)$ be the number of persons in the set X, then we would like to know $n(\sim B)$. That is, we wish to know the number of persons not in the set B. Since there are 1000 persons

$$n(\sim B) = 1000 - n(B)$$

Of course, at the moment we do not know the number of persons in the set B, but we will calculate this shortly. In any case, once we know $n(\sim B)$, the probability $P(A+)$ is just $n(\sim B)/1000$.

Similarly, if we wish to calculate $P(A-)$, we need to know the number of persons who do not have the B antigen or the Rh antigen. That is, we need to know $n\left[\sim(B \cup Rh)\right]$. But

$$n\left[\sim(B \cup Rh)\right] = 1000 - n(B \cup Rh)$$

From (6.7)

$$n\left[\sim(B \cup Rh)\right] = 1000 - n(B) - n(Rh) + n(B \cap Rh)$$

From this brief discussion it should be clear that if we know $n(A)$, $n(B)$, and $n(Rh)$ as well as the cardinal numbers of the intersections of these three sets, then we can calculate the probabilities of each blood type finding a donor. Therefore we will compute these cardinal numbers prior to calculating the probabilities in question.

Now $n(A)$ is the sum of the cardinal numbers of the regions 1, 2, 4, and 5 in Figure 6.14. The cardinal numbers of those regions are the percentages shown in Figure 6.12 multiplied by 10 (to convert to decimal form we divide by 100 and, since there are 1000 individuals, we then multiply by 1000). Thus

$$n(A) = 63 + 6 + 357 + 34 = 460$$

Similarly, B is made up of the regions 2, 3, 5, and 6. Thus

$$n(B) = 6 + 15 + 34 + 85 = 140$$

The set Rh is made up of 4, 5, 6, and 7, so

$$n(Rh) = 357 + 34 + 85 + 374 = 850$$

The set A ∩ B is made up of the regions 2 and 5, so

$$n(A \cap B) = 6 + 34 = 40$$

The set A and Rh is made up of regions 4 and 5, so

$$n(A \cap Rh) = 357 + 34 = 391$$

The set B ∩ Rh is made up of regions 5 and 6, so

$$n(B \cap Rh) = 34 + 85 = 119$$

With this data we now return to the calculation of the probabilities outlined earlier. For example, we noted earlier that

$$P(A+) = n(\sim B)/1000$$

But

$$
\begin{aligned}
n(\sim B) &= 1000 - n(B) \\
&= 1000 - 140 \\
&= 860
\end{aligned}
$$

so

$$P(A+) = 0.86$$

This means that if we select someone at random, the chances are almost 9 out of 10 that he will be able to donate blood to a person with blood type A+.

For A− we can compute

$$P(A-) = n[\sim(B \cup Rh)]/1000$$

But

$$
\begin{aligned}
n[\sim(B \cup Rh)] &= 1000 - n(B) - n(Rh) + n(B \cap Rh) \\
&= 1000 - 140 - 850 + 119 \\
&= 129
\end{aligned}
$$

so

$$P(A-) = 0.129$$

The chances of a randomly selected person being able to donate blood to a person of type A− are very small.

We will do one more example. Consider $P(O+)$. A person of type O+ can receive blood from anyone except those persons with the A or B antigen. Thus the number of people who can donate blood to an O+ are those not in A or B $\{$i.e., $n[\sim(A \cup B)]\}$. But, from (6.7),

$$n[\sim(A \cup B)] = 1000 - n(A) - n(B) + n(A \cap B)$$

This leads to

$$n[\sim(A \cup B)] = 1000 - 460 - 140 + 40$$
$$= 440$$

so

$$P(O+) = 0.44$$

We can notice one rather striking result already. Although O+ is the most prevalent blood type, less than ½ of the people are potential donors to that type. Indeed, A+, which is slightly less prevalent than O+, has almost twice as many potential donors.

We summarize our results in the table in Figure 6.15. The reader should be able to complete the table for himself (See Exercises 1 and 2).

Blood Type	Probability that a Random Selection Produces a Donor	Probability that a Random Selection Produces a Donee
O+	0.44	
A+	0.86	
B+		
AB+		
O−		
A−	0.129	
B−		
AB−		

FIGURE 6.15 Partially completed table showing probabilities of finding blood donors and donees.

EXERCISES FOR SECTION 6.7

*1. Complete the first column of Figure 6.15.

2. Complete the second column of Figure 6.15.

3. Suppose Figure 6.12 were replaced by the following table.

O+	40%	O−	10%
A+	30%	A−	2%
B+	10%	B−	5%
AB+	2%	AB−	1%

    (a) Complete the first column of Figure 6.15.
    (b) Complete the second column of Figure 6.15.

*4. What are the odds that a person with blood type O+ will find a blood donor?

5. What are the odds that a person with blood type A+ will find a blood donor?

6. You are running a computerized dating bureau. In your file of 500 men there are:
    (a) 34 who are less than 5′6″, live outside of town, and do not like sports.
    (b) 91 who are over 5′6″, live outside the town, and do not like sports.
    (c) 89 who are over 5′6″, live in the town, but who do not like sports.
    (d) 102 who are over 5′6″, live outside the town, but who do like sports.
    (e) 63 who are over 5′6″, live in town, and do like sports.
    (f) 34 who are less than 5′6″, live in town, and do not like sports.
    (g) 49 who are less than 5′6″, live outside the town, but who do like sports.
    (h) 38 who are less then 5′6″, live in town, but who do like sports.
    A woman calls and asks you to arrange a date with someone who
        (i) lives in town
        (ii) is over 5′6″ and likes sports
        (iii) either likes sports or lives in town
        (iv) does not like sports

If you simply select at random a man from your file of 500, what is the probability that he will be acceptable in each of the four cases?

6.8 CASE STUDY 6: THE GAME OF CRAPS (A SIMULATION)

The world's most popular dice game is craps. What are the chances of winning at craps? How could we change the rules to improve the chances of winning?

In this case study we will create a computer program in BASIC to play the game of craps. By using this BASIC program to play the game over and over again, we will estimate the probability of winning the game. Later, in Case Study 7, we will develop the necessary mathematical techniques to compute this probability of winning more precisely. For now we will content ourselves with computing the relative frequency of

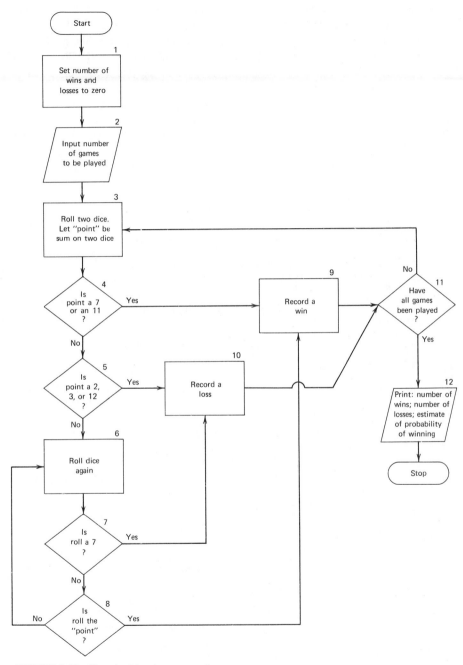

FIGURE 6.16 Flowchart to play game of craps.

winning and using this relative frequency to estimate the probability. (See the discussion of relative frequency and its relation to probability in Section 6.2.) Such a process is called a *simulation* because we are simulating the actual playing of the game.

The rules of the game of craps are described below and recorded in the flowchart in Figure 6.16. This flowchart also describes a procedure for playing a number of games and recording the results. Of course, we will not actually play the game. We will *simulate* the procedure in Figure 6.16 on a computer, hence the term "simulation." First, we turn our attention to the rules of the game.

The rules are as follows. Two six-sided dice are tossed simultaneously [3].[5] The sum of the two numbers that appear is, of course, an integer between 2 and 12. If a 7 or 11 appears, the player wins [4 and 9]. If a 2, 3, or 12 appears, then the player loses [5

```
100    REM   **   SET NØ. ØF WINS AND LØSSES TØ ZERØ   **
200    LET N1=0
300    LET N2=0
400    PRINT "TYPE NØ. ØF GAMES TØ BE PLAYED "
500    INPUT N
600    REM   **   RØLL TWØ DICE AND CØMPUTE PØINT, P   **
700    LET D1=INT(6*RND(0)+1)
800    LET D2=INT(6*RND(0)+1)
900    LET P=D1+D2
1000   REM   **   IF PØINT IS 7 ØR 11, RECØRD A WIN   **
1100   IF P=7 THEN 3100
1200   IF P=11 THEN 3100
1300   REM   **   IF PØINT IS 2,3 ØR 12, RECØRD A LØSS   **
1400   IF P=2 THEN 2800
1500   IF P=3 THEN 2800
1600   IF P=12 THEN 2800
1700   REM   **   IF PØINT IS 4,5,6,8,9 ØR 10, RØLL DICE AGAIN   **
1800   LET D1=INT(6*RND(0)+1)
1900   LET D2=INT(6*RND(0)+1)
2000   LET R=D1+D2
2100   REM   **   IF RØLL IS 7, RECØRD A LØSS   **
2200   IF R=7 THEN 2800
2300   REM   **   IF RØLL IS PØINT, RECØRD A WIN   **
2400   IF R=P THEN 3100
2500   REM   **   IF RØLL IS NEITHER 7 NØR PØINT, RØLL DICE AGAIN   **
2600   GØTØ 1800
2700   REM   **   RECØRD A LØSS   **
2800   LET N1=N1+1
2900   GØTØ 3400
3000   REM   **   RECØRD A WIN   **
3100   LET N2=N"+1
3200   REM   **   IF NØ. ØF WINS PLUS NØ. ØF LØSSES IS LESS THAN NØ.
3300   REM        ØF GAMES TØ BE PLAYED, START ANØTHER GAME          **
3400   IF N1+N2<N THEN 700
3500   REM   **   PRINT FINAL RESULTS   **
3600   PRINT "NØ. ØF WINS =";N2
3700   PRINT "NØ. ØF LØSSES =";N1
3800   PRINT "ESTIMATE FØR PRØBABILITY ØF WINNING IS";N2/N
3900   END
```

FIGURE 6.17 BASIC program to play the game of craps.

[5]The numbers in square brackets refer to the box numbers on the flowchart in Figure 6.16.

and 10]. If any other integer appears (the only possibilities are 4, 5, 6, 8, 9, and 10), then the player neither wins nor loses. Instead, the number that appears is designated as the player's "point" [3]. The player then tosses the two dice again [6]. On the second toss, if the number that appears is a 7, then he loses [7 and 10]. If the number that appears on the second toss is his "point," the player wins [8 and 9]. If any other number appears, once again the player neither wins nor loses. His "point" remains the same (i.e., his "point" is the number that appeared on the first toss), and he tosses the dice again [6]. The player continues to toss the dice until either (1) a 7 appears, or (2) his "point" appears. In the first case, the player loses. In the second, he wins.

We now turn to a description of the simulation process and the flowchart in Figure 6.16. We start by setting the number of wins and losses to zero, since no games have been played [1]. We then specify the number of games we wish to play [2], say 500. We make the first roll of the dice [3] and decide if we have won or lost [4 and 5]. If the result is neither a win nor a loss but a draw, we roll the dice again [6]. Once again we decide whether a win or loss has occurred [7 and 8]. If not, we roll the dice again and again [6].

When either a win or loss occurs, we check to see whether or not we have played all the games we wish to play [11]. If not, we play another game. If we have played the required number of games, we print the results [12].

The computer program in Figure 6.17 follows this flowchart quite closely.

The first problem we encounter in trying to construct the program is: How can we compute the numbers that appear on the faces of the dice?

In order to compute the numbers on the dice in some way that approximates what would happen if actual dice were tossed, we generate two "random" integers between 1 and 6. That is, we generate, using the computer, two integers between 1 and 6 in such a way that any one of the six possible digits is equally likely to occur. In BASIC the function called RND (for random) produces random numbers between 0 and 1. Of course, these numbers are not integers, and one has to be curious about what the word "random" means in this context.

"Random" in the sense just used means that each number between 0 and 1 is equally likely to occur if we generate a large quantity of these numbers. That is, we can expect .65 or .125 or .667342 to appear with equal probability. How can a computer perform such a task? We will not try to answer this question completely. Instead, we will answer it by providing an example of one method for generating such random numbers.

Suppose we wish to generate the integers from 1 to 10 in a random way. We start with any one of the integers, say 8. We then double the integer. If the result is less than 11, we use the number generated through the process of doubling. If the result is 11 or greater, we subtract 11 and use the result of the subtraction. In either case we have generated the next integer. In the example, if we double 8, we get 16. Since this is greater than 11, we subtract 11 getting 5. Thus 5 is the next integer. We then repeat the process. Doubling 5 we get 10. Thus the sequence of random integers thus far is

8, 5, 10

We double 10 getting 20. We then subtract 11, getting 9. Proceeding in this way, we produce the sequence

$$8, 5, 10, 9, 7, 3, 6, 1, 2, 4, \ldots$$

Here we have all 10 integers from 1 to 10 in a scrambled order. If we continue, the sequence will repeat itself. The important point is, however, that the above numbers appear to be random (i.e., there is no discernable pattern to them). Of course, there is a pattern that can be described by our method of computing the integers. Moreover, these numbers behave like random numbers in that in any long sequence each integer is just as likely to occur as any other.

The method just described is not very satisfactory, since the sequence repeats itself after every 10 numbers, and this could not be called random. There are methods of preventing this repetition from occurring so soon. Indeed strings of integers that do not repeat for several hundred thousands of integers are common. Of course, any given number, such as 5, and even combinations such as 5 followed by 3, may be repeated several times in the long string, but the entire string of numbers does not repeat. In addition there is no discernable pattern to the numbers, and they behave as random numbers from a probabilistic point of view.

The function RND in BASIC produces such strings of numbers with strings of exceptional length. In fact, the length of the string is so long that its length is of no concern to the user. The numbers produced are not integers. They are numbers such as .132 or .897, which are between 0 and 1 and do not include 0 and 1 themselves.

Since we want integers between 1 and 6, and RND produces numbers between 0 and 1, we take a random number produced through the use of RND and multiply by 6. This produces a number (not an integer) between 0 and 6, not including 0 and 6. We then add 1 so that we have a number between 1 and 7, not including 1 and 7. If we take the integer part of this number, we will get an integer between 1 and 6, including 1 and 6. Thus the BASIC expression

```
INT(6*RND(0) + 1)
```

produces an integer between 1 and 6. The value zero used with RND is quite arbitrary. If we generate two such random integers and add them, then we have a random toss of two dice. (See statements 700 and 800 and also statements 1800 and 1900.)

Notice that since a computer must have some method of actually calculating the numbers, the numbers produced by RND cannot be truly random. They appear to be random, and they behave as random numbers would in any experiment. For these reasons such numbers are referred to as *pseudo-random numbers*.

The remainder of the program is relatively straightforward and follows the flowchart (Figure 6.16) closely. The reader should run this program for a fairly large number of games. $N = 1000$ is reasonable. The results of running the program to play 1000 games several times are as follows:

	Number of Wins	Number of Losses	Estimate for Probability of Winning
	478	522	.478
	474	526	.474
	521	479	.521
	499	501	.499
Total	1972	2028	.493

Of course, running the program on a different computer would most likely produce different results, since the function RND produces a different sequence on each computer.

From these results the probability of winning seems to be slightly less than one half. That is, the odds are against the person tossing the dice.

Suppose we try to change the rules to give the person tossing the dice a better chance of winning. One way to do this is to eliminate the rule that produces a loss if the first roll is a 12. To accomplish this, we merely delete statement 1600 from the BASIC program, Now, if a 12 appears on the first toss, it becomes the point, and we proceed to the second and succeeding tosses until either a 7 or a 12 appears. On these succeeding tosses, if a 7 appears, the player loses; if a 12 appears, he wins.

If we run this modified program to play several sets of 1000 games, a typical set of results are:

	Number of Wins	Number of Losses	Estimate for Probability of Winning
	477	523	.477
	476	524	.476
	524	476	.524
	505	495	.505
Total	1982	2018	.496

It appears that we have increased the probability of winning, but the odds are still slightly against the person tossing the dice.

The reader should be able to make other changes in the rules and the corresponding changes in the BASIC program to carry out his own experiments. (See Exercise 1.)

EXERCISES FOR SECTION 6.8

1. Estimate the probability of winning a game of dice with the following rules.

	Win on First Toss	Lose on First Toss	Point on First Toss	Win on Later Tosses	Lose on Later Tosses
*(a)	7, 11	2, 3, 12	4, 5, 6, 8, 9, 10	7	Point
(b)	6, 10	2, 3, 12	4, 5, 7, 8, 9, 11	Point	6
*(c)	6, 10	2, 3, 12	4, 5, 7, 8, 9, 11	6	Point
(d)	7, 11	2, 12	3, 4, 5, 6, 8, 9, 10	Point	7
(e)	7, 11	3, 12	2, 4, 5, 6, 8, 9, 10	7	Point
(f)	2, 3, 12	7, 11	4, 5, 6, 8, 9, 10	7	Point
(g)	3, 7	2, 11, 12	4, 5, 6, 8, 9, 10	Point	7
(h)	7, 11	2, 3, 12	4, 5, 6, 8, 9, 10	Point	7, 11
(i)	6, 7, 8	2, 3, 11, 12	4, 5, 9, 10	Point	2, 7, 12
(j)	4, 7, 10	2, 5, 9, 12	3, 6, 8, 11	Point	4, 7, 10
(k)	7, 11	2, 3, 12	4, 5, 6, 8, 9, 10	Point	7, 11
(l)	7, 11	2, 3, 12	4, 5, 6, 8, 9, 10	7, 11	Point

2. Describe the difficulty that is presented in interpreting the rules for the following dice games.

	Win on First Toss	Lose on First Toss	Point on First Toss	Win on Later Tosses	Lose on Later Tosses
*(a)	7	2, 3, 12	4, 5, 6, 8, 9, 10, 11	Point	7, 11
(b)	3, 7	2, 11, 12	4, 5, 6, 8, 9, 10	10	Point
(c)	3, 4, 5, 6, 7	8, 9, 10, 11	12	Point, 12	2, 4, 6, 8, 10
(d)	2, 4, 11	3, 10, 12	5, 6, 8, 9	Point	7

3. Run the BASIC Program in Figure 6.17 for 10 games, 100 games, and 1000 games. Compare and discuss the relative values of the results in estimating the probability of winning the game of craps.

*4. Change the BASIC Program in Figure 6.17 so that while it plays the game of craps, it counts not only the number of wins and losses, but the number of wins and losses on the first toss. Estimate
   (a) The probability of winning on the first toss.
   (b) The probability of losing on the first toss.
   (c) The probability of drawing on the first toss.

5. Use the modified BASIC program in Exercise 4 to estimate the probabilities of winning, losing, and drawing on the first toss in all 10 games described in Exercise 1.

*6. (a) Write a BASIC program that (i) takes two integers as input, (ii) doubles the first integer, (iii) if the result is greater than the second number, subtracts the second number from the first, and (iv) prints the result and uses this result as a new value for the first number. The program should continue printing until the original first number is reached. The program should then stop.

(b) Run the program in part a with 8 as the first number and 11 as the second number. The output of the program should be: 5, 10, 9, 7, 3, 6, 1, 2, 4, 8.

7. (a) Run the BASIC program in Exercise 6a for the following pairs of numbers.
   - (i) 8, 13
   - (ii) 44, 53
   - (iii) 71, 107

   (b) How long would you guess the string of scrambled integers is before it repeats itself?

8. Rewrite the program in Exercise 6a so that it generates a sequence of integers between 1 and 6, inclusive. (*Hint.* Consider the remainder after division by 6.)

7

CONDITIONAL PROBABILITIES AND BAYES' FORMULA

7.1 INTRODUCTION

In the previous chapter we were concerned with finding the probabilities of particular events in the absence of outside influences. But there are many practical cases in which constraining conditions do exist and are important. For example, drilling a well in search of oil has a great deal of uncertainty connected with it. But oil prospectors seldom drill such expensive wells randomly. They are quite likely to conduct certain tests before going to the expense of boring a well. One such test is a seismic test. If the test is positive over some land area, the likelihood of oil being present under the land area in question increases, but it is still far from certain. The oil prospector would like to know the probability of oil being present if a seismic test is positive.

Along these same lines, in the midst of an epidemic of some disease, health authorities would like to detect new cases of the disease as early as possible. Early detection permits early treatment and may retard the outbreak of the epidemic. Suppose both a fever and a rash are symptoms of the disease. Then the medical examiners would like to know: What is the probability that a person has contracted the disease if he has a fever? If he has a rash? If he has both a fever and a rash?

Horse track habitues not only ask: What is the probability that my horse will win, but what is the probability that the horse will win if he has a certain jockey?

The word "if" in all of these questions is significant and changes the probabilities. A positive seismic test increases the likelihood of oil. A fever or a rash or both increases the likelihood that a patient has a disease. A successful jockey increases the likelihood that a horse will win. The "if" implies what is called a *conditional probability*. That is, we seek the probability of an event when certain conditions are imposed on the experiment. The condition may be the result of a seismic test, the result of a physical examination, the announcement of jockey assignments, or some other piece of information.

In this chapter we will examine such conditional probabilities and develop methods of computing them. We also will discuss some simple pictorial ways of keeping track of these probabilities in order to circumvent the sometimes tedious and error-prone algebra.

7.2 CONDITIONAL PROBABILITIES

Suppose we have an urn with four marbles: one red, one green, one white, and one black. We will draw a marble, observe its color, and replace it. We then draw another marble and observe its color. The sample space for this experiment is shown in Figure 6.9 on page 184.

Consider the event

$$X = \{\text{neither marble is white}\}$$

There are nine sample points in this event and

$$P(X) = 9/16$$

But suppose we are told that the first marble is not red. Now what is the probability that neither marble is white?

Let the event Y be

$$Y = \{\text{the first marble is not red}\}$$

Then we wish to find the probability that X occurs given the fact that Y does occur. We express this as

$$P(X \mid Y) \tag{7.1}$$

where the vertical bar is read "given," so that (7.1) is read "the probability of X given Y." To analyze this problem, we have redrawn the sample space of Figure 6.9 in

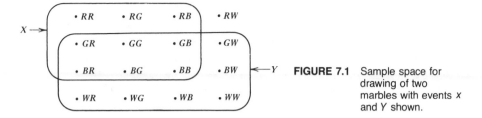

FIGURE 7.1 Sample space for drawing of two marbles with events *x* and *Y* shown.

Figure 7.1. In the latter figure the two events X and Y, as defined above, are shown.

Since the event Y must occur, the only possible sample points are the 12 in the last three rows. Thus, in effect, the sample space is just those 12 points. Of those 12 points the ones favorable to X are the six labeled GR, GG, GB, BR, BG, and BB. Thus the probability that X will occur is 6/12 or 1/2.

$$P(X \mid Y) = 1/2$$

The fact that Y occurs reduces the probability that X will occur from 9/16 to 1/2.

Lest the reader think that our method of attacking this problem is confined to experiments in which the parts of the experiment are independent, we turn to a two-part experiment where the parts are related to each other.

An urn contains six marbles: two red, one black, and three white. We will draw one marble from the urn, but we will not replace it. We then draw a second marble from the urn. Clearly we can get two reds or two whites, but we cannot get two blacks. We will number the red marbles R_1 and R_2. Similarly, we will number the white marbles W_1, W_2, and W_3. Thus, if the first draw results in R_1, the possibilities for the second draw are R_2, B, W_1, W_2, or W_3. Similarly, if the first draw results in B, the second draw can be any one of R_1, R_2, W_1, W_2, or W_3. The sample space is shown in Figure 7.2. Let the

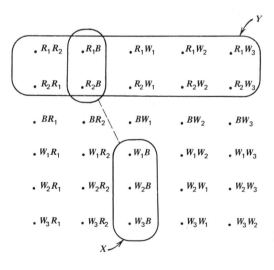

FIGURE 7.2 Sample space for drawing without replacement.

event X be: $\{$the second marble drawn is black$\}$. Let the event Y be: $\{$the first marble drawn is red$\}$. Then $n(U) = 30$ and $n(X) = 5$, so $P(X) = 5/30 = 1/6$. Similarly, $n(Y) = 10$, so $P(Y) = 1/3$. The events X and Y are shown in Figure 7.2. Notice that X is made up of two parts: one containing R_1B and R_2B and the other part consisting of W_1B, W_2B, and W_3B.

Suppose we wish to know $P(X \mid Y)$. That is, we wish to know the probability that the second marble drawn is black if the first marble drawn is red.

Since the first marble drawn is red, there are only 10 possible simple events that can occur. They are the 10 sample points in the first two rows of Figure 7.2. Of those 10 sample points only two are favorable to the event X (second marble drawn is black). Thus $P(X \mid Y) = 2/10 = 1/5$. Notice again that $P(X) = 1/6$, as we calculated earlier. Therefore knowing that the first marble drawn is red has increased the probability that the second marble is black.

We now turn to a general method for computing conditional probabilities. First we observe that in both Figures 7.1 and 7.2 the number of favorable sample points was the number of sample points common to both the events X and Y. That is, the number of favorable sample points was $n(X \cap Y)$. The number of possible sample points was the number of sample points in the event Y. Thus the probability was $n(X \cap Y)/n(Y)$. This is always the case, as we now demonstrate in a more general way.

In Figure 7.3 two events X and Y are shown. $P(X)$ is $n(X)$ divided by $n(U)$ where U is the certain event. If we are given that the event Y occurs, then the only possible sample points must lie in the region indicated as Y in Figure 7.3. Of those points the ones favorable to the event X are shown in the horizontally striped region. But the number of points in the striped region is $n(X \cap Y)$. Since the probability of any event is the number of favorable sample points in the event divided by the number of possible sample points,

$$P(X \mid Y) = \frac{n(X \cap Y)}{n(Y)}$$

Dividing numerator and denominator by $n(U)$ we get

$$P(X \mid Y) = \frac{P(X \cap Y)}{P(Y)} \qquad (7.2)$$

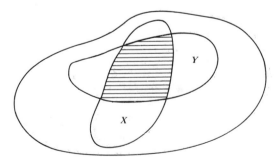

FIGURE 7.3 Sample spaces to compute $P(X \mid Y)$.

We may rewrite this equation as

$$P(X \cap Y) = P(X \mid Y) \cdot P(Y) \qquad (7.3)$$

Thus, to compute the probability of X and Y occurring, we need to know (1) the probability of X given that Y occurs, and (2) the probability of Y occurring. Recall that if X and Y are independent,

$$P(X \cap Y) = P(X) \cdot P(Y)$$

With (7.2) are we now in a position to calculate the probability that a person has a certain disease if he has a fever? It is unlikely that we are able to make this calculation, considering the realities of the situation. It is possible to take a group of people known to have the disease and, by taking their temperatures, find the probability that they have a fever. But this is

$$P(\text{fever} \mid \text{disease})$$

and what we seek is

$$P(\text{disease} \mid \text{fever})$$

and the two are not the same (see Exercises 1a and 2a and compare these with the results above). Similarly, in the oil well problem, we can conduct seismic tests at the sites of known wells and find

$$P(\text{positive seismic test} \mid \text{oil well})$$

but we seek

$$P(\text{oil well} \mid \text{positive seismic test})$$

Thus we need some way to reverse the conditional probabilities, and we turn our attention to that problem in the next section. (See also Exercises 4 to 7.)

EXERCISES FOR SECTION 7.2

(Note. Solutions are given for those exercises marked with an asterisk.)

1. An urn has four marbles: one red, one green, one black, and one white (see Figure 7.1). If two marbles are drawn with replacement, what is the probability that
   *(a) The first marble is not red if neither marble is white?
   (b) Neither marble is white if at least one marble is green?
   (c) At least one marble is green if neither marble is white?
   (d) Neither marble is black if the two marbles are different colors?
   (e) The two marbles are different colors if neither marble is black?

2. An urn contains six marbles: two red, one black, and three white. We draw a marble, observe its color, and do not replace it. We then draw a second marble and observe its color. (See Figure 7.2.) What is the probability that
   *(a) The first marble drawn is red if the second is black?
   (b) Both marbles are the same color if the first marble is not white?
   (c) The first marble is not white if both marbles are the same color?
   (d) The marbles drawn are red and black if the two marbles are not the same color?
   (e) The two marbles are not the same color if the two marbles drawn are red and black?

*3. Use (7.2) to compute $P(X \mid Y)$ when X and Y are independent.

4. Show that

$$P(X \mid Y) \cdot P(Y) = P(Y \mid X) \cdot P(X)$$

*5. Suppose we conduct seismic tests at several oil wells and find that at a well the probability of a positive test is 0.7. If we conduct such seismic tests randomly, well or no well, the probability of a positive test is 0.4. The probability of a well without any tests is 0.2. If we conduct a seismic test and obtain a positive result, what is the probability that a well is present? (*Hint.* See Exercise 4.)

6. A certain disease is reaching epidemic proportions. Thirty percent of the population is now infected. By taking the temperatures of a group of people known to have the disease, we find that the probability that a person has a fever if he has the disease is 0.6. If we take the temperatures of all people, sick or well, we find that 20% have a fever. If a person has a fever, what is the probability that he has the disease? (*Hint.* See Exercise 4.)

7. An investor holds two securities, United Biscuits and National Oxygen. He estimates that the probability that United Biscuits will increase in value is 0.3 and that the probability that National Oxygen will increase in value is 0.2. On past performance, he notices that when the value of National Oxygen increases, then 80% of the time United Biscuits also increases in value. If United Biscuits' value increases, what is the probability that National Oxygen will increase as well? (*Hint.* See Exercise 4.)

8. Recall the blood donor problem in the text of Section 6.7. If a donor is found for someone who is type A+, what is the probability that the donor is
   *(a) A+?
   (b) A−?
   (c) B−?

9. In a survey of 2655 employed persons the following results were obtained:

Education	Less than $5000	Annual Income Between $5000 and $10000	Over $10000
No high school diploma	387	124	36
High school diploma	419	617	404
College degree	109	212	347

If a person is selected at random from this group of 2655 people, what is the probability that
*(a) He has only a high school diploma?

(b) His salary is less than $5000?

(c) He has no college degree?

*(d) He has only a high school diploma if his salary is less than $5000?

(e) His salary is less than $5000 if he has only a high school diploma?

(f) His salary is over $10000 if he has a college degree?

(g) His salary is over $10000 if he has only a high school diploma?

(h) He has a college degree if his salary is between $5000 and $10000?

10. Verify the equation in Exercise 4 using the table in Exercise 9 where

$$X = \text{has only a high school diploma}$$
$$Y = \text{has an annual salary less than } \$5000$$

(*Hint*. See Exercise 9a, 9b, 9d, and 9e.)

7.3 BAYES' PROBABILITIES

Events are called mutually exclusive if the occurrence of one precludes the occurrence of the other. (See Section 6.5.) Suppose we have three events A_1, A_2, and A_3 that are mutually exclusive. Then

$$n(A_1 \cap A_2) = 0$$
$$n(A_2 \cap A_3) = 0$$
$$n(A_1 \cap A_3) = 0$$

If these three events include all possible outcomes of an experiment, they ar said to be a *complete set of mutually exclusive alternatives*. An example is in order.

Consider an urn containing four marbles: one red, one green, one black, and one white. We draw a marble from the urn, observe its color, and replace it. We then draw a second marble and observe its color. We have encountered this experiment before in Section 6.5 and again in Section 7.2. The sample space is shown in Figure 7.1. Let

$$A_1 = \left\{\text{first marble drawn is red}\right\}$$
$$A_2 = \left\{\text{first marble drawn is green}\right\}$$
$$A_3 = \left\{\text{first marble drawn is black or white}\right\}$$

These three events are mutually exclusive — if any one occurs the other two are excluded. The first consists of the first row in Figure 7.1, A_2 is the second row, and A_3 consists of the last two rows. The three events are a complete set, since they account for all possible outcomes.

In a complete set of mutually exclusive alternatives, every simple event (sample point) is accounted for once and only once. The set of simple events themselves each considered as a set with one member is a complete set of mutually exclusive alternatives although there are, in general, more than three parts to this mutually exclusive set.

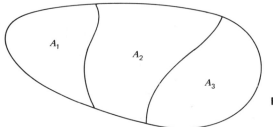

FIGURE 7.4 Venn diagram showing complete set of mutually exclusive alternatives.

If A_1, A_2, and A_3 are a complete set of mutually exclusive alternatives for an experiment with a sample space U, we can represent this by the Venn diagram in Figure 7.4. Notice that A_1, A_2, and A_3 cover all of the set U, but the three sets do not overlap. From the figure we see that

$$n(U) = n(A_1) + n(A_2) + n(A_3)$$

Consider an event X from this experiment, as shown in Figure 7.5. From the figure, it is clear that

$$n(X) = n(X \cap A_1) + n(X \cap A_2) + n(X \cap A_3)$$

Dividing by $n(U)$

$$P(X) = P(X \cap A_1) + P(X \cap A_2) + P(X \cap A_3) \qquad (7.4)$$

Thus the probability that X will occur is the sum of the probabilities of X and each of the alternatives occurring.

Suppose we know the probability of each of the alternatives occurring $\left[\text{i.e., we know } P(A_1), P(A_2), \text{ and } P(A_3)\right]$. Suppose also that we know the probability of X occurring if any one of the alternatives arises $\left[\text{i.e., we know } P(X \mid A_1), P(X \mid A_2), \text{ and } P(X \mid A_3)\right]$. Can we use this information to compute

$$P(A_2 \mid X)$$

This is the probability that the alternative A_2 occurs if the event X occurs. The answer to the question is "yes," but before we show how this is accomplished, we pause to demonstrate why one might wish to do so.

Suppose you are an oil prospector. You would like to find a gusher, but you will be content with a more modest oil deposit. From past experience you know that drilling wells at random will produce a gusher 10% of the time and a moderate supply 20% of the time. The other 70% of the borings are dry. We have then a complete set of mutually exclusive alternatives for well drilling; that is,

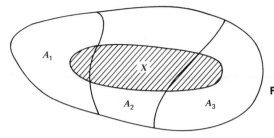

FIGURE 7.5 Venn diagram showing event X and complete set of mutually exclusive alternatives.

$$A_1 = \{\text{gusher or large oil supply}\}$$
$$A_2 = \{\text{moderate oil supply}\}$$
$$A_3 = \{\text{no oil}\}$$

and

$$P(A_1) = 0.1 \qquad P(A_2) = 0.2 \qquad P(A_3) = 0.7$$

The cost of drilling wells is too high to take the risk when the chances of finding oil are as slim as this, so you decide to conduct seismic tests to increase the chances of finding oil. To determine the effectiveness of these tests, you conduct tests over existing wells and over some dry borings. When conducted where a gusher is known to be present, the seismic test produces a positive result 80% of the time; that is,

$$P(X \mid A_1) = 0.8$$

where X is the event "a positive seismic test result." Where modest wells exist, the test is positive 60% of the time, and over dry borings it is positive 30% of the time. Thus

$$P(X \mid A_2) = 0.6 \qquad P(X \mid A_3) = 0.3$$

But what we want to know is: If a seismic test is positive, is there oil present? And how much? In terms of the events described above we would like to calculate $P(A_1 \mid X)$ and $P(A_2 \mid X)$. The first is the probability of a gusher, and the second is the probability of a moderate oil supply.

Let us return then to the computation of $P(A_2 \mid X)$. From (7.2),

$$P(A_2 \mid X) = \frac{P(A_2 \cap X)}{P(X)} \qquad (7.5)$$

But we do not know either of the two probabilities on the right. Consider first the numerator and notice that

$$P(A_2 \cap X) = P(X \cap A_2)$$

From (7.3),

$$P(X \cap A_2) = P(X \mid A_2) \cdot P(A_2)$$

From our assumptions we do know both probabilities on the right in this last equation so we know the numerator in (7.5).

$$P(A_2 \mid X) = \frac{P(X \mid A_2) \cdot P(A_2)}{P(X)} \tag{7.6}$$

The value of $P(X)$ can be calculated from (7.4) if we use (7.3) three more times. In particular, from (7.3),

$$P(X \cap A_1) = P(X \mid A_1) \cdot P(A_1)$$

$$P(X \cap A_2) = P(X \mid A_2) \cdot P(A_2)$$

$$P(X \cap A_3) = P(X \mid A_3) \cdot P(A_3)$$

so from (7.6),

$$P(A_2 \mid X) = \frac{P(X \mid A_2) \cdot P(A_2)}{P(X \mid A_1) \cdot P(A_1) + P(X \mid A_2) \cdot P(A_2) + P(X \mid A_3) \cdot P(A_3)} \tag{7.7}$$

We will use this to solve the oil well problem.

$$P(A_2 \mid X) = \frac{0.6 \times 0.2}{0.8 \times 0.1 + 0.6 \times 0.2 + 0.3 \times 0.7}$$

$$= 0.293$$

If the seismic test is positive, then more than 29% of the time we should expect a moderate oil supply to be present. But what of a gusher?

We could have started out to compute $P(A_1 \mid X)$ instead of $P(A_2 \mid X)$. If we had done so, then instead of (7.7), we would have obtained

$$P(A_1 \mid X) = \frac{P(X \mid A_1) \cdot P(A_1)}{P(X \mid A_1) \cdot P(A_1) + P(X \mid A_2) \cdot P(A_2) + P(X \mid A_3) \cdot P(A_3)}$$

Therefore,

$$P(A_1 \mid X) = \frac{0.8 \times 0.1}{0.8 \times 0.1 + 0.6 \times 0.2 + 0.3 \times 0.7}$$

$$= 0.195$$

A gusher will be present in almost 20% of the cases where a seismic test is positive.

Adding these two results,[1] we see that an oil supply, large or moderate, exists about 50% of the time if we have a positive result from a seismic test. Compared with the 30% chance in the absence of testing, this represents an increase of 20% in the chances of finding oil.

Is the seismic test worth its cost? To answer this question we need to know the cost of conducting the test, the cost of drilling one well, the number of wells planned, the profit from a well, and so on. Given these costs then we ask: Is the cost of the test less than the value of the oil that will be obtained from an increase of 20% in the number of wells? (See Exercise 10.)

Equation 7.7 is called *Bayes' formula* and is an extremely powerful and often used formula.

EXERCISES FOR SECTION 7.3

*1. An urn contains six marbles: two red, one black, and three white. You draw two marbles from the urn without replacing the first marble before the second is drawn. You are blindfolded when the first marble is drawn, but you see that the second marble is red.
(a) What is the probability that the first marble was red?
(b) Someone offers to bet $4 to your $1 that the first marble drawn was not red. Should you accept the bet?

2. Same as Exercise 1 except that the urn contains seven marbles: two red, two black, and three white.

*3. There are two urns. The first urn contains five marbles: one red, two black, and two white. The second urn contains four marbles: one red, two black, and one white. You select an urn (each urn is equally likely to be selected) without knowing whether it is the first urn or the second urn. You draw a marble from the urn and observe its color to be black. What is the probability that the urn selected was the first urn?

4. Same as Exercise 3 except that the marble drawn was observed to be white.

*5. Each person in a group of patients examined at a medical clinic is classified into one of the following four groups.

$$A = \{\text{persons with a rash}\}$$
$$B = \{\text{persons with a fever}\}$$
$$C = \{\text{persons with a headache}\}$$
$$D = \{\text{persons with none of the above three symptoms}\}$$

(a) Are A and B mutually exclusive?
(b) Are B and D mutually exclusive?
(c) Are A, B, and C mutually exclusive?
(d) Describe a complete set of mutually exclusive alternatives using the above four classifications.

6. The patients in the clinic in Exercise 5 also are classified as follows.

[1]We add the results because the events "gusher" and "moderate supply" are mutually exclusive events.

$$M = \{\text{adult males}\}$$
$$F = \{\text{adult females}\}$$
$$C = \{\text{children}\}$$

(a) Are M and F mutually exclusive?

(b) Are F and C mutually exclusive?

(c) Do M, F, and C form a complete set of mutually exclusive alternatives?

7. The patients in the clinic in Exercise 5 also are classified as follows.

$$X = \{\text{had a physical examination within the last year}\}$$
$$Y = \{\text{had a physical examination within the last two years}\}$$
$$Z = \{\text{have never had a physical examination}\}$$

(a) Are X and Y mutually exclusive?

(b) Are X and Z mutually exclusive?

(c) Redefine Z so that Y and Z are a complete set of mutual exclusive alternatives.

*8. If a person has tuberculosis, its early detection is important in order to save the patient's life. Chest X-rays are one method of detecting tuberculosis. If a person has tuberculosis, a chest X-ray will detect that fact 90% of the time. But the X-ray sometimes diagnoses tuberculosis when it is not present. If a patient is healthy, the X-ray will indicate tuberculosis is present 1% of the time. If tuberculosis is present in 5 out of every 10,000 persons, what is the probability that even if an X-ray indicates no tuberculosis that the patient in fact does have the disease?

9. It is essential that flaws in the equipment of a spacecraft be detected if the craft and its occupants are to survive an orbiting mission. Elaborate electronic sensing equipment is installed to detect such flaws even though they rarely occur. Suppose the failure rate of a critical part in a missile is 1/10 of 1%. Suppose also that if the part fails that 2% of the time the electronic sensing equipment does not detect this fact. Finally, suppose that if the part is functioning correctly (not failing), the sensing equipment will say that it does fail 5% of the time. If the sensing equipment does not detect a failure, what is the probability that a failure nonetheless exists?

10. Consider the oil well problem in Section 7.3. Suppose a gusher produces a $1 million profit while a moderate oil supply yields a $500,000 profit. Conducting a seismic test costs $50,000 per well. Is the test profitable if the total number of wells (gushers, moderate and dry) to be drilled is

(a) 10

*(b) 20

(c) 30

11. Two diseases are prevalent in a community. No patient can have both diseases at the same time. Suppose 40% of the population has one or the other of the diseases. The first disease is three times as prevalent as the second one. Both diseases produce a common symptom, say a rash. Of those patients with the first disease, 60% have a rash. Of those with the second disease, 80% have a rash. However, even a person with neither disease has a rash 20% of the time. If a patient is examined and found to have a rash, what is the probability he has

(a) The first disease?

(b) The second disease?

(c) Neither disease?

(d) One of the two diseases?

12. Same as Exercise 11 except that the two diseases are equally prevalent.

13. Same as Exercise 11 except that the rash only appears on patients with one of the two diseases (i.e., none of the persons with neither disease ever has a rash).

7.4 TREES

In mathematics we often find it useful to look at a problem in several ways. If one way is not fruitful, the other may be more revealing. If one method of attacking a problem seems tedious or leads to blunders on our part, another method may be more profitable.

Bayes' formula is a powerful and important tool, but the algebraic manipulations often conceal the underlying process and may be difficult to carry out without committing blunders. In this section, therefore, we turn to a geometrical or graphical representation of Bayes' formula.

Suppose we have an urn with three balls: one red, one black, and one white. We make two draws from the urn without replacing the first ball drawn. The sample space for this experiment is shown in Figure 7.6. This is one graphical or geometrical representation of this experiment. However, others are possible, and we turn to one of the alternative graphical representations now.

● *RB*　　● *RW*

● *BR*　　　● *BW*

● *WR*　　● *WB*　　　**FIGURE 7.6**　Sample space for drawing marbles without replacement.

On the first draw there are three possible outcomes: red, black, and white. We present these graphically as indicated in Figure 7.7. This diagram says that, starting from the empty circle, we can go in one of three directions indicated by the arrows. One arrow leads to a circle containing R. This implies that the result of the first draw is red (R). The other arrows lead to B for black and W for white.

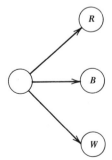

FIGURE 7.7　First stage in tree diagram for drawing marbles.

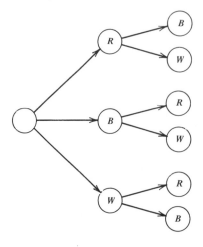

FIGURE 7.8 Complete tree for drawing marbles without replacement.

Starting from each of the outcomes of the first draw, there are two possible outcomes for the second part of the experiment. These are indicated graphically in Figure 7.8.

By following each possible "chain" of arrows, we can determine the six possible outcomes of the experiment. The symbols that we encounter as we travel along the chain *taken in order* represent the outcome of that chain. For example, the uppermost chain in Figure 7.8 encounters R followed by B. Thus this chain indicates the event R followed by B, or a red marble is drawn on the first draw and a black marble on the second. This chain then corresponds to the sample point RB in Figure 7.6. Notice also that each chain corresponds to a sample point in the sample space shown in Figure 7.6.

Diagrams such as those shown in Figures 7.7 and 7.8 are called *trees* because of their treelike structure. (The root of the tree is at the left and the tree appears to be growing horizontally to the right.) The arrows are called the *branches* of the tree. If we start at any circle and follow an arrow to a second circle and an arrow from the second circle to a third, and so on, we construct a *chain* of arrows.

The tree shown in Figure 7.8 corresponds to a two-part experiment (two draws from the urn). Of course, we can have experiments that have two, three, or more parts.

We will label each branch of the tree with the probability that the event at the point of the arrow will occur *given that the event at the tail of the arrow has occurred* (i.e., a conditional probability). That is, if the arrow points from event B to event A, the arrow is labeled with $P(A \mid B)$. For example, the three branches emanating from the empty circle in Figure 7.8 are labeled with 1/3, since the probability of each marble occurring on the first draw is 1/3. All six arrows in the second stage of that figure are labeled with 1/2 since, for example, if R occurs on the first draw, the probability of either a B or a W on the second draw is 1/2. The result of the labeling process is shown in Figure 7.9. These numerical labels placed on the branches are called the *branch weights*.

The probability that we get a red marble on the first draw and a white marble on the second draw in the two draws without replacement experiment is designated by

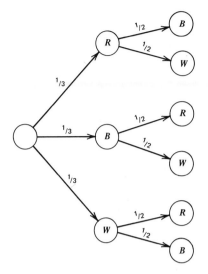

FIGURE 7.9 Tree for drawing marbles with branch weights.

$P(R_1 \cap W_2)$ and read as the "probability that R occurs on the first draw and W on the second." From (7.3),

$$P(R_1 \cap W_2) = P(R_1) \cdot P(W_2 \mid R_1) \tag{7.8}$$

But $P(R_1)$ is the label (branch weight) on the arrow leading from the empty circle to R. Similarly, $P(W_2 \mid R_1)$ is the branch weight on the arrow leading from R to W in the second stage. Thus,

$$P(R_1 \cap W_2) = (1/3) \cdot (1/2) = 1/6$$

Notice that this is the product of the branch weights on the chain.

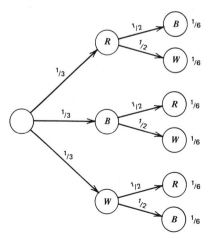

FIGURE 7.10 Tree for drawing marbles with chain probabilities.

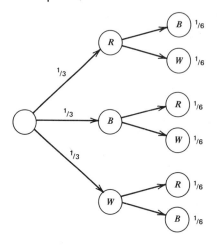

FIGURE 7.11   Tree for drawing marbles where conditional probabilities are unknown.

To find the probability that all of the events on any chain will occur in the order in which they appear, we form the product of the branch weights of the chain. The reader should study the trees and (7.8) to assure himself that this is true of all chains.

With this in mind we will label all chains with their probabilities of occurring by placing the appropriate number at the end of the chain. This number, we repeat, is obtained by multiplying all of the branch weights on the chain. If we do that with the two draw—without replacement—urn problem, we transform Figure 7.9 into Figure 7.10.

In this case all chain probabilities (also sometimes referred to as path probabilities) are identical and equal to 1/6. This is not always the case, but it is always so that the chain probabilities sum to 1 since, in any one particular experiment, one of the paths must be followed to its end.

The use of trees, branch weights, and chain probabilities is a convenient way of computing $P(A \cap B)$. Of course, to compute the branch weights, we need to know the conditional probabilities. Once we know these we can compute the compound probabilities from (7.8). But what if it is the conditional probabilities themselves that we seek? If we know the compound probabilities, we can compute the conditional ones. Returning to the urn problem, let R_1 be red on the first draw, B_2 be black on the second, and so on. If we know $P(R_1)$, $P(B_1)$, $P(W_1)$ and also $P(R_1 \cap B_2)$, $P(R_1 \cap W_2)$, $P(B_1 \cap R_2)$, $P(B_1 \cap W_2)$, $P(W_1 \cap R_2)$, and $P(W_1 \cap B_2)$, we can label the tree as shown in Figure 7.11. How can we determine the branch weights (conditional probabilities) on the second stage? The product of the branch weights on any chain must equal the chain probability. Looking at the topmost chain (R to B), let x be the branch weight on the unlabeled branch. Then

$$(1/3) \cdot x = 1/6$$

or

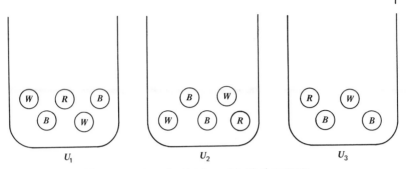

FIGURE 7.12 Three urns, two of which have identical contents.

$$x = 1/2$$

The other chains may be handled in the same way.

The use of branch weights and chain probabilities is just a convenient way of applying (7.3). It is important to note that we have not really added anything new to our knowledge of probabilities by the use of trees and their labels. Everything contained in the tree and everything that can be derived from the tree also can be derived from the algebraic formulas developed in the preceding sections. Trees are merely a convenience that, on occasion, saves a great deal of formula manipulation.

As another example of the convenience of trees, we will discuss how to "flip" a tree. This process will be equivalent to using Bayes' formula and will be somewhat easier to apply than a direct application of (7.7). Some people find it easier to remember the steps in "flipping" a tree than to remember Bayes' formula, and some people find the reverse is true.

Consider three urns. Two of the urns have identical contents: one red marble, two black marbles, and two white marbles. The third urn has one red marble, two black marbles, and one white marble. (See Figure 7.12.) We select an urn, draw a marble from the urn, and observe its color. A tree with branch weights and path probabilities for this experiment is shown in Figure 7.13. Notice that the first two urns are treated as one, since they have identical contents. They are denoted by $U_{1,2}$. These two urns are twice as likely to be selected as urn three (U_3), since they are twice as plentiful.

Suppose we think of the first stage not as selecting an urn but as selecting a marble. Then the second stage consists of selecting an urn. This leads to the tree diagram less the branch weights shown in Figure 7.14.

The chain probabilities are shown in Figure 7.14. They are read directly from the original tree shown in Figure 7.13 as follows. For each chain in Figure 7.13 there is a chain in Figure 7.14. For example, the topmost chain in Figure 7.13 represents the event $U_{1,2} \cap R$. Similarly, the topmost chain in Figure 7.14 represents $R \cap U_{1,2}$. But these are the same event, so they must have the same probability of occurring. Similarly, the chain that passes through U_3 and B in Figure 7.13 represents the compound event $U_3 \cap B$ and has probability 1/6, as indicated by the chain probability. In Figure

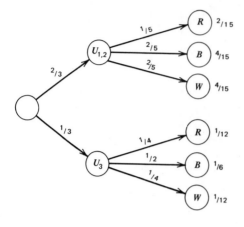

FIGURE 7.13 Tree for drawing from urns of Figure 7.12 with branch weights.

7.14 there is a chain passing through B and U_3. It represents the event $B \cap U_3$ and has the same probability (i.e., 1/6). This value is then placed at the end of the chain passing through B and U_3, as shown in Figure 7.14. Similarly, the chain probabilities for all six chains in Figure 7.14 can be translated from Figure 7.13.

In Figure 7.13 there are two chains leading to R. Thus there are two ways in which a red marble may be drawn. They are mutually exclusive events, since one involves urns 1 and 2 and the other involves urn 3. Therefore the probability that either of these two chains occurs is the sum of the probabilities of the two chains themselves. Thus the probability of a red marble being drawn is 2/15 + 1/12, or 13/60. This is the branch weight for the arrow leading to R in Figure 7.14. Similarly, the probability of a black marble being drawn is

$$4/15 + 1/6 = 26/60$$

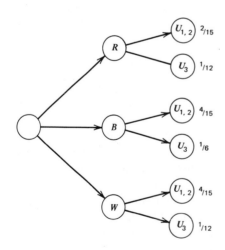

FIGURE 7.14 Tree of Figure 7.13 after it has been "flipped" but without branch weights.

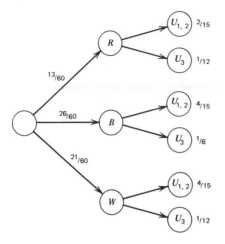

FIGURE 7.15 "Flipped" tree for urn problem with first stage labeled.

and the probability of a white marble is

$$4/15 + 1/12 = 21/60$$

These are the other branch weights for the first stage of Figure 7.14. The result of adding them to Figure 7.14 produces Figure 7.15.

All that remains is to label the branches in the second stage. But we know that the product of the branch weights on any chain must be equal to the chain probability. Thus, if the branch weight on the arrow leading from R to $U_{1,2}$ (topmost branch) is x, then it must be that

$$(13/60) \cdot x = 2/15$$

or

$$x = 8/13$$

This is the label that must be attached to the topmost branch of stage two. Similarly, for the second branch from the top of stage two, we let the branch weight be y. Then it must be that

$$(13/60) \cdot y = 1/12$$

or

$$y = 5/13$$

Continuing in this way, we find the following branch weights.

$$B \text{ to } U_{1,2} \qquad (4/15) \div (26/60) = \quad 8/13$$

$$B \text{ to } U_3 \qquad (1/6) \div (26/60) = \quad 5/13$$

$$W \text{ to } U_{1,2} \qquad (4/15) \div (21/60) = 16/21$$

$$W \text{ to } U_3 \qquad (1/12) \div (21/60) = \quad 5/21$$

Attaching these weights to the appropriate branches of Figure 7.15 produces the final tree diagram shown in Figure 7.16.

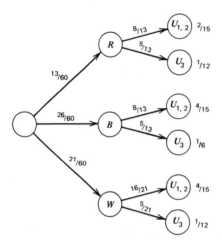

FIGURE 7.16 "Flipped" tree for urn problem completely labeled.

Recall that any branch weight in a tree is the conditional probability that the event at the head of the arrow occurs given that the event at the tail of the arrow has occurred. Thus, 8/13, the branch weight on the arrow leading from B to $U_{1,2}$ is $P(U_{1,2} \mid B)$. This agrees with the result calculated from Bayes' formula. If we let $U_{1,2}$ and U_3 be the set of mutually exclusive alternatives, then Bayes' formula produces

$$P(U_{1,2} \mid B) = \frac{P(B \mid U_{1,2}) \cdot P(U_{1,2})}{P(B \mid U_{1,2}) \cdot P(U_{1,2}) + P(B \mid U_3) \cdot P(U_3)}$$

Using the probabilities given above,

$$P(U_{1,2} \mid B) = \frac{(2/5) \times (2/3)}{(2/5) \times (2/3) + (1/2) \times (1/3)}$$

$$= 8/13$$

We now summarize the rules for flipping a two-stage tree.

1. The events in stage one of the original tree become the events in stage two of the flipped tree and vice versa (see Figure 7.17).

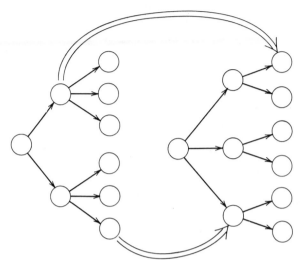

FIGURE 7.17   Diagram indicating interchange of stages in flipping a tree (Step 1).

2. For each chain in the original tree identify the corresponding chain (same set of events) in the flipped tree and assign the same chain probability to it. (See Figure 7.18.) In this way *all* chains in the flipped tree are labeled.

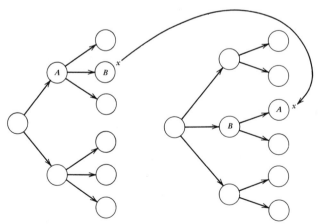

FIGURE 7.18   Diagram indicating labeling of all chain probabilities in flipped tree (Step 2).

3. In the original tree find all chains terminating in a given event, say B. Sum the chain probabilities of all these chains. Assign this sum to the branch leading to the event B in the flipped tree (See Sigure 7.19). In this way *all* branches in the first stage of the flipped tree can be labeled.

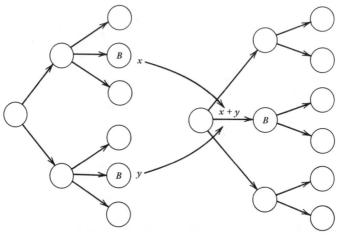

FIGURE 7.19 Diagram indicating labeling procedure of branches in first stage of flipped tree (Step 3.)

4. To find the branch weight on any second stage branch of the flipped tree, divide the chain probability by the first stage branch weight (see Figure 7.20). This completes the flipped tree.

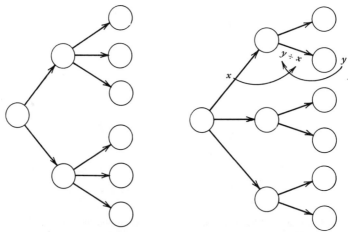

FIGURE 7.20 Diagram indicating computation of branch weights in second stage of flipped tree. (Step 4.)

EXERCISES FOR SECTION 7.4

1. Draw and label trees for the following experiments.

   *(a) An urn contains three marbles: two red and one white. A marble is drawn from the urn, its color is observed, but it is not replaced. A second marble is drawn from the urn. (*Note*. There should be only two outcomes for the first stage: red and white.)

   (b) An urn contains four marbles: two red, one green, and one white. A marble is drawn from the urn, its color observed, but it is not replaced. A second marble is drawn from the urn. (*Note*. There should be only three outcomes for the first stage: red, green, and white.)

   (c) There are two urns. The first urn contains seven marbles: three red, three white, and one black. The second contains 10 marbles: five red, three white, and two black. An urn is selected. A marble is selected from the urn and its color is observed.

   (d) Consider the same two urns described in part c. A marble is selected from the first urn. Then a marble is selected from the second urn.

2. Complete the labeling on the following trees.

(a)

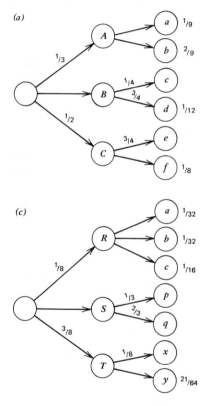

(b)

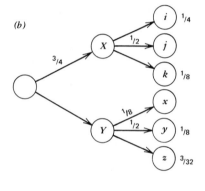

(c)

*3. In a dozen eggs, three eggs are known to be spoiled. Two eggs are examined in succession. The examined eggs are not replaced.

   (a) Draw a tree diagram describing this two-stage experiment. There are two possible outcomes at each stage: a spoiled egg or an unspoiled egg. Label the tree.

(b) What is the probability that both of the eggs that are examined are spoiled?

(c) That precisely one is spoiled?

4. In a dozen light bulbs, four are known to be defective. Three bulbs are examined in succession. The examined bulbs are not replaced.

(a) Draw a tree diagram describing this three-stage experiment. There are two possible outcomes at each stage: defective or good. Label the tree.

(b) What is the probability that all three bulbs that are examined are defective?

(c) What is the probability that precisely two of the bulbs that are examined are defective?

*5. There are three types of urns: A, B, and C. Out of 10 urns three are of type A, four of type B, and three of type C. Type A urns contain two red marbles, three white marbles, four black marbles, and three yellow marbles. Type B urns contain no red marbles, four white marbles, one black marble, and five yellow marbles. Type C urns have six red marbles, one black marble, one white marble, and no yellow marbles. An urn is selected at random and a marble drawn from the urn so selected.

(a) Draw a tree describing this experiment and label the branches and chains.

(b) "Flip" the tree so that the first stage represents the color of the marble selected and the second stage represents the type of urn selected.

(c) What is the probability that the marble is white if the urn is type B?

(d) What is the probability that the urn was type B if the marble was white?

6. There are three types of urns: X, Y, and Z. Out of 10 urns there are two of type X, three of type Y and five of type Z. Type X urns contain four red marbles, five white marbles, and one black marble. Type Y urns contain two red marbles, two white marbles, and six black marbles. Type Z urns contain no red marbles, seven white marbles, and three black marbles. An urn is selected at random and a marble is drawn from the urn so selected.

(a) Draw a tree diagram describing this experiment and label the branches and chains.

(b) "Flip" the tree so that the first stage represents the color of the marble and the second stage represents the type of urn.

(c) What is the probability that the marble is red if the urn is of type Z?

(d) What is the probability that the urn is of type Z if the marble is red?

*7. If we drill for oil in a certain area the probability of a gusher is 0.2, the probability of a moderate oil supply is 0.3 and the probability of no oil is 0.5. A gravity test can be conducted over a well or a dry boring. If the test is conducted over a gusher, it registers a highly favorable result 80% of the time, a favorable result 10% of the time, and a negative result 10% of the time. If this same test is conducted over a moderately well-supplied well, it registers highly favorable 20% of the time, favorable 60% of the time, and negative 20% of the time. Over a dry well the gravity test is highly favorable 5%, favorable 25%, and negative 70%.

(a) Draw a tree diagram describing the two-stage experiment: (i) drill a well and discover whether it is a gusher, a moderate supply, or dry; (ii) conduct and record a gravity test. Label the tree.

(b) "Flip" the tree so that stage one is the result of the gravity test and stage two is the type of well.

(c) If a gravity test is favorable, what is the probability that the well will be dry?

(d) If a gravity test is favorable or highly favorable, what is the probability that the well will be a gusher?

8. Same as Exercise 7 except that the probabilities of a gusher, a moderate oil supply, and a dry well in the area are 0.15, 0.4, and 0.45, respectively.

9. Of the registered voters in a certain district, 40% are Republicans, 25% are Democrats, and 35% are Independents. Republicans vote for their party's candidate 80% of the time. Democrats vote Democratic 90% of the time. Independents vote Democratic 60% of the time.
   (a) Draw a tree diagram in which the first stage is the selection of a party affiliation: Republican, Democrat, or Independent; and the second stage is the party that receives the vote: Republican or Democratic.
   (b) "Flip" the tree so that the first stage is the party receiving the vote and the second stage is the affiliation.
   (c) If the vote is cast for a Democrat, what is the probability that the voter was an Independent?
   (d) If the vote is cast for a Republican, what is the probability that the voter was a Republican?

*10. You are entered in a tournament in which you play three matches and must win two *consecutive* matches in order to win an award. You must alternate playing matches against opponent X and opponent Y. However, you may elect the opponent, X or Y, against whom your first match is to be played. You estimate your probability of winning against X is 1/3 and your probability of winning against Y is 2/3. Who should you choose to play first, X or Y? (*Hint.* Draw trees describing the two alternatives.)

11. Same as Exercise 10 except that you estimate the probability of winning against X is 1/2 and the probability of winning against Y is 2/5.

7.5 A BASIC PROGRAM FOR BAYES' FORMULA

We now have two methods for reversing conditional probabilities—using (7.7) or flipping a tree.[2] In this section we will develop and use a BASIC program to accomplish this task. The program that we develop will evaluate (7.7).

Suppose once more that we wish to compute $P(X \mid A_2)$ where A_2 is one of a complete set of mutually exclusive events. To do so we need the probabilities of each of the mutually exclusive events. In our previous discussion we have confined ourselves to sets of three mutually exclusive events: A_1, A_2, and A_3. We now broaden our outlook to take into account a set of m events (i.e., $A_1, A_2, \ldots, A_m$). Of course, if $m = 3$, this reduces to the previous case.

Thus we need the values of

$$P(A_1), P(A_2), \ldots, P(A_m) \tag{7.9}$$

We also need the conditional probabilities of each mutually exclusive event, given that the event X has occurred. That is, we need to know the values of

$$P(A_1 \mid X), P(A_2 \mid X), \ldots, P(A_m \mid X) \tag{7.10}$$

---

[2]Recall that these two methods are really two different ways to do the same thing.

The flow chart in Figure 7.21 describes the procedure we will follow. The number of events in the mutually exclusive set, m, is required (Box 1) and then both (7.9) and (7.10) are required (Boxes 2 and 3). For a set of m events, Bayes' formula, (7.7), becomes

$$P(A_2 \mid X) = \frac{P(X \mid A_2) \cdot P(A_2)}{P(X \mid A_1) \cdot P(A_1) + P(X \mid A_2) \cdot P(A_2) + \ldots + P(X \mid A_m) \cdot P(A_m)}$$

$$(7.11)$$

Box 4 in Figure 7.21 computes the denominator on the right and calls it D. This value, which is $P(X)$, is printed in (Box 5). In Box 6, $P(A_1 \mid X), P(A_2 \mid X, \ldots, P(A_m \mid X)$ are computed from (7.11) and the analogous formulas. For example,

$$P(A_1 \mid X) = \frac{P(X \mid A_1) \cdot P(A_1)}{P(X \mid A_1) \cdot P(A_1) + P(X \mid A_2) \cdot P(A_2) + \ldots + P(X \mid A_m) \cdot P(A_m)}$$

$$(7.12)$$

Finally, the results are printed (Box 7).

A BASIC program that follows this flowchart is shown in Figure 7.22. The correspondence between flowchart and program is so close that we will not comment on the program other than to explain the way in which $P(A_1)$, $P(A_2)$, $\ldots$, $P(A_m)$ and $P(X \mid A_1), P(X \mid A_2), \ldots, P(X \mid A_m)$ are represented in BASIC.

To achieve this representation in a reasonable way, we introduce the use of a *subscripted variable* in BASIC. In our usual mathematical notation we call an expression such as A_4 a subscripted variable. The name of the variable is A, and the subscript[3] is 4. One very common use of subscripts is to sequentially number a set of variables all of which have something in common. For example, $A_1, A_2, \ldots, A_m$ all represent events in a mutually exclusive set, so we call them all by the same name, A, and distinguish one from the others with subscripts 1, 2, $\ldots$, m. There are m variables in this set.

In BASIC there is no way to write numbers below a line, since a teletype or a key punch types all on one line. Therefore we need a different device to represent a subscripted variable. To that end we enclose the subscript in parentheses so that A_4 becomes A(4). Similarly, A_{10} becomes A(10) in BASIC. Subscripts need not be constants. They may be variables themselves. For example, A_j becomes A(J). Of course, in BASIC, J must have a value assigned to it before the expression A(J) is encountered; notice that A(J) does appear in statement 2000 in Figure 7.22. The preceding statement assigns a value to J. In this case J is first 1, then 2, and so on, so that each time statement 2000 is encountered, J has one of the values from 1 to M and each time it has a different value.

---

[3]"Sub" for below and "script" for written: "written below" the line. The 4, the subscript, is written below the line on which the variable, A, appears.

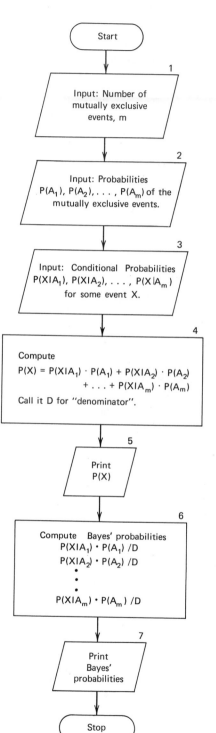

FIGURE 7.21 Flowchart for using Bayes' formula, (7.11).

```
100    PRINT "NØ. ØF MUTUALLY EXCLUSIVE EVENTS IS",
200    INPUT M
300    PRINT
400    REM  **  DATA: PRØBABILITIES, P(A(J)), ØF SET ØF
500    REM        MUTUALLY EXCLUSIVE EVENTS                    **
600    FØR J=1 TØ M
700    PRINT "PRØBABILITY ØF EVENT";J;" IS",
800    INPUT A[J]
900    NEXT J
1000   PRINT
1100   REM  **  DATA: CØNDITIØNAL PRØBABILITIES, P(X GIVEN A(J))  **
1200   FØR J=1 TØ M
1300   PRINT "PRØBABILITY ØF EVENT X GIVEN THAT EVENT";J;"ØCCURS IS";
1400   INPUT P[J]
1500   NEXT J
1600   PRINT
1700   REM  **  CØMPUTE DENØMINATØR ØF EQ. (7.11)  **
1800   LET D=0
1900   FØR J=1 TØ M
2000   LET D=D+P[J]*A[J]
2100   NEXT J
2200   REM  **  DENØMINATØR ØF EQ. (7.11) IS Ø(X)  **
2300   PRINT "PRØBABILITY THAT EVENT X ØCCURS IS";D
2400   PRINT
2500   REM  **  CØMPUTE BAYES PRØBABILITY FØR A(J) USING
2550   REM        (7.11), )7.12), ... .                      **
2600   FØR J=1 TØ M
2700   LET B[J]=P[J]*A[J]/D
2800   PRINT "PRØBABILITY ØF EVENT ";J;" GIVEN X IS";B[J]
2900   NEXT J
3000   END
```

FIGURE 7.22 BASIC program to compute conditional probabilities using Bayes' formula, (7.11).

In our program the probability $P(A_1)$ is represented by A(1), and in general

$$P(A_j) = A(J) \qquad J = 1, 2, \ldots, M$$

Similarly, the BASIC program contains a second subscripted variable P(1), P(2), . . . , P(M), which represents

$$P(X \mid A_1) = P(1)$$
$$P(X \mid A_2) = P(2)$$

$$\cdot$$
$$\cdot$$
$$\cdot$$

$$P(X \mid A_m) = P(M)$$

Therefore the term P(J) * A(J) in statement 2000 is $P(X \mid A_j) \cdot P(A_j)$.

Two points should be emphasized.

1. The BASIC variables A1 and A(1) are different variables. The first is a non-subscripted variable named "A1." The second is the first element of a subscripted variable named "A."

2. Names of subscripted variables must be a single letter. Thus A, P, and Z are legitimate names for subscripted variables, but A6, P2, and Z3 are not. The latter set are legitimate variable names in BASIC, but only if the variable has no subscripts.

We now look at the FOR-NEXT loop in statements 1800 to 2100. D is the denominator of (7.17). It is set to zero. Then, with $J = 1$, it becomes

$$D = 0 + P(1) * A(1)$$
$$= 0 + P(X \mid A_1) \cdot P(A_1)$$

J is set equal to 2, and statement 2000 is repeated so

$$D = 0 + P(1) * A(1) + P(2) * A(2)$$
$$= 0 + P(X \mid A_1) \cdot P(A_1) + P(X \mid A_2) \cdot P(A_2)$$

Continuing, when at last $J = M$, statement 2000 produces

$$D = P(X \mid A_1) \cdot P(A_1) + P(X \mid A_2) \cdot P(A_2) + \ldots + P(X \mid A_m) \cdot P(A_m)$$

as we wish it to be.

We close our discussion of the program in Figure 7.22 by noting that

$$B(1) = P(A_1 \mid X)$$
$$B(2) = P(A_2 \mid X)$$
$$\cdot$$
$$\cdot$$
$$\cdot$$
$$B(M) = P(A_m \mid X)$$

The B stands for Bayes.

The results of running this program for the oil drilling problem in Section 7.3 are shown in Figure 7.23. Recall that in that problem there were three mutually exclusive events.

$$A_1 = \{\text{gusher}\}$$
$$A_2 = \{\text{moderate oil supply}\}$$
$$A_3 = \{\text{no oil}\}$$

and

$$P(A_1) = 0.1 \qquad P(A_2) = 0.2 \qquad P(A_3) = 0.7$$

The event X is

```
NØ. ØF MUTUALLY EXCLUSIVE EVENTS IS              ?3

PRØBABILITY ØF EVENT 1        IS ?0.1
PRØBABILITY ØF EVENT 2        IS ?0.2
PRØBABILITY ØF EVENT 3        IS ?0.7

PRØBABILITY ØF EVENT X GIVEN THAT EVENT 1      ØCCURS IS?0.8
PRØBABILITY ØF EVENT X GIVEN THAT EVENT 2      ØCCURS IS?0.6
PRØBABILITY ØF EVENT X GIVEN THAT EVENT 3      ØCCURS IS?0.3

PRØBABILITY THAT EVENT X ØCCURS IS .41

PRØBABILITY ØF EVENT  1       GIVEN X IS .195122
PRØBABILITY ØF EVENT  2       GIVEN X IS .292683
PRØBABILITY ØF EVENT  3       GIVEN X IS .512195
```

FIGURE 7.23 Results of running BASIC program in Figure 7.22 for oil-drilling problem.

$$X = \{\text{positive seismic test}\}$$

and

$$P(X \mid A_1) = 0.8 \qquad P(X \mid A_2) = 0.6 \qquad P(X \mid A_3) = 0.3$$

Figure 7.23 indicates that

$$P(X) = 0.41$$

and

$$P(A_1 \mid X) = 0.195 \qquad P(A_2 \mid X) = 0.293 \qquad P(A_3 \mid X) = 0.512$$

which agrees with our hand calculations near the close of Section 7.3.

EXERCISES FOR SECTION 7.5

*1. Use the BASIC program in Figure 7.22 to find the solution to Exercise 8, Section 7.3.

2. Use the BASIC program in Figure 7.22 to find the solution to Exercise 9, Section 7.3.

*3. Suppose we hope to drill an oil well in an area in which oil is not exceptionally plentiful. In fact, in this given area for any given well,

P(more than 10 million barrels)   = 0.05
P(between 5 and 10 million barrels)   = 0.10
P(between 3 and 5 million barrels)   = 0.20

P(between 1 and 3 million barrels)   = 0.15
P(less than 1 million barrels)   = 0.50

If we conduct a seismic test we get a positive indication of oil being present with the following probabilities.

More than 10 million barrels   0.9
Between 5 and 10 million barrels   0.7
Between 3 and 5 million barrels   0.6
Between 1 and 3 million barrels   0.4
Less than 1 million barrels   0.1

Use the BASIC program in Figure 7.22 to compute:
(a) The probability that a positive result is obtained from a seismic test conducted in this area.
(b) The probabilities of the various amounts of oil being present if a positive result is obtained from a seismic test.
(c) If a gusher is anything over 5 million barrels, what is the probability of a gusher if a positive seismic result is obtained?

4. Same as Exercise 3 except the area selected is known to be rich in oil so that the probabilities of the various amounts of oil are reversed; that is,

P (more than 10 million barrels)   = 0.50
P(between 5 and 10 million barrels)   = 0.15
P(between 3 and 5 million barrels)   = 0.20
P(between 1 and 3 million barrels)   = 0.10
P(less than 1 million barrels)   = 0.05

*5. Since $A_1, A_2, \ldots, A_m$ are a complete set of mutually exclusive events,

$$P(A_1) + P(A_2) + \ldots + P(A_m) = 1$$

Modify the program in Figure 7.22 so that after $A(1), \ldots, A(M)$ are specified, they are summed. If the sum does not equal 1, a warning message is printed, and the program stops. If the sum does equal 1, the program continues as before.

6. All of the probabilities used in the BASIC program in Figure 7.22 must be between 0 and 1. Modify that program so that if any probability violates this requirement, a warning message is printed, and the program stops.

7.6 CASE STUDY 7: THE GAME OF CRAPS REVISITED

We turn now to an analysis of the game of craps that we simulated in the previous chapter. We start our analysis by constructing the sample space for one toss of two dice. That sample space is shown in Figure 7.24. Each row corresponds to a number on the first die, and each column corresponds to a number on the second die. The numbers in the table correspond to the *sum* of the numbers appearing on the two dice. Thus, for example, in the third row and fourth column, the first die has a 3 and the second die has

Number on second die

	1	2	3	4	5	6
1	● 2	● 3	● 4	● 5	● 6	● 7
2	● 3	● 4	● 5	● 6	● 7	● 8
3	● 4	● 5	● 6	● 7	● 8	● 9
4	● 5	● 6	● 7	● 8	● 9	● 10
5	● 6	● 7	● 8	● 9	● 10	● 11
6	● 7	● 8	● 9	● 10	● 11	● 12

Number on first die

FIGURE 7.24 Sample space for tossing two dice.

a 4. Therefore the number appearing in the table in the third row and fourth column is $3 + 4 = 7$.

Each of the sample points is equally likely to occur. Since there are 36 points, the probability of any one point occurring is 1/36. We now calculate the probability of any given sum appearing based on the table in Figure 7.24.

First note that the sums range from 2 to 12. There is only one sample point that produces a 2 (both dice are 1). Thus $P(2) = 1/36$. There are two sample points that produce a 3 (first die is 1, second die is 2 or first die is 2, second die is 1). Thus $P(3) = 2/36 = 1/18$. Similarly, we calculate the probabilities of each number appearing as the sum of the two dice. They are:

$$P(2) = 1/36 \qquad P(8) = 5/36$$
$$P(3) = 1/18 \qquad P(9) = 1/9$$
$$P(4) = 1/12 \qquad P(10) = 1/12$$
$$P(5) = 1/9 \qquad P(11) = 1/18$$
$$P(6) = 5/36 \qquad P(12) = 1/36$$
$$P(7) = 1/6$$

All of these events are mutually exclusive, so that $P(3 \cup 6) = P(3) + P(6)$.

Recall that on the first toss the player wins if either a 7 or a 11 occurs. Therefore the probability of winning on the first toss is

$$P(7 \cup 11) = P(7) + P(11)$$
$$= 1/6 + 1/18$$
$$= 2/9$$

That is, on the average, we should expect to win 2 out of 9 times on the first toss.

The player loses on the first toss if either a 2 or a 3 or a 12 appears. Thus the probability of losing on the first toss is

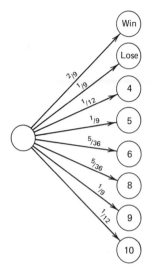

FIGURE 7.25 First stage in game of craps tree.

$$P(2 \cup 3 \cup 12) = P(2) + P(3) + P(12)$$
$$= 1/36 + 1/18 + 1/36$$
$$= 1/9$$

Thus we expect to lose on the first toss 1 out of 9 times.

To continue our analysis, we construct a tree to describe the results of the first toss. It is shown in Figure 7.25. The branches are appropriately labeled. This is the first stage in the game. The upper two circles end the game. However, in the event that we reach any of the six lower circles, we continue to the next stage of the process. That is, the dice are tossed again until either a 7 appears or the "point" appears.

Let us consider what happens if the first toss of the dice results in a 4. We are then at the third circle from the top in Figure 7.25. We toss the dice one or more additional times until either a 4 appears or a 7 appears. All other results are ignored. If the 4 appears, the player wins. If the 7 appears, he loses. Therefore the second stage can be represented by the diagram in Figure 7.26a. The *D* indicates a draw and implies yet another toss (i.e., a third stage). This third stage is shown in Figure 7.26b. The second

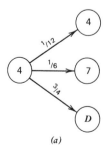

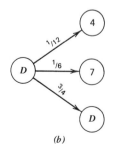

FIGURE 7.26 Second and third stages of part of the game of craps.

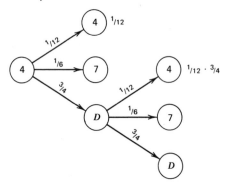

FIGURE 7.27 Second and third stages of part of game of craps combined.

and third stages together are shown in Figure 7.27. Each chain ending in a 4 represents a win. Of course, the stages can go on indefinitely. In Figure 7.28 we show stages two, three, and four. Notice the chain probabilities for the three chains leading to a win. They are $1/12$, $(1/12) \cdot (3/4)$, and $(1/12) \cdot (3/4)^2$. The chain in the next, or fifth, stage leading to a 4 will have a chain probability of $(1/12) \cdot (3/4)^3$.

All chains are mutually exclusive events. Hence the probability of winning on the second or third or fourth toss is the sum of the probabilities of winning on each individual toss. Thus, if the point is 4, the probability of winning is

$$\frac{1}{12} + \frac{1}{12} \times \frac{3}{4} + \frac{1}{12} \times \left(\frac{3}{4}\right)^2 + \frac{1}{12} \times \left(\frac{3}{4}\right)^3 + \ldots$$

or

$$\frac{1}{12} \left[1 + (3/4) + (3/4)^2 + (3/4)^3 + \ldots \right] \tag{7.13}$$

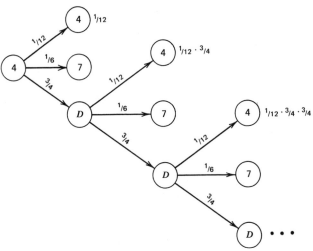

FIGURE 7.28 Second, third, and fourth stages of part of game of craps.

The term in square brackets is a geometric series and was encountered in Section 3.3. There we found that

$$1 + M + M^2 + \ldots + M^{k-1} = \frac{1 - M^k}{1 - M}$$

provided $M \neq 1$. $\Big[$See (3.30)$\Big]$. If $M = 3/4$ this becomes

$$1 + (3/4) + (3/4)^2 + \ldots + (3/4)^{k-1} = \frac{1 - (3/4)^k}{1 - (3/4)} \tag{7.14}$$

In (7.13) the series does not stop at $k - 1$ but continues indefinitely (i.e., k gets larger and larger). What happens to the right side of (7.14) as k gets large? The only term whose value changes at all is the term $(3/4)^k$, and it gets smaller and smaller. So small, in fact, that eventually we can ignore it. Thus the sum becomes very close to

$$\frac{1}{1 - (3/4)}$$

or 4. Therefore the probability expressed in (7.13) becomes

$$\frac{1}{12}\Big[1 + (3/4) + (3/4)^2 + (3/4)^3 + \ldots\Big] = \frac{1}{12} \times 4$$

$$= \frac{1}{3}$$

This is then the probability of winning if the point is a 4, no matter how many tosses are required. Since we must either win or lose, the probability of losing if the point is a 4 is 2/3.

In a like way we find

$$P(W \mid 5) = 2/5$$
$$P(W \mid 6) = 5/11$$
$$P(W \mid 8) = 5/11$$
$$P(W \mid 9) = 2/5$$
$$P(W \mid 10) = 1/3$$

This allows us to complete the tree in two stages, as shown in Figure 7.29. The W indicates a win, and the L indicates a loss. Notice that these two stages are different than the ones we referred to earlier. We have collapsed stages two, three, four, and so on into one stage—the second.

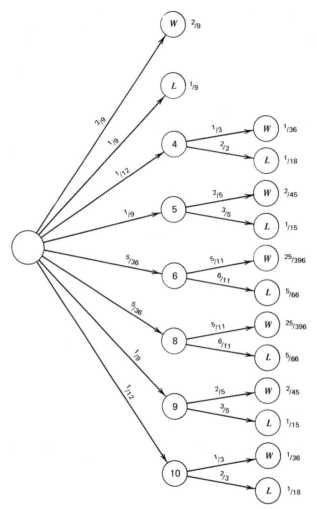

FIGURE 7.29 Complete tree diagram for the game of craps.

The chain probabilities are also shown in Figure 7.29. They are, as usual, the products of the probabilities on the branches of the chain. Thus the probability of getting a 4 on the first toss and then losing is 1/18, which is the product of 1/12 and 2/3.

There are seven chains that lead to a win. Each of these chains represents a mutually exclusive event. For example, the topmost W represents the event {either a 7 or a 11 occurs on the first toss}. The W on the third chain from the top represents the event {a 4 occurs on the first toss and another 4 appears before a 7 appears on succeeding tosses}. These cannot both occur and, thus, the events are mutually exclusive.

Since all W chains are mutually exclusive, the probability of following some chain that will end in a W is the sum of the probabilities of the individual chains. Therefore the probability of winning is the sum of the chain probabilities for the chains ending in a W. The probability of winning the game of craps is

$$(2/9) + (1/36) + (2/45) + (25/396) + (25/396) + (2/45) + (1/36)$$

or

$$244/495 = 0.49292929292 \ldots$$

The probability of winning is just slightly less than $1/2$. Out of 495 games we should expect that on the average we will win 244 and lose 251.

We now summarize our results. We have assumed that on any toss of the dice, each of the sample points in Figure 7.24 is equally likely. We also have assumed that the game eventually ends. With these assumptions we find that, on the average, out of any group of 495 games, there will be 244 wins and 251 losses. Did your experiment with the computer program in Case Study 6 bear this out? Can you explain why?

The above analysis is not at all simple, and there are many pitfalls into which the unwary can fall. Suppose we had asked the more complicated question: How many times of the 495 games should we expect to win in three or less tosses? We would find it difficult to answer this question, although a similar analysis could certainly be carried out. Try to think of other interesting questions whose answers you would like to have. Another example is: If the point is 5, how many tosses would you expect to make before either winning or losing?

You should consider such questions and try to outline how you could write a computer program that would help you to answer them. Of course, we must always keep in mind that the results of computer experiments are only approximations to the probabilities that we seek.

EXERCISES FOR SECTION 7.6

1. At the close of Section 6.8 we modified the game of craps so that a 12 on the first toss did not lose, but resulted in a draw, and the 12 became the "point." Using a tree diagram, compute the probability of winning this modified game. Compare the result with the simulation results in Section 6.8.

2. Draw a tree diagram similar to Figure 7.29 for the following dice games. Compute the probability of winning each game.

	Win on First Toss	Lose on First Toss	Point on First Toss	Win on Later tosses	Lose on Later tosses
*(a)	7, 11	2, 3, 12	4, 5, 6, 8, 9, 10	7	Point
(b)	6, 10	2, 3, 12	4, 5, 7, 8, 9, 11	Point	6
*(c)	6, 10	2, 3, 12	4, 5, 7, 8, 9, 11	6	Point
(d)	7, 11	2, 12	3, 4, 5, 6, 8, 9, 10	Point	7
(e)	7, 11	3, 12	2, 4, 5, 6, 8, 9, 10	7	Point
(f)	2, 3, 12	7, 11	4, 5, 6, 8, 9, 10	7	Point
(g)	3, 7	2, 11, 12	4, 5, 6, 8, 9, 10	Point	7
(h)	7, 11	2, 3, 12	4, 5, 6, 8, 9, 10	Point	7, 11
(i)	6, 7, 8	2, 3, 11, 12	4, 5, 9, 10	Point	2, 7, 12
(j)	4, 7, 10	2, 5, 9, 12	3, 6, 8, 11	Point	4, 7, 10
(k)	7, 11	2, 3, 12	4, 5, 6, 8, 9, 10	Point	7, 11
(l)	7, 11	2, 3, 12	4, 5, 6, 8, 9, 10	7, 11	Point

3. A game is played with two pyramid-shaped dice. Each of the four sides of each die are numbered 1, 2, 3, and 4 (See Figure 7.30). When the dice are tossed, one face is down. The "score" on a toss is the sum of the numbers on the two faces that are down.

*(a) What are the possible scores?

(b) What are the probabilities of each of the scores?

FIGURE 7.30 Two pyramid-shaped dice each with four sides.

4. In the dice game described in Exercise 3, the rules are: On the first toss a 5 wins, while a 4 or 6 loses. Any other "score" is a draw and the score is called the "point." On the second and succeeding tosses a 5 loses, and the "point" wins.

*(a) What is the probability of winning on the first toss?

(b) What is the probability of losing on the first toss?

*(c) If the "point" is a 3, what is the probability of winning on the second or succeeding tosses?

(d) What is the probability of winning the game?

5. In the game of craps (see Section 7.6):

*(a) What is the probability that if the player wins he did so on the first toss?

(b) What is the probability that if the player wins his point was 4?

(c) What is the probability that if the player loses his point was 6?

8
BINOMIAL PROBABILITIES

8.1 BINOMIAL TRIALS

We now turn our attention to a special, but important, type of probabilistic experiment. As before, we will conduct the experiment over and over. Each time that we carry out the experiment, we say that we have carried out one *trial*. If, for example, the experiment consists of drawing a marble from an urn, observing its color, and returning the marble to the urn, then each marble drawn is a trial.

The particular experiments we will consider can be arranged so that there is a set of *two* mutually exclusive alternatives that cover all of the sample points. We call these two alternatives *success* and *failure*. This requirement is not as restrictive as it might seem at first. Suppose the experiment consists of drawing a marble from an urn containing three black marbles, two white marbles, and four red marbles. If we define success to be drawing a white marble and failure to be selecting a black or a red marble, then success and failure are mutually exclusive and cover all possible outcomes.

We will place three other restrictions on our experiment.

1. There must be a fixed number of trials. Thus we must decide to draw 10 marbles or 6 marbles or 67 marbles. Remember, each draw is a trial.
2. The event {success on the first trial} must be independent of the event {success on the second trial}. This is certainly true of our urn example, but it would not be true if we did not replace the marble after drawing.
3. The probability of success on each trial must be the same as the probability of success on any other trial. In the urn example the probability of success, drawing a white marble, is 2/9 on each and every trial, so this requirement is satisfied. Once again, had we not replaced the marbles after they were drawn, this would not be so.

Add to this the requirement stated earlier.

4. Each trial must result in either a success or a failure.

An experiment that satisfies these four requirements is called a *binomial trials* experiment. The "bi" arises because there are two outcomes. Recall there may be many different outcomes, but we categorize them into two distinct and mutually exclusive groups.

We will compute the probability of obtaining s successes in n trials given the probability of success on one trial. In terms of the urn problem, suppose that we draw three marbles from the urn, replacing each marble drawn before drawing another. What is the probability that precisely two of the marbles drawn are white? We also will be concerned with finding the probability that two or more marbles drawn are white.

8.2 COMPUTER SOLUTIONS OF BINOMIAL TRIALS

Before turning to a mathematical analysis of binomial trials experiments, we will construct a BASIC program that estimates the probability of obtaining 0, 1, 2, 3, . . . successes in n trials. For definiteness we may think of this as drawing n marbles from an urn where each marble drawn is replaced before the next drawing. If we draw a white marble we record a success. Otherwise we record a failure. We perform this experiment (drawing n marbles) over and over (perhaps 1000 or more times). Each time we do so we get 0, 1, 2, . . . , or n successes (white marbles). We record the number of times we get each of these numbers of successes, divide by the number of experiments, and thereby obtain the relative frequency of each of 0, 1, 2, . . . , n successes in n trials. We use these relative frequencies as estimates of the probabilities (see Section 6.2).

This is in much the same spirit as our simulation of the game of craps (Section 6.8). Once again we will generate a random number and use it to decide upon a success or a failure. In each experiment we will generate not one but n random numbers and count the number of successes. A flowchart of the simulation process is shown in Figure 8.1.

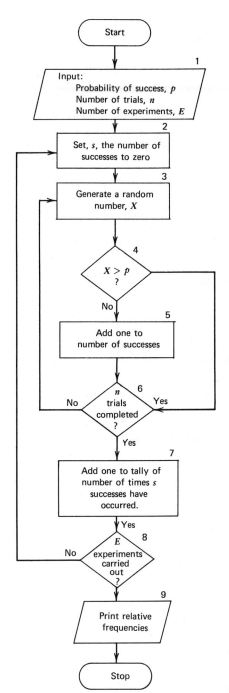

FIGURE 8.1 Flow chart of simulation of binomial trials.

In the flowchart, boxes 2 to 6 describe one experiment and, for the moment, we concentrate our attention on that one experiment. The number of successes in the experiment will be recorded by the variable s. Since at the start of the experiment, before any trials are carried out, we have no successes, s is set equal to zero (Box 2). We generate a random number (Box 3). If this random number is less than or equal to p, we add one to the number of successes (Boxes 4 and 5). If not, we do not change the number of successes. These three boxes—3, 4, and 5—represent one trial in the experiment. After each trial, we increase s if the trial resulted in a success. We continue for n trials (Box 6). When n trials have been carried out, the variable s will be an integer between 0 and n. Moreover, s will be equal to the number of successes obtained on this one experiment. This last statement is true for all of the E experiments.

To compute the relative frequencies we must keep a tally of the number of experiments that result in no successes, the number that result in one success, the number that result in two successes, and so on. When we leave Box 6 with a "yes," we will have finished an experiment that resulted in s successes. Hence we add one to the

```
100    REM  **  DATA: VALUES ØF P, N AND E  **
200    PRINT "PRØBABILITY ØF SUCCESS ØN ØNE TRIAL",
300    INPUT P
400    PRINT "NUMBER ØF TRIALS",
500    INPUT N
600    PRINT "NUMBER ØF EXPERIMENTS",
700    INPUT E
800    REM  **  SET TALLIES ØF 0, 1, 2, ...., N SUCCESSES TØ ZERØ
900    REM      NØTE THAT T(J) CØNTAINS NØ. ØF TIMES J-1 SUCCESSES
1000   REM      ØCCURS                                            **
1100   FØR J=1 TØ N+1
1200   LET T[J]=0
1300   NEXT J
1400   REM  **  CØNDUCT E EXPERIMENTS  **
1500   FØR K=1 TØ E
1600   REM  **  SET NUMBER ØF SUCCESSES TØ ZERØ  **
1700   LET S=0
1800   REM  **  CARRY ØUT N TRIALS FØR EACH EXPERIMENT  **
1900   FØR J=1 TØ N
2000   LET X=RND(0)
2100   REM  **  IF RANDØM NUMBER EXCEEDS PRØBABILITY ØF
2200   REM      SUCCESS, DØ NØT RECØRD SUCCESS              **
2300   IF X>P THEN 2500
2400   LET S=S+1
2500   NEXT J
2600   REM  **  INCREASE TALLY FØR S SUCCESSES BY ØNE.
2700   REM      RECALL THAT T(S+1) CØNTAINS NØ. ØF TIMES S
2800   REM      SUCCESSES ØCCURS.                          **
2900   LET T[S+1]=T[S+1]+1
3000   NEXT K
3100   PRINT
3200   REM  **  PRINT RELATIVE FREQUENCIES ØF 0,1,2,....,N SUCCESSES  **
3300   PRINT "PRØBABILITY ØF"
3400   FØR J=0 TØ N
3500   PRINT J;" SUCCESSES = ";T[J+1]/E
3600   NEXT J
3700   END
```

FIGURE 8.2 BASIC program to simulate binomial trials.

appropriate tally (Box 7). If there are experiments yet to be carried out (Box 8), we begin a new experiment (Boxes 2 to 6). When all E experiments have been conducted, we will print the relative frequencies [i.e., the various tallies for the number of successes divided by the number of experiments (Box 9)].

A BASIC program that follows this flowchart is given in Figure 8.2. We will use a subscripted variable to keep the tally of the number of successes. In BASIC subscripts are enclosed in parentheses, so that T_1 is written T(1) and X_k is written X(K).

We will use T (for tally) to record the tallies of $0, 1, 2, \ldots, n$ successes. It would be convenient to let T_0 be the tally for zero successes, T_1 be the tally for one success, and so on. Unfortunately, some versions of BASIC do not allow zero subscripts. Hence we let T_1 be the tally for zero successes, T_2 be the tally for one success, and so on until T_{n+1} is the tally for n successes (see REM statements 900 and 1000).

The program starts by reading the values of P, n, and E (statements 100 to 700). Then all tallies are set to zero (statements 800 to 1300). The FOR-NEXT loop from statement 1500 to statement 3000 controls each experiment and corresponds to Boxes 2 to 8 in Figure 8.1. The FOR-NEXT loop from statement 1900 to statement 2500 carries out n trials in each experiment and corresponds to Boxes 3 through 6.

The tally increase described in Box 7 is accomplished in statement 2900. Since there are s successes, we wish to increase the tally for that number of successes by one. Recall that T(1) is the tally for zero successes, T(2) is the tally for one success, and T(S + 1) is the tally for S successes.

Finally, we note that the relative frequency of zero successes is the tally of zero successes, T(1), divided by the total number of experiments, E. Hence, in statement 3500, we print T(J + 1)/E for J = 0, 1, . . . , N.

An example of the use of this program is shown in Figure 8.3. This particular example is the urn problem discussed in the previous section (i.e., drawing a marble

```
PRØBABILITY ØF SUCCESS ØN ØNE TRIAL          ?0.222222
NUMBER ØF TRIALS                 ?3
NUMBER ØF EXPERIMENTS            ?1000

PRØBABILITY ØF
  0      SUCCESSES =  .499
  1      SUCCESSES =  .381
  2      SUCCESSES =  .113
  3      SUCCESSES =  .007


PRØBABILITY ØF SUCCESS ØN ØNE TRIAL          ?0.222222
NUMBER ØF TRIALS                 ?3
NUMBER ØF EXPERIMENTS            ?2000

PRØBABILITY ØF
  0      SUCCESSES =  .4965
  1      SUCCESSES =  .388
  2      SUCCESSES =  .103
  3      SUCCESSES =  .0125
```

FIGURE 8.3 Results of computer simulation of binomial trials: three trials.

```
PRØBABILITY ØF SUCCESS ØN ØNE TRIAL          ?0.5
NUMBER ØF TRIALS                 ?6
NUMBER ØF EXPERIMENTS            ?10000

PRØBABILITY ØF
  0      SUCCESSES =   •0138
  1      SUCCESSES =   •0949
  2      SUCCESSES =   •2331
  3      SUCCESSES =   •3137
  4      SUCCESSES =   •2331
  5      SUCCESSES =   •095
  6      SUCCESSES =   •0164
```

FIGURE 8.4 Results of computer simulation of binomial trials: six trials.

from an urn containing three black marbles, two white marbles, and four red marbles). A white marble is a success. Therefore the probability of success on each draw is $2/9 = 0.222222$. An experiment consists of three draws with replacement. In Figure 8.3 the program was run twice, once with 1000 experiments and once with 2000 experiments. The results differ slightly. Which do you think is more accurate? Why?

Finally we run the program once more for six trials (e.g., drawing six marbles in succession from an urn), where the probability of success on any one trial is 1/2. The results are shown in Figure 8.4.

The probabilities increase to a maximum at three successes. On either side of three successes the decreases in the probabilities are more or less equal (i.e., the probability of two successes is almost equal to the probability of four successes).

EXERCISES FOR SECTION 8.2

1. Run the BASIC program in Figure 8.2 and use the results to estimate the probability of 0, 1, 2, . . . successes for the following cases.
   *(a) Five trials with probability of success on one trial of 0.25.
    (b) Five trials with probability of success on one trial of 0.75.
   *(c) Eight trials with probability of success on one trial of 0.30.
    (d) Eight trials with probability of success on one trial of 0.70.

*2. Consider an experiment where the probability of success is 1/2. Suppose we carry out the experiment six times. Use Figure 8.4 to estimate the probability that the number of successes is between 2 and 4 (i.e., the number is 2, 3, or 4).

3. Using the experiment described in Exercise 2 and Figure 8.4, estimate the probability that the number of successes is an odd number.

*4. Modify the flowchart in Figure 8.1 so that it produces the probabilities of 0, 1, 2, . . . , n or more successes on n trials. (*Note.* The probability of zero or more successes is always 1.)

5. Modify the BASIC program in Figure 8.2 so that it produces the probabilities of 0, 1, 2, . . . , n or more successes on n trials. (*Hint.* See Exercise 4.)

8.3 A FORMULA FOR BINOMIAL TRIALS PROBABILITIES

We now turn to the mathematical analysis of binomial trials experiments. In the previous section we estimated the probabilities of zero, one, two, or three successes in three trials using a BASIC program (Figure 8.3). In this section we will compute those probabilities more precisely.

Let us first construct the tree for the process described for the urn problem of Section 8.1. The letter S stands for success (a white marble is drawn) and the letter F stands for failure (a white marble is not drawn). The tree is shown in Figure 8.5. Each stage in the tree corresponds to one trial. If two white marbles are drawn, there must be precisely two S's on the chain. There are three such chains, and all three have chain probabilities of $(2/9)^2(7/9)$. Since there are three such chains and all chains represent mutually exclusive events, the probability of obtaining two white marbles is

$$3 \times (2/9)^2 \times (7/9) = 84/729 = .115 \qquad (8.1)$$

Compare this result with the estimates in Figure 8.3.

If we ask for the probability that two or more white marbles are drawn, we must accept precisely two white marbles or precisely three white marbles. There is only one

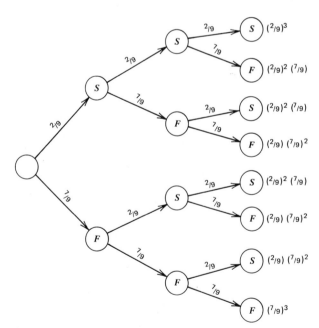

FIGURE 8.5 Tree diagram for drawing three marbles from an urn that contains two white marbles and seven nonwhite marbles.

chain with three white marbles, and its chain probability is $(2/9)^3$. Thus the probability of two or more white marbles is

$$3 \times (2/9)^2 \times (7/9) + (2/9)^3 = 92/729 = 0.126 \qquad (8.2)$$

Once again this should be compared with the results in Figure 8.3 where the last two probabilities should be summed for comparison with (8.2).

Now consider a more general urn problem. Suppose the urn contains 100 marbles and that w of them are white. Once more we call drawing a white marble a success. The probability of success will be $w/100$. If we let $p = w/100$ then the probability of failure will be $1 - p$. The tree for this experiment is shown in Figure 8.6.

The event "precisely two successes occur" is represented by the three chains, which contain two S's and one F. Each of these chains has a probability of $p^2(1 - p)$ of occurring. Thus the probability of precisely two successes is $3p^2(1 - p)$. The probability of two or more successes is $3p^2(1 - p) + p^3$. For $p = 2/9$ this reduces to (8.2).

We could have more than three trials in this urn experiment if we wished. We could construct the tree for any number of stages in a straightforward way. However, we can

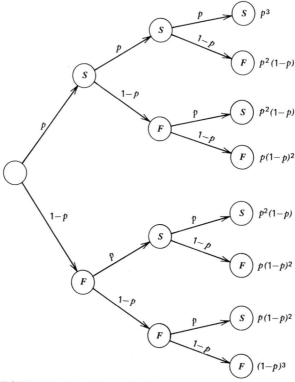

FIGURE 8.6 Tree diagram for binomial trials experiment with three stages.

FIGURE 8.7 Five possible strings of five symbols: one *S* and four *F*'s.

visualize the tree without actually drawing it, and we can deduce the results that we are seeking from our visualization.

Suppose that we have *n* trials in the experiment (i.e., *n* stages in the tree). In the urn example this means that we draw *n* marbles from the urn, each time replacing the marble drawn before drawing again. We wish to know the probability of precisely *s* successes.

Any chain with *s* successes will contain *s* of the *S*'s and $n - s$ of the *F*'s. Each branch leading to an *S* has a branch weight of *p*, and each branch leading to an *F* has a branch weight of $1 - p$. Therefore the chain probability of a chain with *s* successes is $p^s(1 - p)^{n-s}$. The earlier example was for the case $n = 3$ and $s = 2$.

It remains to determine how many chains have precisely *s* *S*'s and $n - s$ *F*'s. We turn to that now. Any such chain can be represented by a string of *n* letters, each one being either an *S* or an *F*. Of these *n* letters precisely *s* must be *S*'s and precisely $n - s$ must be *F*'s. Before attempting this general case of *s* *S*'s out of a string of *n* symbols, we look at the simpler case of placing 1, 2, or 3 *S*'s in a string of five symbols. The method we use for counting the number of ways of doing this will be identical with the one we will use later in the more general case.

We number the letters in the string consecutively from 1 to 5. The first *S* can be placed in any one of these five numbered positions. Suppose, for example, that we place the first *S* in position 3. Then the third trial results in a success. We could have chosen any one of the other four positions for the first *S*. Thus there are five different ways of placing the first *S*. If we have only one *S* to place, there are five possible strings, and they are shown in Figure 8.7.

We now have answered the question of how many possible strings there are for the case of one success in five trials. The answer is five.

But suppose there were two *S*'s to be placed in the string of five symbols. If the first *S* is placed in position 1, then there are four choices for the placement of the second *S*. They are positions 2, 3, 4, or 5. The possible strings are shown in Figure 8.8. On the

FIGURE 8.8 Four possible strings of five symbols: two *S*'s and three *F*'s, if first *S* is in position 1.

other hand, if the first *S* were placed in position 2, there are also four choices for placement of the second *S*: 1, 3, 4, or 5. In this case the possible strings are shown in Figure 8.9.

The same argument can be applied to the other three choices for placement of the first *S*. Those choices, to reiterate, are positions 3, 4, and 5. That is, for *each* of the five

FIGURE 8.9 Four possible strings of five symbols: two *S*'s and three *F*'s, if the first *S* is in position 2.

choices for the first *S*, there are four choices for the second *S*. The total number of choices for the first two *S*'s is

$$5 \times 4$$

Of these 20 choices, eight are given in Figures 8.8 and 8.9.

Finally, suppose we wish to place three *S*'s in our string of five symbols. As we noted above, there are 5×4 ways to place the first two of these three *S*'s. We consider *one* of these ways (i.e., positions 1 and 2). Then there are three choices for placement of the third *S*. We could choose position 3, position 4, or position 5. The resulting strings from each of these choices are shown in Figure 8.10.

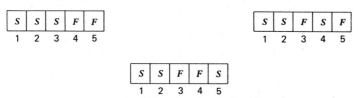

FIGURE 8.10 Three possible strings of five symbols: three *S*'s and two *F*'s, if the first two *S*'s are in positions 1 and 2.

For *each* of the 5×4 choices for the first two S's, we have three choices for placing the third S. The total number of choices for placing three S's therefore is

$$5 \times 4 \times 3$$

In summary, the number of strings of s S's in five symbols is given by

s	Number of Strings
1	5
2	5×4
3	$5 \times 4 \times 3$

Had we started with a string of six symbols, then we would have obtained

s	Number of Strings
1	6
2	6×5
3	$6 \times 5 \times 4$

We now turn to a string of n symbols. The first S can be placed in any one of the n positions. Thus there are n strings containing one S and $n - 1$ F's. Following the same logic as we did above, once the first S has been placed, there are $n - 1$ choices for the second S's position. Moreover, this statement applies to each and every one of the n choices for the first S. Hence the number of ways we can place the first two S's is

$$n \times (n - 1)$$

When two S's have been placed, there are $n - 2$ vacancies available for the third S. Therefore the number of ways we can place three S's is

$$n \times (n - 1) \times (n - 2)$$

Continuing in this way, we obtain

s	Number of Strings
1	n
2	$n \times (n - 1)$
3	$n \times (n - 1) \times (n - 2)$
4	$n \times (n - 1) \times (n - 2) \times (n - 3)$

and so on.

Notice that the last factor in the product is n minus an integer. That integer always is one less than s. In other words, we should subtract $s - 1$ from n to find the last factor. But

$$n - (s - 1) = n - s + 1$$

so the number of ways we can place s S's in a string of n symbols is

$$n \times (n - 1) \times (n - 2) \times \ldots \times (n - s + 1) \tag{8.3}$$

We pause to interpret this last formula. It means start with n and successively subtract 1 from the latest number and multiply. Thus we subtract 1 from n, producing $n - 1$, and multiply to get

$$n \times (n - 1)$$

We subtract 1 again to get $n - 2$ and multiply once more to get

$$n \times (n - 1) \times (n - 2)$$

Then we get $n - 3$ and multiply

$$n \times (n - 1) \times (n - 2) \times (n - 3)$$

But when are we to stop? We stop when the number reaches $n - s + 1$. For example, if $s = 6$, since $n - s + 1 = n - 6 + 1 = n - 5$, we stop when we have

$$n \times (n - 1) \times (n - 2) \times (n - 3) \times (n - 4) \times (n - 5)$$

A BASIC program to evaluate (8.3) is shown in Figure 8.11. The DATA statement, of course, means that the value for A (A for "answer") printed is for $n = 5$ and $s = 3$. The reader should study this program and the expression in (8.3) and assure himself of their direct correspondence.

An interesting and important result arises if $s = n$ in (8.3). In this case we get

$$n \times (n - 1) \times (n - 2) \times \ldots \times 2 \times 1$$

This is the product of all the positive integers from 1 to n. It arises so often in mathematics that we give it a special name and a special symbol. We write

$$n \times (n - 1) \times (n - 2) \times \ldots \times 2 \times 1 = n! \tag{8.4}$$

```
100    READ N,S
200    DATA 5,3
300    LET A=1
400    FØR J=N TØ N-S+1 STEP -1
500    LET A=A*J
600    NEXT J
700    PRINT A
800    END
```

FIGURE 8.11 BASIC program to evaluate (8.3) for $n = 5$, $s = 3$.

The symbol $n!$ is called n *factorial*. We encountered this expression earlier at the close of Section 2.4. As we have defined it, $n!$ only makes sense if n is a positive integer. For example,

$$1! = 1$$
$$2! = 2 \times 1 = 2$$
$$3! = 3 \times 2 \times 1 = 6$$
$$4! = 4 \times 3 \times 2 \times 1 = 24$$

By convention and for our later convenience, we define

$$0! = 1$$

The factorial can be defined for other values of n, but we will not do so here.

We now use the factorial symbol to abbreviate (8.3). Suppose we multiply and divide (8.3) by $n - s$.

$$\frac{n \times (n - 1) \times (n - 2) \times \ldots \times (n - s + 1) \times (n - s)}{(n - s)}$$

Notice that $n - s$ is one less than $n - s + 1$, so that the pattern in the numerator is continued (i.e., each factor is one less than its predecessor). We next multiply both numerator and denominator by $n - s - 1$, which is one less than $n - s$

$$\frac{n \times (n - 1) \times (n - 2) \times \ldots \times (n - s + 1) \times (n - s) \times (n - s - 1)}{(n - s) \times (n - s - 1)}$$

We continue in this way, reducing the factor used by one each time, until the factor becomes 1. At this point we have

$$\frac{n \times (n - 1) \times (n - 2) \times \ldots \times (n - s + 1) \times (n - s) \times \ldots \times 2 \times 1}{(n - s) \times (n - s - 1) \times \ldots \times 2 \times 1}$$

But the numerator is just $n!$ and the denominator is $(n - s)!$, so (8.3) can be written

$$\frac{n!}{(n - s)!} \qquad (8.5)$$

For $n = 5$ and $s = 3$ this becomes

$$\frac{5 \times 4 \times 3 \times 2 \times 1}{2 \times 1} = 5 \times 4 \times 3 = 60$$

which agrees with our earlier result for the number of strings of five symbols with three S's and two F's.

The expression in (8.5) is sometimes referred to as "the number of *permutations* of n things taken s at a time," and it is written as $P(n, s)$ or as $_nP_s$.

It appears that our task—counting the number of chains that have precisely s S's and $n - s$ F's—has been accomplished, and that the number of chains is given by (8.3) or equivalently by (8.5). Unfortunately, that is not the case. Not all of the possible arrangements included in those counted by (8.5) are distinct. To see this we return to our example where $n = 5$ and $s = 2$. The first case in Figure 8.8 is identical with the first case in Figure 8.9. Both yield the string $SSFFF$. This string indicates a success on the first two trials and a failure on the last three trials. If we use (8.5) to represent the number of strings we will have counted this particular string twice in the number of arrangements. We must have some way of removing these duplications.

In order to avoid these unwanted duplications we will write down the choices of the positions for placement of the S's. Recall that for the strings in Figure 8.8, the first S was placed in position one. Therefore we represent the four strings in that figure by

$$1, 2$$
$$1, 3$$
$$1, 4$$
$$1, 5$$

The first pair—1, 2—means the first S was placed in position one and the second was placed in position two. Figure 8.9 can be represented by

$$2, 1$$
$$2, 3$$
$$2, 4$$
$$2, 5$$

Up until now we have counted all of these as different strings. But 1, 2 and 2, 1 both correspond to the string

$$SSFFF$$

On the basis of our earlier analysis we decided that there were $5 \times 4 = 20$ different strings with two S's. However, two of them (1, 2 and 2, 1) are duplicates. Similarly, 1, 3 and 3, 1 are duplicates, as are 3, 4 and 4, 3. In fact, for any pair, there is another pair that duplicates it (see Exercise 1). What appeared to be 20 pairs is, in fact, only 10 pairs. We must *divide* the result given by (8.5) by 2.

Let us look at one more example. Suppose $s = 3$ (i.e., there are three S's in the string). One choice for the three locations of the S's is 5, 2, 3. If $n = 5$ this string is

FSSFS

But many other choices produce this same string. One choice is 2, 3, 5. Indeed there are six possible choices, all of which produce the string *FSSFS*. They are

2, 3, 5

2, 5, 3

3, 2, 5

3, 5, 2

5, 2, 3

5, 3, 2

Notice in this list that any one of the three integers may be selected to be the first. Once we choose the first integer, we have two choices for the second. The third integer then allows us no freedom of choice. Since there are three choices for the first integer, and each of these allows two choices for the second integer, there are 3×2 choices in total. Notice that there are $6 = 3 \times 2$ arrangements in the above list.

In general we will have chosen s integers that represent the positions of the S's in the string of n symbols. We wish to know the number of ways in which these s integers can be rearranged (permuted among themselves). Each of these rearrangements of the integers will produce the same string of F's and S's.

Given s integers, we can choose any one of them for the first integer. Once we have chosen the first integer, there are $s - 1$ choices for the second integer. Then there are $s - 2$ choices for the third integer and so on until there is only one choice for the last integer. Since, for each choice of the first integer, we have $s - 1$ choices for the second integer, there are $s(s - 1)$ choices for the first two integers. Continuing in this way we find that there are $s(s - 1)(s - 2) \ldots (2)(1)$ choices for the s integers. Each of these choices produces the same string of F's and S's. Using the definition of factorial given in (8.4), notice that this number is

$$s! \tag{8.6}$$

Expression 8.5 gives the number of ways in which we can choose s integers from the integers 1 to n. For each such choice the number of other choices that produce the same string of F's and S's is given by (8.6). Thus the number of *distinct* strings of F's and S's is given by (8.5) divided by (8.6).

This leads us to the conclusion that the number of strings of n symbols, each string containing precisely s S's and $n - s$ F's, is given by

$$C(n, s) = \frac{n!}{(n - s)! \, s!} \tag{8.7}$$

This often is referred to as the "number of *combinations* of n things taken s at a time." It is sometimes written as $_nC_s$.

Each of these chains has a probability of $p^s(1 - p)^{n-s}$ of occurring, since that was the chain probability calculated at the beginning of this section. Since each chain is a mutually exclusive event, the probability of any one of the chains occurring is the sum of the probabilities of the individual chains. Thus the probability of a chain with precisely s S's is

$$P(n,s;p) = \frac{n!}{(n - s)! \, s!} \, p^s(1 - p)^{n-s} \tag{8.8}$$

This is the probability of s successes in n trials where the probability of success in one trial is p. It is called a *binomial probability* because of its relationship to binomial trials.[1]

As an example we return to the urn problem at the beginning of this section. The probability of two successes (a success is drawing a white marble) on three trials is, from (8.8),

$$P(3,2;\ 2/9) = \frac{3 \times 2 \times 1}{(1)(2 \times 1)} \, (2/9)^2(7/9)$$

$$= 3 \times (2/9)^2(7/9)$$

which agrees with our previous result, (8.1).

For three successes on five trials (8.8) yields

$$P(5,3;2/9) = \frac{5 \times 4 \times 3 \times 2 \times 1}{(2 \times 1)(3 \times 2 \times 1)} \, (2/9)^3(7/9)^2 = .066386$$

Finally we look at the probability of s *or more* successes in n trials. This is the probability of s successes, $s + 1$ successes, or $s + 2$ successes, and so on up to and including n successes. But s successes and $s + 1$ successes are mutually exclusive events, so the probability of s successes or $s + 1$ successes is

$$P(n,s;p) + P(n,s + 1;p)$$

The probability of s or more successes is

$$P(n,s;p) + P(n,s + 1;p) + \ldots + P(n,n - 1;p) + P(n,n;p)$$

We will call this sum $P^*(n,s;p)$ to distinguish it from $P(n,s;p)$. The latter is the probability of precisely s successes, while the former is the probability of s or more successes. Indeed,

---

[1]The coefficients in (8.8), which are the $C(n,s)$ given by (8.7), are called the *binomial coefficients*. They appear elsewhere in mathematics, most notably in the expansion of $(a + b)^n$.

$$P*(n,s;p) = P(n,s;p) + P(n,s+1;p) + \ldots + P(n,n;p) \qquad (8.9)$$

In the urn example suppose we wish to determine the probability of two or more successes on three draws where the probability of success (a white marble) on one trial is 2/9. Then (8.9) yields

$$
\begin{aligned}
P*(3,2;2/9) &= P(3,2;2/9) + P(3,3;2/9) \\
&= \frac{3!}{1!\,2!}(2/9)^2(7/9) + \frac{3!}{0!\,3!}(2/9)^3(7/9)^0 \\
&= 3 \times (2/9)^2 \times (7/9) + 1 \times (2/9)^3 \\
&= 0.126
\end{aligned}
$$

which agrees with (8.2).

EXERCISES FOR SECTION 8.3

*1. Write down the 20 pairs of numbers that represent the 5×4 choices for placement of two S's in five positions. Recall that any one of 1, 2, 3, 4, or 5 can be chosen for the first S. Once the first position is selected, there are four choices for the second S. Match the pairs that correspond to the same string and, thereby, show that there are exactly 10 different strings possible.

2. Write down the 12 pairs of numbers that represent the 4×3 choices for placement of two S's in four positions. Match the pairs that correspond to the same string and, thereby, show that there are exactly six different strings possible.

3. (a) Draw a tree diagram, similar to Figure 8.6, for a four-stage binomial experiment.
   *(b) How many chains contain two successes?
   (c) What is the probability of each of the chains in part b?
   *(d) What is the probability of two successes in four trials?

*4. In the BASIC program in Figure 8.11 it is necessary that $N \geq S$. Insert statements immediately after statement 100 to check that this is true. If it is not, print an error message and stop.

5. Modify the BASIC program in Figure 8.11 so that the STEP can be eliminated from the FOR statement (i.e., the STEP should be + 1).

6. Suppose we have B different books. We wish to place C of them on a shelf beside each other. How many different arrangements are possible if:
   *(a) $B = 6$, $C = 4$?
   (b) $B = 7$, $C = 3$?
   (c) $B = 5$, $C = 5$?

7. Suppose in Exercise 6 the order of the C books is immaterial. How many different arrangements are possible?

8. What is the probability of exactly three successes on five trials if the probability of success on one trial is:

*(a) 1/5?

(b) 1/2?

(c) 7/10?

9. What is the probability of three or more successes on five trials if the probability of success on one trial is:

   *(a) 1/5?

   (b) 1/2?

   (c) 7/10?

10. Of a group of securities, 60% increase in value over a three-year period. If you buy six of the securities today, what is the probability that three years from now

    *(a) At least four of them will have increased in value?

    (b) All of them will have increased in value?

    (c) No more than four of them will have decreased in value?

11. In a certain community 30% of the people switch their brand of toothpaste each month. If you question 10 people, what is the probability that:

    *(a) Seven will have switched brands?

    (b) Five will have switched brands?

    (c) Three will have switched brands?

    (d) None will have switched brands?

12. Five six-sided dice are tossed. What is the probability that:

    *(a) There are three or more ones?

    (b) There are three or more even numbers?

    (c) There are precisely three ones?

    (d) There are precisely three even numbers?

13. In a very large group of applicants for a set of jobs 60% are men and 40% are women.[2] If there are 10 identical jobs to be filled from this pool of applicants, what is the probability of

    *(a) Seven or more men being selected?

    (b) Six men and four women being selected?

    (c) Four or more women being selected?

*14. In the job situation described in Exercise 13, if there are seven or more men selected for the 10 jobs, would you say there was evidence of discrimination shown? Against whom? Why?

15. In the job situation described in Exercise 13, if there are five or more women selected for the 10 jobs, would you say there was evidence of discrimination? Against whom? Why?

---

[2]The observant reader will note that condition 3 for a binomial trials experiment is not true in this example. That is, the probability of success is not constant from trial to trial because if a man is selected then the total number of men is reduced by one while the number of women is unchanged. This is equivalent to drawing marbles from an urn *without* replacement. However, we will assume that the number of applicants is so large that the change in the probabilities from trial to trial is very small—small enough to be neglected. Hence we will treat this problem as a binomial trials experiment.

8.4 A COMPUTER PROGRAM FOR BINOMIAL PROBABILITIES

Obviously the calculation of $P(n,s; p)$ and $P^*(n,s; p)$ are tedious and error prone. Therefore we turn to the computer to assist us in such calculations.

A program in BASIC that takes as input the number of trials, N; the number of successes, S; and the probability of success on one trial, P; and then prints the probability of precisely S successes on N trials is shown in Figure 8.12.

Some comments on this program are in order. First we note that in (8.8) for very large n, the numerator $n!$ will become very large. In fact, it may become so large that it will be impossible to compute it even in a large computer. (See Exercise 8.)

Thus we recall that (8.3) and (8.5) are identical; that is,

$$n \times (n - 1) \times (n - 2) \times \ldots \times (n - s + 1) = \frac{n!}{(n - s)!}$$

Therefore we can rewrite

$$\frac{n!}{(n - s)! \, s!}$$

as

$$\frac{n(n - 1) \ldots (n - s + 1)}{s(s - 1) \ldots (1)}$$

We first compute n/s. We then compute $(n - 1)/(s - 1)$ and multiply n/s by this ratio. We then have

$$\frac{n(n - 1)}{s(s - 1)}$$

Next we compute $(n - 2)/(s - 2)$ and multiply the last ratio by this number. We continue until we compute $(n - s + 1)/1$. These are the calculations performed in the

```
100   PRINT "TYPE NUMBER ØF TRIALS"
200   INPUT N
300   PRINT "TYPE NUMBER ØF SUCCESSES"
400   INPUT S
500   PRINT "TYPE PRØBABILITY ØF SUCCESS ØN ØNE TRIAL"
600   INPUT P
700   LET M=1
800   FØR I=S TØ 1 STEP -1
900   LET M=M*((N-S+I)/I)
1000  NEXT I
1100  LET M=M*P↑S*(1-P)↑(N-S)
1200  PRINT
1300  PRINT "PRØBABILITY ØF";S;"SUCCESSES ØN";N;"TRIALS IS";M
1400  END
```

FIGURE 8.12 BASIC program to compute $P'(n,s;p)$.

```
TYPE NUMBER ØF TRIALS
?15
TYPE NUMBER ØF SUCCESSES
?6
TYPE PRØBABILITY ØF SUCCESS ØN ØNE TRIAL
?0.4

PRØBABILITY ØF 6    SUCCESSES ØN 15    TRIALS IS .206598




TYPE NUMBER ØF TRIALS
?100
TYPE NUMBER ØF SUCCESSES
?90
TYPE PRØBABILITY ØF SUCCESS ØN ØNE TRIAL
?0.7

PRØBABILITY ØF 90    SUCCESSES ØN 100    TRIALS IS 1.17042E-06
```

FIGURE 8.13 Typical results from running program in Figure 8.12.

FOR-NEXT loop in statements 800, 900, and 1000. Statement 1100 then multiplies by $P^S(1 - P)^{N-S}$.

The results of running this program for several cases are shown in Figure 8.13.

The reader is warned that the program may fail when N is very large (on the order of 1000) if P is also large (0.9 or so). The behavior of the program is very dependent on the particular computer being used. Changes in the program can be made to avoid such difficulties, but they are of little practical use to us. Failure of the program usually means that the probability being calculated for $P(n,s;\ p)$ is so small that it may effectively be taken to be zero. Notice, for example, in the last example shown in Figure 8.13 that the probability of 90 successes in 100 trials is very small indeed, (0.00000117).

Tables of the values of $P(n,s;p)$ are given in several places. One such table can be found in Appendix B. This table includes values of n from 2 to 10 with s varying from 0 to n in each case. The values of p are .1, .2, .3, .4, .5, .6, .7, .8, and .9. This table was generated with a slight modification of the BASIC program in Figure 8.12. The modified program is not shown.

A BASIC program for computing $P^*(n,s;p)$, as defined in (8.9), is shown in Figure 8.14. It is quite similar to the program in Figure 8.12. In fact, starting with Figure 8.12, we make two changes and four additions to obtain Figure 8.14. Statement 400 is changed to rename the input X instead of S. Statement 1300 is changed to print different text and also to print the value of T instead of M. The four added statements are 610, 620, 1110, and 1120. They sum the various probabilities as indicated in (8.9).

A table of the values of $P^*(n,s;p)$ for $n = 2, 3, \ldots, 9, 10$ and for $p = 0.1, 0.2, \ldots, 0.8, 0.9$ is given in Appendix C. That table was produced by a BASIC program that is a slight modification of Figure 8.14.

```
100   PRINT "TYPE NUMBER ØF TRIALS"
200   INPUT N
300   PRINT "TYPE NUMBER ØF SUCCESSES"
400   INPUT X
500   PRINT "TYPE PRØBABILITY ØF SUCCESS ØN ØNE TRIAL"
600   INPUT P
610   LET T=0
620   FØR S=X TØ N
700   LET M=1
800   FØR I=S TØ 1 STEP -1
900   LET M=M*((N-S+1)/I)
1000   NEXT I
1100   LET M=M*P↑S*(1-P)↑(N-S)
1110   LET T=T+M
1120   NEXT S
1200   PRINT
1300   PRINT "PRØBABILITY ØF";X;"ØR MØRE SUCCESSES ØN";N;"TRIALS =";T
1400   END
```

FIGURE 8.14 BASIC program to compute $P^*(n,s;p)$.

EXERCISES FOR SECTION 8.4

1. Of all the voters in a district 20% change their party registration every four years. If there are 20 people on a given election roll in the district, what is the probability that four years from now
   *(a) At least five will have changed their registration?
   (b) No more than 13 will have the same party registration?
   (c) No more than five will have changed their registration?

2. In a certain community 35% of the people change the brand of deodorant they purchase every six months. If you survey 100 people what is the probability that in the last six months
   *(a) Fifty or more people will have changed their deodorant brand?
   (b) Precisely 50 people will have changed their deodorant brand?
   (c) At least one third of the people surveyed will have changed their deodorant brand?
   (d) At most 40% of the people surveyed will have changed their deodorant brand?

3. Find the value of q if
   *(a) $P(10,5;q) = 0.2$
   (b) $P^*(10,5;q) = 0.2$
   (c) $P(8,2;q) = 0.1$
   (d) $P^*(8,2;q) = 0.1$

*4. A pharmaceutical firm is testing a new drug reported to cure a certain disease. The firm claims that if the drug is administered to 100 patients known to have the disease, in 60% of the tests, 83 or more persons will be cured. What is the probability that any one patient will be cured by the drug?

5. Same as Exercise 4 except that with tests of 50 patients the drug cures 40 or more patients in 75% of the tests.

6. Write a BASIC program that reads two integers n and s and computes the number of combinations of n objects taken s at a time from (8.7). Test the program with $n = 7$ and $s = 3$. The correct result is 35.

7. Write a BASIC program that reads two integers n and s and computes the number of permutations of n objects taken s at a time from (8.5). Test the program with $n = 7$ and $s = 3$. The correct result is 210.

8. Write a BASIC program that computes and prints 1!, 2!, 3! and so on until the result becomes too large for the computer to handle. What is the largest value of $n!$ that can be computed in your computer?

8.5 INFORMATION THEORY REVISITED

In Section 6.6 we discussed a problem in information theory. The problem considered was that of sending a signal, either 0 or 1. In order to increase the reliability of the message we sent three signals. If we wished to send a 1, we sent three 1's. Similarly, if we wished to send a 0, we sent three 0's. At the receiving end we could receive three 1's, two 1's and one 0, two 0's and one 1, or three 0's. In the first two cases the message was interpreted as a 1 and in the latter two cases the message was interpreted as a 0. That is, we used a majority rule.

We can interpret this as a binomial trials experiment. Each line is considered to be a trial. The line can fail (transmit an erroneous message) or succeed (transmit a correct message). An erroneous message is received only if two or three lines fail. That is, failure of the system occurs if there are two or more failures in the three trials. Thus, if p is the probability of the failure of a single line, we wish to know the probability of two or more failures on three trials. From (8.9) this is

$$P^*(3,2;p) = P(3,2;p) + P(3,3;p)$$
$$= \frac{3!}{1!\,2!}\,p^2(1-p) + \frac{3!}{0!\,3!}\,p^3$$

Using $p = 0.1$, as we did in Section 6.6, this produces

$$= 0.027 + 0.001 = 0.028$$

which is the result obtained earlier. There we had to count the number of possible signals and compute the probability of each signal being transmitted.

If we had tried to use five lines in Section 6.6 the computations would become exceedingly difficult. With the use of binomial probabilities, however, the computation of the reliability of a five-line system is straightforward.

Suppose then we use five lines and once again use a majority rule to determine the message. Then there are five trials, and failure occurs only if three or more lines fail. Thus failure of the system occurs with probability

$$P^*(5,3;p) = P(5,3;p) + P(5,4;p) + P(5,5;p)$$

Again using $p = 0.1$, this becomes

$$\frac{5!}{2! \; 3!} (.1)^3(.9)^2 + \frac{5!}{1! \; 4!} (.1)^4(.9) + \frac{5!}{0! \; 5!} (.1)^5$$

or

$$.0081 \quad + \quad .00045 \quad + \quad .00001$$

The probability of failure then is 0.00856.

Thus the failure rate in the five-line system is less than 1%. Compare this with the three-line system where the failure rate was just a little less than 3%. As was to be expected then, the five-line system is much more reliable.

EXERCISES FOR SECTION 8.5

1. In the five-line system, what is the probability of an error being received if the error rate on a single line is
   *(a) 20%?
    (b) 30%?
    (c) 60%?

2. Suppose we use a seven-line system with majority rule. What is the probability of an error being received if the probability of an error on a single line is
   *(a) 0.1?
    (b) 0.2?
    (c) 0.01?

*3. Suppose in the five-line system we would like the probability of an error being received to be 0.01. What should be the probability of an error on a single line?

4. Suppose in a seven-line system with majority rule (see Exercise 2), we would like the probability of an error being received to be 0.01. What should be the probability of an error on a single line?

5. A special committee of the United Nations is voting on whether to send a resolution on military intervention to the General Assembly. There are 13 members of the committee, and they will vote by secret ballot. A political observer of the United Nations estimates that the probability that any one of the committee members will vote to send the resolution to the General Assembly is 0.4.
   (a) If a simple majority is required to send the resolution to the General Assembly, what is the probability that the resolution will indeed be sent?
   (b) If a two-thirds majority is required, what is the probability that the resolution will be sent?

6. Same as Exercise 5 except that the probability that any one of the committee members will vote to send the resolution to the General Assembly is 0.7.

8.6   CASE STUDY 8:   A LEGAL PROBLEM

We conclude our discussion of probability with a problem that brings to bear all of the aspects of probability that we have encountered. In particular it will draw on trees, "flipped" trees, Bayes' formula, and binomial probabilities. We posed this problem in Section 6.1 at the outset of our study of probability. We restate the problem here.

An employee is suing his employer on the basis that the machine that he was operating was malfunctioning and, therefore, he sustained injuries through the negligence of the company. The machine produces light bulbs. From observations both before and after the accident it has been verified that the machine functions properly 95% of the time. When the machine functions properly, 9 out of 10 light bulbs (on the average) are without defects. On the other hand, when the machine is malfunctioning, only 5 out of 10 bulbs are acceptable.

Now it turns out that when the accident occurred, a quality control inspector was examining the output of the machine. He reports that he tested three bulbs and found two to be defective. What is the probability that the machine was malfunctioning at the time?

We consider the event that precisely two of three bulbs are defective. If the machine is functioning properly, then the probability of any one bulb being defective is 0.1. Thus the probability of two out of three bulbs being defective is

$$P(3,2;0.1) = \frac{3!}{2! \ 1!} (0.1)^2(0.9) \tag{8.10}$$
$$= 0.027$$

On the other hand, if the machine is malfunctioning, the probability of any one given bulb being defective is 0.5. Thus the probability of two out of three bulbs being defective is

$$P(3,2;0.5) = \frac{3!}{2! \ 1!} (0.5)^2(0.5) \tag{8.11}$$
$$= 0.375$$

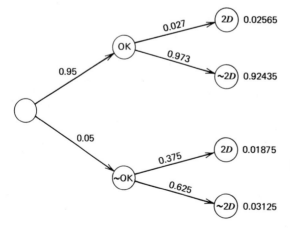

FIGURE 8.15 Tree diagram for selecting status of machine and then selecting the results of quality control test.

The events {machine malfunctioning} and {machine functioning properly} are a complete set of mutually exclusive alternatives. We have just computed the probability of two out of three bulbs being defective for each of these alternatives. Consider the process in which we first decide whether or not the machine is functioning properly and then decide whether or not two out of three bulbs are defective. The tree for this process is shown in Figure 8.15. The symbol *"OK"* designates that the machine is functioning properly. Notice that the branch leading to that symbol is labeled with 0.95, since the machine operates properly 95% of the time. The symbol *"2D"* represents two defective bulbs. The branch labels for the second stage were computed earlier. In particular in (8.10) we computed $P(2D \mid OK)$ and in (8.11) we found $P(2D \mid \sim OK)$. The other branch labels are obtained by subtracting these probabilities from 1. The chain probabilities also are shown in Figure 8.15.

What we wish to compute is $P(\sim OK \mid 2D)$, since we know that $2D$ (two bulbs were defective) occurred, and we want to know the probability of the machine malfunctioning. To compute this probability, we could use Bayes' formula. Instead, we perform an equivalent calculation that arises from "flipping" the original tree. We thus construct a tree where the first stage represents the event that two bulbs are or are not defective. The second stage is a choice of the condition of the machine (OK or notOK). The "flipped" tree is shown in Figure 8.16. The chain probabilities shown in Figure 8.16 can be read from the original tree in Figure 8.16 directly. We need to calculate the branch label on the branch going from $2D$ to $\sim OK$.

To compute this branch label we first need the branch label on the branch leading to $2D$. That branch label is the sum of the chain probabilities on the chains leading to $2D$ in the first tree, or

$$0.02565 + 0.01875 = 0.0444$$

This is the probability that precisely two out of three bulbs are defective, regardless of the condition of the machine.

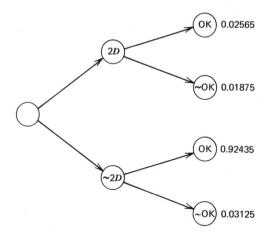

FIGURE 8.16 "Flipped" tree for legal problem with chain probabilities.

The product of the branch labels must equal the chain probability. Thus whatever the branch label is on the branch leading from $2D$ to $\sim OK$ when it is multiplied by 0.0444, the result must be 0.01875. Therefore the branch label we seek is

$$\frac{0.01875}{0.0444} = 0.4223$$

Our conclusion is that the probability that the machine was malfunctioning at the time of the accident is 0.4223. In other words, the likelihood is that the machine was in proper running order.

EXERCISES FOR SECTION 8.6

*1. Suppose in the situation described in Section 8.6 that the quality control expert had tested four light bulbs and found three of them to be defective. What is the probability that the machine had been malfunctioning at the time?

2. Same as Exercise 1 except that three out of five bulbs were found to be defective.

3. Suppose in the situation described in Section 8.6 that the machine is known to function properly 90% of the time. When functioning properly, nine out of 10 bulbs are without defects; when malfunctioning, five out of 10 bulbs are acceptable. What is the probability that the machine was malfunctioning at the time of the accident if the quality control expert found
   *(a) two out of three defective bulbs?
   (b) three out of four defective bulbs?
   (c) three out of five defective bulbs?

4. Suppose in the situation described in Section 8.6 the machine functions properly 95% of the time. The quality control expert finds two out of three light bulbs to be defective. What is the probability that the machine is malfunctioning if
   (a) When the machine is functioning properly 19 out of 20 bulbs are acceptable, while only six out of 10 are acceptable if the machine malfunctions?
   (b) When the machine is functioning properly eight out of 10 bulbs are acceptable, while only seven out of 10 are acceptable if the machine is malfunctioning?

*5. Suppose an archeologist wants to determine whether a given site that has been uncovered recently is the ruins of a village from Tribe 1 or from Tribe 2. The investigators that found the site report that they have discovered three artifacts and that two of these three are composed of a particular type of red clay. From the results of previous diggings and analyses the archeologist knows that in the sites that belonged to Tribe 1, one out of 10 artifacts contained red clay. Similarly, Tribe 2's sites produced five out of 10 artifacts with red clay content. Finally, in the general area where the site was found, of all the sites that have been positively identified, 95% of them have been identified as Tribe 1 and the other 5% are known to be Tribe 2. What is the probability that the site in question is Tribe 2?

6. Same as Problem 5 except that another artifact is uncovered at the site in question and is found to contain red clay. That is, there have been four artifacts uncovered and three contain red clay.

7. Suppose an archeologist wants to determine whether a given site that has been uncovered recently is the ruins of a village from Tribe 1, Tribe 2, or Tribe 3. The team of investigators that found the site reports that they have discovered three artifacts and that two of these three are composed of a particular type of red clay. From the results of previous diggings and analyses, the archeologist knows that in the sites that belonged to Tribe 1, one out of 10 of the artifacts contained red clay. Similarly, Tribe 2's sites produced five out of 10 artifacts with red clay content. For Tribe 3 there were eight out of 10 artifacts with red clay. Finally, in the general area where the site was found, of all of the known sites that have been positively identified, 50% of them have been identified as Tribe 1, 20% have been identified as Tribe 2, and 30% have been identified as Tribe 3. What tribe is the most likely to have been associated with the new site? What is the probability that this tribe should be associated with the site?

9
VECTORS
AND MATRICES

9.1 VECTORS

S uppose we have the test scores of three students: Jones, Smith, and White; on each of four tests. The tests are named A, B, C, and D. For example, suppose Jones has scores of

$$80, 50, 75, 88$$

A convenient way of writing this is

$$\mathbf{j} = (80, 50, 75, 88) \tag{9.1}$$

We are to understand from the form in which this is written that these are Jones' test scores (the $\mathbf{j}$ stands for Jones). Moreover, they are the scores of the tests in the order A, B, C, and D.

273

Equation 9.1 represents a *row vector*. A row vector is a row of numbers, separated by commas and enclosed in parentheses. The name of the row vector shown on the right of (9.1) is **j**.

We could write row vectors that represent the test scores of the other two students. For example,

$$\mathbf{s} = (60, \ 67, \ 75, \ 100) \tag{9.2}$$

$$\mathbf{w} = (80, \ 33, \ 50, \ 75) \tag{9.3}$$

These mean that, for example, Smith scored 75 on the third test or Test C. On the other hand, White scored 80 on Test A and 75 on Test D.

There are four different numbers in each of the row vectors shown here. These numbers, which are separated from each other by commas, are called the *components* or the *elements* of the row vector. Thus the components of **j** are 80, 50, 75, and 88. The components are numbered consecutively starting with 1. Thus, again using (9.1), the first component is 75 and the fourth component is 88.

Notice that a row vector with one component is a number. For example, writing a row vector as

$$\mathbf{x} = (3)$$

is really the same as saying that a variable x has the value 3. We can add two numbers (e.g., $3 + 4$), so if we add two one-component row vectors we would like to obtain a one-component row vector that is the sum of the two numbers represented by the two row vectors. Thus, if

$$\mathbf{y} = (4)$$

is another one-component row vector, then we would like

$$\mathbf{x} + \mathbf{y} = (7)$$

The most straightforward way to accomplish this is to define the addition of two one-component row vectors to be a one-component row vector whose lone component is the sum of the components of the two vectors being added. We now generalize this to row vectors with more than one component in a way that is consistent with the one-component case.

We define the *addition of two row vectors* as follows. If **a** and **b** are two row vectors, each with the same number of components,[1] then **a** + **b** is a row vector with the same number of components as both **a** and **b**. Moreover, each component of **a** + **b** is the sum of the corresponding components of **a** and **b**.

---

[1]The requirement that the two vectors have the same number of components is crucial. Indeed, if **a** and **b** have different numbers of components, the sum **a** + **b** is not defined (i.e., does not exist).

Consider the sum of the vectors **j** and **s** given in (9.1) and (9.2). According to the above definition, the sum is again a row vector with four components, since both **j** and **s** have four components. The first component of the sum is the sum of the first components of **j** and **s**. The first components of **j** and **s** are 80 and 60, respectively. Therefore the first component of the sum is $80 + 60 = 140$. Similarly, we compute the other three components as follows.

$$(80, 50, 75, 88) + (60, 67, 75, 100) = (80+60, 50+67, 75+75, 88+100)$$
$$= (140, 117, 150, 188)$$

Why would we ever want to find the sum of two vectors? To answer this, we carry our rules for arithmetic of vectors one step further. A single number is called a *scalar*. If we multiply a row vector by a scalar, we multiply each component of the row vector by the scalar. For example, if a vector **x** is

$$\mathbf{x} = (3, -4, 6)$$

then **x** is a row vector with three components. If we multiply **x** by the scalar -2, then we get another row vector with three components. Each component of the product is -2 times the corresponding component of **x**. This means that -2 times **x** is

$$-2 \cdot \mathbf{x} = (-2 \times 3, -2 \times -4, -2 \times 6) = (-6, 8, -12)$$

Now suppose we multiply the sum of the vectors **j** and **s** as shown above by the scalar $1/2$. The result is

$$(1/2) \cdot (\mathbf{j} + \mathbf{3}) = (70, 58.5, 75, 94)$$

These components are the average scores of the first two persons on each of the four tests.

Thus addition of vectors together with multiplication by a scalar has provided us with a convenient way to find the averages of the test scores.

We could, of course, add the third vector **w** given in (9.3) to the sum **j** + **s** and then multiply this result by 1/3. We then should obtain the average of all three students on each of the four tests.

So far we have written one row vector for each subject. Each row vector contained the scores from four tests. Suppose we wish to look at the scores of all students on one test (e.g., Test *A*). Those scores are 80, 60, and 80 for Jones, Smith, and White, respectively. Since we used a separate row for each student earlier, we will continue to do so. We write a column of numbers representing Test *A* as follows.

$$\mathbf{a} = \begin{pmatrix} 80 \\ 60 \\ 80 \end{pmatrix} \qquad (9.4)$$

Since this is a column of numbers surrounded by parentheses, we call it a *column vector*. This column vector has three components. They are numbered consecutively from the top, starting with one. Thus the first component is 80, the second component is 60, and the third component is 80.

We would write column vectors for each of the other three tests as follows.

$$\mathbf{b} = \begin{pmatrix} 50 \\ 67 \\ 33 \end{pmatrix} \qquad \mathbf{c} = \begin{pmatrix} 75 \\ 75 \\ 50 \end{pmatrix} \qquad \mathbf{d} = \begin{pmatrix} 88 \\ 100 \\ 75 \end{pmatrix} \tag{9.5}$$

We repeat that each column vector represents the test with which its name is associated. The scores are those of the three students taken in order: Jones, Smith, and White.

Now, if we add **a** and **b**, we get

$$\mathbf{a} + \mathbf{b} = \begin{pmatrix} 80 + 50 \\ 60 + 67 \\ 80 + 33 \end{pmatrix} = \begin{pmatrix} 130 \\ 127 \\ 113 \end{pmatrix}$$

where we have added component by component just as we did with row vectors. Thus the first component of **a** + **b** is the sum of the first components of **a** and **b**.

If we now multiply **a** + **b** by the scalar ½ we get

$$(1/2) \cdot (\mathbf{a} + \mathbf{b}) = \begin{pmatrix} 65 \\ 62.5 \\ 56.5 \end{pmatrix}$$

These numbers represent the average scores on the first two tests of each of the three students. If we added **a** + **b** + **c** + **d** and multiplied by ¼, we would obtain the average scores on all four tests of each of the three students.

We have carefully avoided discussing the sum of a row vector and a column vector. Such addition is not defined. Recall from footnote 1 that the sum of two row vectors that do not have the same number of components also is not defined.

In general, the addition of two vectors (row or column) is only defined if the vectors have the same number of components. For example,

$$(3, 2, -1) + (1, -4, 3) = (4, -2, 2)$$

and

$$(6, 4) + (7, -5) = (13, -1)$$

On the other hand,

$$(3, 2, -1) + (6, 4)$$

is not defined, since the two row vectors do not have the same number of components. The first vector has three components and the second has only two.

The same rule applies to column vectors. That is,

$$\begin{pmatrix} 8 \\ -2 \\ 7 \end{pmatrix} + \begin{pmatrix} 4 \\ 6 \\ -7 \end{pmatrix} = \begin{pmatrix} 12 \\ 4 \\ 0 \end{pmatrix}$$

However,

$$\begin{pmatrix} 4 \\ -2 \\ 6 \end{pmatrix} + \begin{pmatrix} 2 \\ 3 \end{pmatrix}$$

is not defined since again the column vectors do not have the same number of components.

The sum of a row vector and a column vector is not defined, even if the two vectors have the same number of components. For example, while

$$(2, 1) + (-1, 6) = (1, 7)$$

and

$$\begin{pmatrix} 2 \\ 1 \end{pmatrix} + \begin{pmatrix} -1 \\ 6 \end{pmatrix} = \begin{pmatrix} 1 \\ 7 \end{pmatrix}$$

there is no meaning ascribed to

$$(2, 1) + \begin{pmatrix} -1 \\ 6 \end{pmatrix}$$

nor to

$$\begin{pmatrix} 2 \\ 1 \end{pmatrix} + (-1, 6)$$

Clearly, it would be helpful to have a way to distinguish row vectors from column vectors so we could avoid inadvertently adding two different types of vectors. To that end we will use boldface letters such as **v** to represent column vectors, and we will use similar letters with a superscript T for row vectors. The T stands for the word *transpose,* indicating that we have changed the orientation from a column-one to a row-one.

Of course, choosing to use the T to denote a row vector and no superscript to denote a column vector is quite arbitrary. It could just as well have been the other way around. Nevertheless, we will write a *row vector with p components* as

$$\mathbf{v}^T = (v_1, v_2, \ldots, v_p) \tag{9.6}$$

where v_1 represents the first component of $\mathbf{v}^T$, v_2 is the second component, and so on until v_p is the pth or last component. If we have another row vector, $\mathbf{u}^T$, with p components,

$$\mathbf{u}^T = (u_1, u_2, \ldots, u_p)$$

then the *sum of two row vectors* $\mathbf{u}^T$ and $\mathbf{v}^T$ is defined to be

$$\begin{aligned} \mathbf{u}^T + \mathbf{v}^T &= (u_1, u_2, \ldots, u_p) + (v_1, v_2, \ldots, v_p) \\ &= (u_1 + v_1, u_2 + v_2, \ldots, u_p + v_p) \end{aligned} \tag{9.7}$$

In the same way we will write a *column vector with n components as*

$$\mathbf{a} = \begin{pmatrix} a_1 \\ a_2 \\ \cdot \\ \cdot \\ \cdot \\ a_n \end{pmatrix} \tag{9.8}$$

If we have a second column vector also with n components,

$$\mathbf{b} = \begin{pmatrix} b_1 \\ b_2 \\ \cdot \\ \cdot \\ \cdot \\ b_n \end{pmatrix}$$

Then we define the *sum of two column vectors* $\mathbf{a}$ and $\mathbf{b}$ to be

$$\mathbf{a} + \mathbf{b} = \begin{pmatrix} a_1 \\ a_2 \\ \cdot \\ \cdot \\ \cdot \\ a_n \end{pmatrix} + \begin{pmatrix} b_1 \\ b_2 \\ \cdot \\ \cdot \\ \cdot \\ b_n \end{pmatrix} = \begin{pmatrix} a_1 + b_1 \\ a_2 + b_2 \\ \cdot \\ \cdot \\ \cdot \\ a_n + b_n \end{pmatrix} \tag{9.9}$$

It is important to recall that addition of a row vector and a column vector is not defined. Moreover, addition of two row vectors or two column vectors is not defined unless the vectors have the same number of components.

Finally, we define the *product of a scalar k and a row vector* $\mathbf{v}^T$ where $\mathbf{v}^T$ is given by (9.6) to be

$$k \cdot \mathbf{v}^T = (kv_1, kv_2, \ldots, kv_p) \tag{9.10}$$

and the *product of a scalar k and a column vector* $\mathbf{a}$ where $\mathbf{a}$ is given by (9.8) as

$$k \cdot \mathbf{a} = \begin{pmatrix} ka_1 \\ ka_2 \\ \cdot \\ \cdot \\ \cdot \\ ka_n \end{pmatrix} \tag{9.11}$$

We say two vectors are *equal* if each component of the two vectors are equal. That is, if $\mathbf{a}$ and $\mathbf{b}$ are the column vectors on the previous page, if

$$\mathbf{a} = \mathbf{b}$$

it follows that

$$a_1 = b_1$$
$$a_2 = b_2$$
$$\cdot$$
$$\cdot$$
$$\cdot$$
$$a_n = b_n$$

EXERCISES FOR SECTION 9.1

(Note. Solutions are given for those exercises marked with an asterisk.)

1. Find the sums of the following pairs of vectors
   *(a) $(2, -1, 4)$ $(-2, 2, 1)$
   (b) $(1, 0, 1)$ and $(0, 2, 1)$
   (c) $\begin{pmatrix} -2 \\ 3 \\ -3 \end{pmatrix}$ and $\begin{pmatrix} 1 \\ -1 \\ -1 \end{pmatrix}$

(d) $\begin{pmatrix} 6 \\ -6 \end{pmatrix}$ and $\begin{pmatrix} -6 \\ 6 \end{pmatrix}$

2. For the four row vectors

   $\mathbf{a}^T = (2, -1, 4)$ $\mathbf{b}^T = (6, 3)$
   $\mathbf{c}^T = (4, 4, 0)$ $\mathbf{d}^T = (0, 0)$
   find where possible

   *(a) $\mathbf{a}^T - \mathbf{b}^T$ (b) $\mathbf{a}^T + \mathbf{b}^T$
   *(c) $\mathbf{a}^T - \mathbf{c}^T$ (d) $\mathbf{a}^T + \mathbf{c}^T$
    (e) $\mathbf{b}^T + \mathbf{d}^T$ (f) $\mathbf{c}^T - \mathbf{d}^T$
   *(g) $2 \cdot \mathbf{a}^T$ (h) $-3\mathbf{b}^T$
    (i) $5\mathbf{d}^T$ (j) $0 \cdot \mathbf{c}^T$

3. For the four column vectors

   $\mathbf{e} = \begin{pmatrix} 1 \\ 0 \end{pmatrix}$ $\mathbf{g} = \begin{pmatrix} 6 \\ -3 \\ 2 \end{pmatrix}$

   $\mathbf{f} = \begin{pmatrix} 0 \\ 1 \end{pmatrix}$ $\mathbf{h} = \begin{pmatrix} 0 \\ 2 \\ 3 \end{pmatrix}$

   Find where possible
   *(a) $\mathbf{e} + \mathbf{f}$ (b) $\mathbf{e} - \mathbf{f}$
    (c) $\mathbf{f} - \mathbf{e}$ *(d) $\mathbf{e} + \mathbf{g}$
    (e) $\mathbf{h} + \mathbf{g}$ (f) $-\mathbf{g}$
   *(g) $3 \cdot \mathbf{h}$ (h) $-2\mathbf{e} + 3\mathbf{f}$
    (i) $3\mathbf{f} + 2\mathbf{g}$ (j) $2\mathbf{h} + 2\mathbf{g}$

4. For the row vectors

   $\mathbf{a}^T = (1, 0)$ $\mathbf{b}^T = (0, 1)$
   $\mathbf{c}^T = (-5, 4)$ $\mathbf{d}^T = (5, -4)$
   $\mathbf{e}^T = (0, 0)$ $\mathbf{f}^T = (1, 1)$
   which of the following equations are true?
   *(a) $\mathbf{a}^T + \mathbf{b}^T = \mathbf{f}^T$ (b) $\mathbf{a}^T = \mathbf{b}^T$
   *(c) $\mathbf{a}^T - \mathbf{b}^T = \mathbf{e}^T$ (d) $\mathbf{c}^T + \mathbf{d}^T = \mathbf{e}^T$
    (e) $5\mathbf{a}^T - 4\mathbf{b}^T = \mathbf{d}^T$ (f) $-5\mathbf{a}^T + 4\mathbf{b}^T = -\mathbf{c}^T$

*5. If

$$\mathbf{r} = \begin{pmatrix} 2 \\ -1 \end{pmatrix} \quad \mathbf{s} = \begin{pmatrix} -3 \\ 2 \end{pmatrix} \quad \mathbf{t} = \begin{pmatrix} -1 \\ 0 \end{pmatrix}$$

is

$$\mathbf{r} + \mathbf{s} = \mathbf{t}?$$

6. If

$$\mathbf{r} = \begin{pmatrix} 3 \\ 2 \\ 1 \end{pmatrix} \quad \mathbf{s} = \begin{pmatrix} 1 \\ 2 \\ 3 \end{pmatrix} \quad \mathbf{t} = \begin{pmatrix} 1 \\ 1 \\ 1 \end{pmatrix}$$

is

$$\mathbf{r} + \mathbf{s} = 4\mathbf{t}?$$

7. If

$$\mathbf{u} = \begin{pmatrix} 3 \\ -1 \\ 2 \end{pmatrix} \qquad \mathbf{v} = \begin{pmatrix} 1 \\ -2 \\ 2 \end{pmatrix} \qquad \mathbf{w} = \begin{pmatrix} -1 \\ 2 \\ 2 \end{pmatrix}$$

compute
*(a) $\mathbf{u} + \mathbf{v} - 2\mathbf{w}$
 (b) $2\mathbf{u} - \mathbf{v} - \mathbf{w}$
 (c) $2(\mathbf{u} - \mathbf{v}) + \mathbf{w}$
 (d) $-3\mathbf{u} + 2\mathbf{v} - \mathbf{w}$

8. Find x and y if
*(a) $\begin{pmatrix} x \\ 4 \end{pmatrix} = \begin{pmatrix} 3 \\ x + y \end{pmatrix}$

 (b) $\begin{pmatrix} 9 \\ x \end{pmatrix} = y \begin{pmatrix} 3 \\ 5 \end{pmatrix}$

 (c) $\begin{pmatrix} x \\ y \end{pmatrix} = x \begin{pmatrix} x \\ 2 \end{pmatrix}$

 (d) $(5, 4, 2) = (x, x + y, -2y)$
 (e) $2 \cdot (x, -1) = (y, y)$

*9. You own stock in the following companies: 50 shares of National Oxygen, 20 shares of United Biscuits, and 10 shares of Consolidated Power. All were purchased on the last day of trading in 1973 at prices of 31, 72, and 104, respectively. The prices of these stocks on the last day of trading in 1974, 1975, 1976 are given in the following table.

	1974	1975	1976
National Oxygen	33½	32	30
United Biscuits	69	76	78
Consolidated Power	102	97	97

(a) Write three column vectors, $\mathbf{n}$, $\mathbf{u}$, and $\mathbf{c}$, each with four components and each of which represents successive prices at year end of one of the three stocks.
(b) Using the three vectors in part a, define a column vector $\mathbf{t}$ whose components represent the total value of your investment at the end of each of the four years from 1973 to 1976.
(c) Using the results of parts a and b, find the value of your investment at the end of each of the four years.

10. Use the data in Exercise 9 to complete this exercise.
(a) Write four row vectors $\mathbf{f}^T$, $\mathbf{s}^T$, $\mathbf{t}^T$, and $\mathbf{x}^T$, each with three components and each of which represents the prices of the three stocks in a given year.
(b) Using the four vectors in part a, define a row vector $\mathbf{a}^T$ whose components represent the average price of each stock over the four year period.
(c) Using the results of parts a and b, find the average price of each of the three stocks.

9.2 VECTOR PRODUCTS

Let us return now to the test score problem described in Section 9.1. We will enlarge on that problem and introduce the concept of the product of two vectors.

Suppose we have a column vector **f** with four components. The components of this vector **f** are the number of questions on each of the four tests administered to the students. Recall that there were three students. We could have written **f** as a row vector, but we will find it more convenient to write **f** as a column vector.

For example, let us take **f** to be

$$\mathbf{f} = \begin{pmatrix} 5 \\ 6 \\ 4 \\ 8 \end{pmatrix} \tag{9.12}$$

This means that there were five questions on the first test, six on the second, and so on.

Suppose now that we wish to calculate the total number of correct responses by each subject on all four tests. The sum of all four components of **f** is the total number of responses (correct and incorrect) on all four tests. Before we proceed with the calculation of the total number of correct responses on all tests, let us see what we would have to do just to find the number of correct responses on one test.

If a person had 75% correct responses on a test of eight questions, then the number of correct responses is six, or $(75/100) \times 8$. In general we would take the percentage of correct responses, multiply that result by the total number of responses, and divide by 100.

We now turn to the case of several tests. Consider the first person, Jones, in our original problem. His test scores were

$$\mathbf{j}^T = (80, 50, 75, 88)$$

We will interpret these as percentages and, for our later convenience, convert them to decimal form. We accomplish this by multiplying $\mathbf{j}^T$ by the scalar $1/100$. The resulting row vector we will call $\mathbf{x}^T$. Thus

$$\mathbf{x}^T = \frac{1}{100}\mathbf{j}^T = (0.8, 0.5, 0.75, 0.88)$$

On the first test, which had five responses according to the column vector **f**, the number of correct responses was

$$0.8 \times 5 = 4$$

since Jones scored 80% on five questions. On the second test, which had six questions, Jones' score was 50%. The number of correct responses was

$$0.5 \times 6 = 3$$

Similarly, the number of correct responses on the third and fourth tests were

$$0.75 \times 4 = 3$$

and

$$0.88 \times 8 = 7.04$$

The residual 0.04 arises because the percentages were assumed to be rounded to the nearest 1%. This is, of course, a common practice. In any case, the last result indicates that there were actually seven correct responses.

On all four tests the total number of correct responses of the first student, Jones, is

$$4 + 3 + 3 + 7 = 17$$

How could we define vector multiplication to help us achieve such a result?

Notice that the first test score in decimal form (0.8) is multiplied by the number of responses on the first test (five), the second test score (0.5) is multiplied by the second test's number of responses (six), and so on. After all of these multiplications have been performed, the results of the four multiplications are added. This is an indication of how we might usefully define multiplication of vectors.

With this as background we define the product of a row vector $\mathbf{x}^T$ and a column vector $\mathbf{f}$ to be the sum of the products of the corresponding components of each vector. Then

$$
\begin{aligned}
\mathbf{x}^T \cdot \mathbf{f} &= 0.8 \times 5 + 0.5 \times 6 + 0.75 \times 4 + 0.88 \times 8 \\
&= \quad 4 \quad + \quad 3 \quad + \quad 3 \quad + \quad 7.04 \\
&= 17.04
\end{aligned}
$$

which is precisely the desired result. Rounded to the nearest integer it produces 17, the number of correct responses by the first student on all four tests.

Moreover, it should be clear that if we added a fifth test, we would get one more score (a fifth component of $\mathbf{x}^T$) and one more question number (a fifth component of $\mathbf{f}$). If we multiplied these two components, we would have the number of correct responses on the fifth test. If we then added this to 17, we would have the number of correct responses on all five tests. Adding a sixth, seventh, or even more tests would produce similar results.

Thus the definition of the product of a row vector and a column vector will produce the total number of correct responses for any number of tests. It is for this reason that the definition is as it is. We will come back to a more general definition after we examine a few more examples.

As another example, let us take the second student, Smith, whose test scores were

$$\mathbf{s}^T = (60, 67, 75, 100)$$

We convert these to decimal:

$$\mathbf{y}^T = (0.6, 0.67, 0.75, 1)$$

Now

$$
\begin{aligned}
\mathbf{y}^T \cdot \mathbf{f} &= 0.60 \times 5 + 0.67 \times 6 + 0.75 \times 4 + 1 \times 8 \\
&= \quad 3 \quad + \quad 4.02 \quad + \quad 3 \quad + \quad 8 \\
&= 18.02
\end{aligned}
$$

So the second student made 18 correct responses.

Let us look now for a convenient way of keeping track of the arithmetic when we multiply vectors. To that end we write

$$\mathbf{y}^T \cdot \mathbf{f} = (0.6, 0.67, 0.75, 1) \begin{pmatrix} 5 \\ 6 \\ 4 \\ 8 \end{pmatrix} \leftarrow$$

If we place our left index finder under the first component of the row vector (see ↑) and our right index finger under the first component of the column vector (see ←), we should multiply the two numbers to which we are pointing. The result in this case is 3. We save this product.

We then move our left index finger one place to the right and our right index finger one place down so that the fingers point as follows.

$$(0.6, 0.67, 0.75, 1) \begin{pmatrix} 5 \\ 6 \\ 4 \\ 8 \end{pmatrix} \leftarrow$$

We again multiply the two numbers to which our fingers point ($0.67 \times 6 = 4.02$) and add this to the previous result. We now have $3 + 4.02 = 7.02$.

We continue in this way, moving our left index finger one place to the right and our right index finger one place down until we have exhausted the vectors. Try the method to compute the product of $\mathbf{y}^T$ and $\mathbf{f}$. The result, as noted earlier, is 18.02.

Our definition of the product of two vectors requires that the two vectors have the same number of components. Notice that the first vector must be a row vector and the second one must be a column vector. The order is important. Recall, in addition, that both vectors had to be of the same type (i.e., either both column vectors or both row vectors).

We will define the product of two vectors in general as follows. Suppose we have a row vector $\mathbf{u}^T$ with n components.

$$\mathbf{u}^T = (u_1, u_2, \ldots, u_n)$$

and a column vector $\mathbf{v}$ also with n components

$$\mathbf{v} = \begin{pmatrix} v_1 \\ v_2 \\ \cdot \\ \cdot \\ v_n \end{pmatrix}$$

We define the product $\mathbf{u}^T\mathbf{v}$ to be a scalar quantity equal to

$$\mathbf{u}^T\mathbf{v} = u_1v_1 + u_2v_2 + \ldots + u_nv_n \tag{9.13}$$

Notice that if we write

$$\mathbf{u}^T\mathbf{v} = (u_1, u_2, \ldots, u_n)\begin{pmatrix} v_1 \\ v_2 \\ \cdot \\ \cdot \\ v_n \end{pmatrix}$$

the process of (1) using the left index finger on the row vector and moving that finger to the right one component at a time, (2) using the right index finger on the column vector and moving that finger down one component at a time at the same speed that the left finger moves, (3) multiplying at each move the two numbers to which the two index fingers point, and (4) adding each product to the previous sum (initially the sum is zero) produces the product

$$u_1v_1 + u_2v_2 + \ldots + u_nv_n$$

We note that (1) the product of two row vectors is not defined, (2) the product of two column vectors is not defined, and (3) the product of a column vector and a row vector is not defined. The last of these could be defined, but we will not do so here. Only the product of a row vector and a column vector (note the order—row first, column second) will be defined. Even in this last case, the product is only defined if the row and column vectors have the same number of components.

EXERCISES FOR SECTION 9.2

1. Find the product of

$$\mathbf{a}^T = (3, 2, -1)$$

with each of the following

*(a)
$$\mathbf{x} = \begin{pmatrix} 1 \\ 0 \\ 3 \end{pmatrix}$$

(b)
$$\mathbf{y} = \begin{pmatrix} 2 \\ -1 \\ 3 \end{pmatrix}$$

(c)
$$\mathbf{s} = \begin{pmatrix} 0 \\ 0 \\ 1 \end{pmatrix}$$

(d)
$$\mathbf{t} = \begin{pmatrix} -1 \\ 3 \\ 3 \end{pmatrix}$$

2. Find the product of each of the following with

$$\mathbf{b} = \begin{pmatrix} -1 \\ 4 \\ 2 \end{pmatrix}$$

*(a) $\mathbf{d}^T = (1, 1, 1)$

(b) $\mathbf{e}^T = (-1, -1, -1)$

(c) $\mathbf{f}^T = (3, 1, 2)$

(d) $\mathbf{g}^T = (-1, 4, 2)$

*3. Using $\mathbf{a}^T$ and $\mathbf{b}$ as given in Exercises 1 and 2, compute

$$(1/3)\mathbf{a}^T\mathbf{b}$$

4. Using the vectors given in Exercises 1 and 2, compute
   (a) $\mathbf{d}^T\mathbf{x}$ (b) $\mathbf{e}^T\mathbf{y}$
   (c) $\mathbf{f}^T\mathbf{s}$ (d) $\mathbf{g}^T\mathbf{t}$

5. If

$$\mathbf{u}^T = (1, x, -3) \qquad \mathbf{v} = \begin{pmatrix} 2 \\ -5 \\ 4 \end{pmatrix}$$

find the values of x for which
   *(a) $\mathbf{u}^T\mathbf{v} = 0$ (b) $\mathbf{u}^T\mathbf{v} = -10$ (c) $\mathbf{u}^T\mathbf{v} = 10$

*6. Using the vector $\mathbf{f}$ in (9.12), compute the total number of correct responses by each of Smith and White [see (9.2) and (9.3)].

7. If there were 15 questions on the first test, 12 questions on each of the second and third tests, and 16 questions on the fourth test, compute the total number of correct responses given by
   (a) Jones.
   (b) Smith.
   (c) White.

*8. Suppose you own 100 shares of common stock in United Biscuits valued at $112, 200 shares in National Oxygen valued at $87, and 150 shares in Consolidated Power valued at $92.

(a) Write a row vector where each component is the number of shares of a different stock. Arrange the stocks in alphabetical order.

(b) Write a column vector where each component is the current price of each stock in the same order as part a.

(c) Use the row and column vectors from parts a and b to compute the total value of the three stocks.

9. Do parts a, b, and c of Exercise 8 if you own 120 shares of National Oxygen valued at $90, 100 shares of Consolidated Power valued at $75, and 150 shares of United Biscuits valued at $102.

*10. A manufacturer is preparing a nut mixture for sale. One pound of the mixture contains 6 ounces of cashews, no peanuts, 4 ounces of hazelnuts, and 6 ounces of Brazil nuts. The cost per pound of the various nut types is: cashews, 24¢; peanuts, 16¢; hazel nuts, 16¢; and Brazil nuts, 32¢.

(a) Write a row vector where each component is the number of ounces of the different nut types in each pound of the mix.

(b) Write a column vector where each component is the price of each of the nut types.

(c) Use the results of parts a and b to compute the cost of each pound of the mix.

11. Same as Exercise 10 except the mix contains no cashews, 8 ounces of peanuts, 6 ounces of hazelnuts, and 2 ounces of Brazil nuts.

9.3 MATRICES

In the test score problem described in Section 9.1 there were four tests, each of which was administered to three students. For any one test the scores achieved by the three individuals were represented by the components of a column vector. But the vector for each test had a different name. Since there were four tests, there were four names: **a**, **b**, **c**, and **d**. If we were to increase the number of tests to, say, 10, we would need 10 names—an unpleasant thought.

Similarly, for any one person, the scores that individuals obtained on the four tests were represented by a row vector.

Each of these vectors had a different name. There were three students: Jones, Smith, and White; so there were three row vectors: $\mathbf{j}^T$, $\mathbf{s}^T$, and $\mathbf{w}^T$. Again, an increase in the number of people to 20 would raise difficulties for us.

But suppose we write these three row vectors on successive lines.

$\mathbf{j}^T$:	80	50	75	88
$\mathbf{s}^T$:	60	67	75	100
$\mathbf{w}^T$:	80	33	50	75

The four columns in this arrangement are the four column vectors **a**, **b**, **c**, **d** referred to earlier. Now, however, all of the scores are represented in a table with three rows and four columns. The score of the second student on the third test is the number in the second row, third column of the table (i.e., 75). If we wish to include the scores of additional people we simply write down more rows and number them consecutively. Similarly, if we wish to include the scores on more tests, we add columns to the table. In any case, we need only one name, the name of the table: M. Of course, we must remember that row 2 corresponds to Smith, but we can make a note of this correspondence at the start, forget it during any calculations, and then reconstruct the correspondence later if we have need of it.

We will refer to such a table as a *matrix* (plural: matrices). We will write it as a rectangular array of numbers surrounded by parentheses. For example, the above table can be written

$$M = \begin{pmatrix} 80 & 50 & 75 & 88 \\ 60 & 67 & 75 & 100 \\ 80 & 33 & 50 & 75 \end{pmatrix} \tag{9.14}$$

M is said to be a 3×4 (read "3 by 4") matrix because it has three rows and four columns (note: rows first, columns second). Each number in the array is called a *component* or element of the array. The components are denoted by m_{ij}, which denotes the number in the ith row and the jth column. For example,

$$m_{23} = 75$$

$$m_{12} = 50$$

$$m_{32} = 33$$

In general, $m_{ij} \neq m_{ji}$.

A row vector is a matrix. Indeed, it is a matrix with only one row. If the row vector has p components, as does $\mathbf{v}^T$ in (9.6), it is a $1 \times p$ matrix. Similarly, a column vector is a matrix. For example, the column vector $\mathbf{a}$ in (9.8) is an $n \times 1$ matrix. Because vectors are matrices, the rules we develop for the addition and multiplication of matrices should be consistent with our earlier definitions of these arithmetic operations for vectors.

When we multiplied a vector by a scalar, we multiplied each component of the vector by the scalar. We will do the same for matrices. Using the test score matrix, M, as an example, if we multiply M by 0.01, we get a new matrix, D.

$$D = 0.01 \times M = \begin{pmatrix} 0.8 & 0.5 & 0.75 & 0.88 \\ 0.6 & 0.67 & 0.75 & 1. \\ 0.8 & 0.33 & 0.5 & 0.75 \end{pmatrix} \tag{9.15}$$

The components of this new matrix are the various scores in decimal form. That is, the third student responded correctly on .33 of the total number of questions on test 2 (the component in the third row and second column is 0.33). In this case, at least, the rule[2] for multiplication by a scalar produces meaningful results.

Before turning to the product of a vector and a matrix and the addition and multiplication of matrices, we will define a matrix in more general terms. A *matrix* is a rectangular array of numbers; that is,

$$
A = \begin{pmatrix}
a_{11} & a_{12} & a_{13} & \cdots & a_{1p} \\
a_{21} & a_{22} & a_{23} & \cdots & a_{2p} \\
\cdot & & & & \cdot \\
\cdot & & & & \cdot \\
\cdot & & & & \cdot \\
a_{n1} & a_{n2} & a_{n3} & \cdots & a_{np}
\end{pmatrix}
\tag{9.16}
$$

The element in the kth row and the jth column is denoted by a_{kj}. If there are n rows and p columns in A, A is said to be an $n \times p$ matrix.

Two matrices are said to be *equal* if, and only if, all of their components are identical. Implicit in this definition is the fact that two equal matrices must have the same numbers of rows and columns.

We close this section with a discussion of the transpose of a matrix. We used the word transpose in Section 9.1 in our discussion of vectors. If we used the components of a column vector to form a row vector, we called the row vector the transpose of the column vector. Indeed, row vectors were labeled with a T to denote that they were the transpose of a column vector.

If we take any matrix and interchange its rows and columns, we call the new matrix the *transpose* of the original matrix. Thus, if

$$
A = \begin{pmatrix} 6 & 0 & -2 \\ 1 & 3 & 4 \end{pmatrix}
$$

then its transpose is

$$
A^T = \begin{pmatrix} 6 & 1 \\ 0 & 3 \\ -2 & 4 \end{pmatrix}
$$

Notice that the first row of A is the first column of A^T, and that the second column of A is the second row of A^T.

If the matrix is a column vector

$$
\mathbf{x} = \begin{pmatrix} 2 \\ 3 \\ -1 \end{pmatrix}
$$

---

[2] The "rule" is that when a matrix is multiplied by a scalar, each element of the matrix is multiplied by the scalar.

then

$$\mathbf{x}^T = (2, 3, -1)$$

so that our definition of the transpose of a matrix is consistent with our earlier use of transpose with vectors.

If A is $n \times p$, A^T is $p \times n$. The transpose of the transpose is the original matrix again; that is,

$$(A^T)^T = A$$

EXERCISES FOR SECTION 9.3

1. For the matrices

$$A = \begin{pmatrix} 2 & 1 & 0 \\ 1 & 3 & 1 \end{pmatrix} \qquad B = (1/2) \begin{pmatrix} 4 & 2 \\ 2 & 6 \\ 0 & 2 \end{pmatrix}$$

$$C = 2 \begin{pmatrix} 3 & 0 \\ 0 & 2 \end{pmatrix} \qquad D = \begin{pmatrix} 2 & 1 \\ 1 & 3 \\ 0 & 1 \end{pmatrix}$$

$$E = \begin{pmatrix} 1 & 2 \\ 3 & 1 \\ 1 & 0 \end{pmatrix} \qquad F = \begin{pmatrix} 6 & 0 \\ 0 & 4 \end{pmatrix}$$

*(a) Find two pairs of equal matrices.

(b) Find two matrices that are the transpose of each other.

*2. If

$$\begin{pmatrix} x & 6 \\ 2 & y \end{pmatrix} = \begin{pmatrix} 1 & a \\ b & 3 \end{pmatrix}$$

Find x, y, a, and b.

3. If

$$\begin{pmatrix} r & 0 & 6 \\ 2 & 1 & -2 \\ -3 & s & 1 \end{pmatrix} = \begin{pmatrix} -5 & 0 & t \\ u & 1 & -2 \\ -3 & 4 & 1 \end{pmatrix}$$

Find r, s, t, and u.

*4. If

$$A = \begin{pmatrix} 3 & 6 & -2 \\ 5 & -6 & 1 \end{pmatrix}$$

(a) Find A^T.

(b) Find $(A^T)^T$.

5. If

$$B = \begin{pmatrix} 2 & 3 & -2 \\ 0 & -2 & -3 \\ 1 & 0 & 4 \\ 4 & 5 & 0 \end{pmatrix}$$

(a) Find B^T.
(b) Find $(B^T)^T$.

*6. If

$$A = \begin{pmatrix} 2 & -1 & a \\ 3 & b & 0 \end{pmatrix}$$

and

$$A^T = \begin{pmatrix} 2 & d \\ c & 6 \\ -2 & e \end{pmatrix}$$

Find a, b, c, d, and e.

7. If

$$B = \begin{pmatrix} 4 & -2 & d \\ a & 4 & -4 \\ b & 2 & e \\ c & 0 & 8 \end{pmatrix}$$

and

$$B^T = \begin{pmatrix} 4 & -1 & 1 & 0 \\ f & 4 & g & h \\ 7 & k & 3 & 8 \end{pmatrix}$$

Find a, b, c, d, e, f, g, h, and k.

9.4 MULTIPLICATION OF A MATRIX AND A VECTOR

The components of the column vector **f** given in (9.12) represented the number of questions on each of the four tests. When we formed the product of the row vector $\mathbf{x}^T$ with **f**, we obtained a scalar.[3] That scalar was the number of correct responses on all four tests by Jones.

[3]Recall that $\mathbf{x}^T$ contained Jones' scores on the four tests in decimal form. The row vector $\mathbf{j}^T$ contained those same scores in percentage form.

But each row of D, the matrix defined in (9.15), is a row vector of percentage scores for some individual. The first row of D is just $\mathbf{x}^T$. Hence each row of M, when multiplied by $\mathbf{f}$, should produce a scalar. Moreover, that scalar should be the total number of correct responses by the corresponding student. If we use all three rows of D we should obtain the number of correct responses by all three students.

This leads us to consider the multiplication of a matrix D by a column vector $\mathbf{f}$ as follows: (1) each row of the matrix is considered as a row vector; (2) each row vector so obtained is multiplied by the column vector; and (3) the scalars so obtained are arranged as a column vector with one component for each row of the original matrix D.

Using D and $\mathbf{f}$ again as examples, we write $D \cdot \mathbf{f}$ as

$$\begin{pmatrix} 0.80 & 0.50 & 0.75 & 0.88 \\ 0.60 & 0.67 & 0.75 & 1.00 \\ 0.80 & 0.33 & 0.50 & 0.75 \end{pmatrix} \begin{pmatrix} 5 \\ 6 \\ 4 \\ 8 \end{pmatrix}$$

The result should be a column vector with three components, since M has three rows. The first component of the result is obtained by multiplying the first row of D by $\mathbf{f}$. This produces

$$0.80 \times 5 + 0.50 \times 6 + 0.75 \times 4 + 0.88 \times 8 = 17.04$$

The second component of the result if obtained by multiplying the second row of D by $\mathbf{f}$. Schematically, this last multiplication can be represented by Figure 9.1. The results from the three vector multiplications produce the column vector

$$\begin{pmatrix} 17.04 \\ 18.02 \\ 13.98 \end{pmatrix} \tag{9.17}$$

This column vector is $D \cdot \mathbf{f}$.

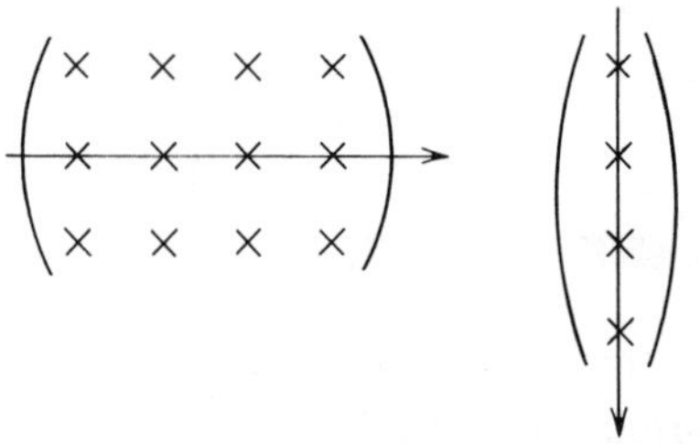

FIGURE 9.1 Schematic representation of part of the product of a matrix and a vector.

Notice that we place the column vector $\mathbf{f}$ on the right of the matrix D and that the number of columns in the matrix must be the same as the number of elements (or rows) of the column vector. In the example, the matrix D has four columns and the vector $\mathbf{f}$ has four elements. In summary:

Multiplication of a matrix by a column vector is only defined if: (1) the matrix is the left-most entry in the product, and (2) the number of columns in the matrix is the same as the number of rows in the column vector. The multiplication proceeds as follows: each row of the matrix is considered as a row vector and is multiplied by the column vector. The scalars so produced are arranged in a column vector with each element corresponding to one row of the original matrix.

Now that we can multiply a matrix on its right by a column vector, what can be said about multiplication of a matrix by a row vector?

To this end, suppose that more than three students are tested. However, we will suppose that all of the students achieve one of the three possible sets of scores.

$$(0.80 \quad 0.50 \quad 0.75 \quad 0.88)$$

or

$$(0.60 \quad 0.67 \quad 0.75 \quad 1.00)$$

or

$$(0.80 \quad 0.33 \quad 0.50 \quad 0.75)$$

These, of course, are the same scores in the matrix D, but now a set of scores may apply to more than one person. If a student scores 0.80 on the first test, he must score either 0.50 or 0.33 on the second test. On the other hand, if the student scores 0.60 on the first test, then he scores 0.67 on the second test, 0.75 on the third, and 1.00 on the fourth. This is not an unreasonable assumption in many cases. Consider, for example, aptitude tests where the results of one part of the test are highly correlated with scores on other parts. Of course, in most realistic cases, there are many more possible sets of scores. We use three possible sets here for illustrative purposes only.

Let $\mathbf{n}^T$ be a row vector with three components. Each component of $\mathbf{n}^T$ represents the number of students who achieve the same scores on all four tests. For example, if three subjects each score 80%, 50%, 75%, and 88% on the four tests, then n_1 (the first component of $\mathbf{n}^T$) is 3. As an example, we will let

$$\mathbf{n}^T = (3, 2, 4)$$

Thus two students scored 0.60, 0.67, 0.75, and 1.00, in that order, on the four tests.

Now we define a vector $\mathbf{e}$ that has all of its components equal to 1. The column vector $\mathbf{e}$ with three components is

$$\mathbf{e} = \begin{pmatrix} 1 \\ 1 \\ 1 \end{pmatrix}$$

The product of $\mathbf{n}^T$ and $\mathbf{e}$ is

$$\mathbf{n}^T\mathbf{e} = (3,\ 2,\ 4)\begin{pmatrix}1\\1\\1\end{pmatrix} = 3 + 2 + 4 = 9$$

Thus $\mathbf{n}^T\mathbf{e}$ is the total number of students tested. But what about multiplication of $\mathbf{n}^T$ and a matrix? We turn to that now; however, we first look at the columns of the matrix in question.

The first column of the matrix D is the column vector $\mathbf{a}$ given in (9.4) and multiplied by 0.01 (i.e., converted to decimal form). The column vector $\mathbf{a}$ contained the scores of all persons on the first test. Thus

$$\mathbf{p} = 0.01\mathbf{a} = \begin{pmatrix}0.80\\0.60\\0.80\end{pmatrix}$$

contains the decimal scores of all persons on test one. If we multiply $\mathbf{n}^T$ by $\mathbf{p}$ we will get a scalar that is the sum of the decimal scores of all nine persons on the first test. If we then divide by 9 (the number of persons taking the tests) we will obtain the average on the first test of all persons. Thus

$$\mathbf{n}^T\mathbf{p} = (3,\ 2,\ 4)\begin{pmatrix}0.80\\0.60\\0.80\end{pmatrix} = 2.40 + 1.20 + 3.20 = 6.80$$

or

$$(\mathbf{n}^T\mathbf{p})/(\mathbf{n}^T\mathbf{e}) = 6.80/9 = 0.7556$$

We have not divided vectors. We formed two vector products. Each of these products produced a scalar. We then divided scalars. The parentheses indicate this but, even without the parentheses, this would be the only possible interpretation we could give to something like $\mathbf{n}^T\mathbf{p}/\mathbf{n}^T\mathbf{e}$, since division of a vector by a vector is not defined. Care must be exercised then that we do not try to divide common factors from a vector quotient. For example, $\mathbf{n}^T\mathbf{p}/\mathbf{n}^T\mathbf{e}$ is not equal to $\mathbf{p}/\mathbf{e}$. Indeed, the latter expression has no meaning.

Now each column of the matrix M is one of the column vectors $\mathbf{a}$, $\mathbf{b}$, $\mathbf{c}$, or $\mathbf{d}$ defined in (9.4) and (9.5). Each of these vectors corresponds to the scores on a specific test. Hence each column of D is one of these vectors in decimal form. Therefore we could multiply each by $\mathbf{n}^T$ and then divide by 9 just as we did with $\mathbf{p}$. In each case we would obtain the average score of all nine students on the test in question.

We now consider the four column vectors $\mathbf{a}$, $\mathbf{b}$, $\mathbf{c}$, and $\mathbf{d}$ taken as a whole. In decimal form they are represented by the matrix D. Therefore, if we multiply each column of D

by the row vector $\mathbf{n}^T$ and divide by 9, we will obtain the average scores on the four tests.

This leads us to define the product of the row vector $\mathbf{n}^T$ and the matrix D as follows: (1) each column of the matrix is considered as a column vector, (2) the row vector $\mathbf{n}^T$ is multiplied by each column vector, and (3) the scalars so obtained are arranged as a row vector with one component for each column of the original matrix, D.

Using $\mathbf{n}^T$ and D we write

$$(3, 2, 4)\begin{pmatrix} 0.80 & 0.50 & 0.75 & 0.88 \\ 0.60 & 0.67 & 0.75 & 1.00 \\ 0.80 & 0.33 & 0.50 & 0.75 \end{pmatrix}$$

The result should be a row vector with four components, since M has four columns. The first component of the result is obtained by multiplying $\mathbf{n}^T$ and the first column of D. This produces

$$3 \times 0.80 + 2 \times 0.60 + 4 \times 0.80 = 6.80$$

The second component of the result is obtained by multiplying $\mathbf{n}^T$ by the second column of D. Schematically, this can be represented by Figure 9.2.

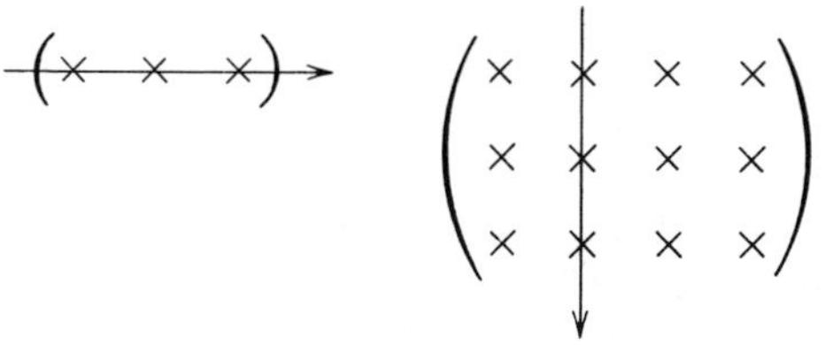

FIGURE 9.2 Schematic representation of part of the product of a row vector and a matrix.

The results of the four vector multiplications are:

$$\mathbf{n}^T D = (6.80, 4.16, 5.75, 7.64) \qquad (9.18)$$

Moreover, if we divide this result by $\mathbf{n}^T \mathbf{e} = 9$, we get

$$(\mathbf{n}^T D)/(\mathbf{n}^T \mathbf{e}) = (0.7556, 0.4622, 0.6389, 0.8489)$$

Notice that $\mathbf{n}^T \mathbf{e}$ is a scalar, so when we divide $\mathbf{n}^T D$, which is a row vector, by $\mathbf{n}^T \mathbf{e}$, we are dividing a vector by a scalar. Our previous rules tell us that we should divide each component of the vector by the scalar, which is what we have done.

The last vector above lists the averages of all of the nine persons on each of the four tests in decimal form.

We summarize the discussion of the multiplication of a row vector by a matrix as follows.

Multiplication of a row vector by a matrix is only defined if: (1) the matrix is the right-most entry in the product, and (2) the number of rows of the matrix is the same as the number of columns in the row vector. The multiplication proceeds as follows: each column of the matrix is considered as a column vector and is multiplied by the row vector. The scalars produced are arranged in a row with each element corresponding to one column of the original matrix.

Note the correspondence between this definition and the definition given earlier for multiplication of a matrix by a column vector.

Now $\mathbf{n}^T D$, as given in (9.18), is a row vector whose four components are the total scores of all nine persons on the four tests. Recall that $\mathbf{f}$, as given in (9.12), is a column vector whose four components are the number of questions on each of the four tests. Thus we can multiply the row vector $\mathbf{n}^T D$ by the column vector $\mathbf{f}$, and we should obtain a single scalar that is the number of correct responses by all nine persons on all four tests. Thus we turn our attention to

$$(\mathbf{n}^T D)\mathbf{f} = (6.80,\ 4.16,\ 5.75,\ 7.64) \begin{pmatrix} 5 \\ 6 \\ 4 \\ 8 \end{pmatrix}$$
$$= 34 + 24.96 + 23 + 61.12$$
$$= 143.08$$

But suppose instead we consider the product of the row vector $\mathbf{n}^T$ and the column vector $D\mathbf{f}$. Once again, this is a scalar. Moreover, $\mathbf{n}^T$ is a row vector whose three components are the number of students achieving identical scores on all four tests, and $D\mathbf{f}$ is a column vector whose three components are the number of correct responses by each student in each of the three groupings. Hence $\mathbf{n}^T$ multiplied by $D\mathbf{f}$ should also produce the total number of correct responses by all nine students (i.e., the value just computed, 143.08).

$$\mathbf{n}^T(D\mathbf{f}) = (3,\ 2,\ 4) \begin{pmatrix} 17.04 \\ 18.02 \\ 13.98 \end{pmatrix}$$
$$= 51.12 + 36.04 + 55.92$$
$$= 143.08$$

Thus the order of the multiplication is immaterial, as we would hope.

Notice that the scalar (143.08, which rounds to 143) is the total number of correct responses by all nine students on all four tests.

EXERCISES FOR SECTION 9.4

1. Multiply the matrix

$$A = \begin{pmatrix} 2 & 1 & -2 \\ -1 & 1 & 4 \\ 2 & 3 & 1 \end{pmatrix}$$

by each of the following vectors.

*(a) $\mathbf{x}^T = (2, 1, 0)$

(b) $\mathbf{x} = \begin{pmatrix} 2 \\ 1 \\ 0 \end{pmatrix}$

(c) $\mathbf{y}^T = (1, 1, 1)$

(d) $\mathbf{y} = \begin{pmatrix} 1 \\ 1 \\ 1 \end{pmatrix}$

(e) $\mathbf{z}^T = (-2, 3, 1)$

(f) $\mathbf{z} = \begin{pmatrix} -2 \\ 3 \\ 1 \end{pmatrix}$

2. Multiply the matrix

$$T = \begin{pmatrix} 4 & 0 & 5 \\ 0 & 1 & -6 \\ 3 & 0 & 4 \end{pmatrix}$$

by each of the following vectors.

*(a) $\mathbf{x} = \begin{pmatrix} 4 \\ -18 \\ -3 \end{pmatrix}$

(b) $\mathbf{y} = \begin{pmatrix} 0 \\ 1 \\ 0 \end{pmatrix}$

(c) $\mathbf{z} = \begin{pmatrix} -5 \\ 24 \\ 4 \end{pmatrix}$

*(d) $\mathbf{u}^T = (4, 0, -5)$

(e) $\mathbf{v}^T = (-18, 1, 24)$

(f) $\mathbf{s}^T = (-3, 0, 4)$

3. Multiply the matrix

$$S = \begin{pmatrix} 4 & 0 & -5 \\ -18 & 1 & 24 \\ -3 & 0 & 4 \end{pmatrix}$$

by each of the following vectors.

*(a) $\mathbf{a} = \begin{pmatrix} 4 \\ 0 \\ 3 \end{pmatrix}$

(b) $\mathbf{b} = \begin{pmatrix} 0 \\ 1 \\ 0 \end{pmatrix}$

(c) $\mathbf{c} = \begin{pmatrix} 5 \\ -6 \\ 4 \end{pmatrix}$

*(d) $\mathbf{d}^T = (4, 0, 5)$

(e) $\mathbf{e}^T = (0, 1, -6)$

(f) $\mathbf{f}^T = (3, 0, 4)$

4. Given the two matrices

$$A = \begin{pmatrix} 2 & 1 & -3 \\ 6 & -4 & 5 \end{pmatrix} \qquad B = \begin{pmatrix} 0 & 1 \\ -1 & -1 \\ 1 & 1 \end{pmatrix}$$

and the vectors

$$\mathbf{x}^T = (2, 1, 0) \qquad \mathbf{y}^T = (5, 1) \qquad \mathbf{z}^T = (0, 1, 3)$$

Compute where possible

*(a) $\mathbf{x}^T B$ (b) $B^T \mathbf{x}$

(c) $\mathbf{y}^T A$ (d) $\mathbf{y}^T B$

(e) $\mathbf{z}^T A$ (f) $A\mathbf{z}$

(g) $\mathbf{y}^T B^T$ (h) $A^T \mathbf{z}$

5. You own stock in the following companies: 50 shares of National Oxygen, 20 shares of United Biscuits, and 10 shares of Consolidated Power. All were purchased on the last day of trading in 1973 at prices of 31, 72, and 104, respectively. The prices of these stocks on the last day of trading in 1974, 1975, and 1976 are given in the following table.

	1974	1975	1976
National Oxygen	33½	32	30
United Biscuits	69	76	78
Consolidated Power	102	97	97

(a) Write a 4×3 matrix M that displays the prices of all three stocks at the end of each of the four years.

(b) Write a column vector $\mathbf{n}$ that displays the number of shares of each stock that you own.

(c) Using the results of parts a and b, compute the product of M and $\mathbf{n}$.

(d) What does the result of part c represent?

6. An attitude survey has been administered to 100 persons. The survey contains six statements, each of which can be rated in one of five ways: much in favor, somewhat in favor, no opinion, somewhat opposed, or much opposed. The number of responses in each case has been tabulated as follows.

Question	Much in Favor	Somewhat in Favor	No Opinion	Somewhat Opposed	Much Opposed
1	22	19	29	13	17
2	31	7	32	17	13
3	42	19	19	20	0
4	0	33	34	25	8
5	22	23	18	19	18
6	16	6	58	2	18

(a) Write a 6×5 matrix Q that contains this data.

(b) What operation should be carried out on Q so that the result is the percentage of responses in each category?

(c) If the five responses are assigned values from 1 to 5 (1 = much in favor and 5 = much opposed), write a column vector **r** so that $(1/100) \cdot Q\mathbf{r}$ contains the "average" response to each of the six statements.

(d) What vector should Q be multiplied by to produce the total number of responses on all six questions in each of the five categories?

(e) Using the results of part c, find the average response to each of the six questions.

(f) Using the results of part d, find the total number of responses in the five categories.

9.5 ADDITION OF MATRICES

Each decade the Census Bureau publishes the populations of all of the states of the United States by race. For example, the populations of four particular states in the 1970 census are shown in Figure 9.3.

	White	Negro	Indian	Japanese	Chinese
New York	15885.9	2164.6	28.4	20.4	81.4
California	17855.2	1398.5	91.0	213.3	170.1
Alabama	2535.8	903.0	2.4	1.1	0.6
Illinois	9616.3	1422.4	11.4	17.3	14.5

FIGURE 9.3 1970 census figures for four states (data in thousands).

We could think of this table of numbers as a matrix with four rows and five columns as follows.

$$\begin{pmatrix} 15885.9 & 2164.6 & 28.4 & 20.4 & 81.4 \\ 17855.2 & 1398.5 & 91.0 & 213.3 & 170.1 \\ 2535.8 & 903.0 & 2.4 & 1.1 & 0.6 \\ 9616.3 & 1422.4 & 11.4 & 17.3 & 14.5 \end{pmatrix} \qquad (9.19)$$

Each row corresponds to a state and each column corresponds to a race. We must, of course, have some way of keeping track of which state goes with each row and which column is associated with each race. That is a bookkeeping problem that, while trivial, is important and cannot be ignored.

The corresponding population figures for 1960 are shown in Figure 9.4.

	White	Negro	Indian	Japanese	Chinese
New York	15302.0	1417.5	16.5	8.7	37.6
California	14541.4	883.9	39.0	157.3	95.6
Alabama	2284.4	980.2	1.3	0.5	0.3
Illinois	9017.8	1037.5	4.7	14.1	7.0

FIGURE 9.4 1960 census figures for four states (data in thousands).

Suppose we wished to know the increase from 1960 to 1970 in the number of Japanese in California. Then we would subtract the 1960 figure (157.3) from the 1970 figure (213.3) and obtain an increase of 56.0. This means that there were 56,000 more Japanese in California in 1970 than there were in 1960.

Similarly, for any given state and race, if we wished to know the population increase between 1960 and 1970, we would simply subtract the corresponding entries in the tables.

If we think of both tables as matrices then, if we subtract corresponding entries (or components) of the two matrices, we obtain a matrix whose elements are the change in the various populations over the decade 1960 to 1970.

Thus, if we subtract the matrix,

$$
\begin{pmatrix}
15302.0 & 1417.5 & 16.5 & 8.7 & 37.6 \\
14541.4 & 883.9 & 39.0 & 157.3 & 95.6 \\
2284.4 & 980.2 & 1.3 & 0.5 & 0.3 \\
9017.8 & 1037.5 & 4.7 & 14.1 & 7.0
\end{pmatrix} \tag{9.20}
$$

from (9.19), we obtain

$$
\begin{pmatrix}
583.9 & 747.1 & 11.9 & 11.7 & 43.8 \\
3313.8 & 514.6 & 52.0 & 56.0 & 74.5 \\
251.4 & -77.2 & 1.1 & 0.6 & 0.3 \\
589.5 & 384.9 & 6.7 & 3.2 & 7.5
\end{pmatrix} \tag{9.21}
$$

This matrix can be represented more explicitly by Figure 9.5.

	White	Negro	Indian	Japanese	Chinese
New York	583.9	747.1	11.9	11.7	43.8
California	3313.8	514.6	52.0	56.0	74.5
Alabama	251.4	−77.2	1.1	0.6	0.3
Illinois	589.5	384.9	6.7	3.2	7.5

FIGURE 9.5   Increase in race populations from the 1960 census to the 1970 census.

In order to obtain the increase in the different race populations in these states, we have found it convenient to subtract two matrices element by element. Of course, when we added or subtracted vectors (either row or column), we performed the operation of addition or subtraction element by element, so we should not be surprised when we find it convenient that matrices are added and subtracted the same way. After all, as we pointed out earlier, vectors are simply special cases of matrices (i.e., a vector is a matrix with only one row or with only one column).

In any case, our population problem, together with our previous definitions for vectors, lead us to define the *addition of two matrices* as follows.

1. The sum (or difference) of two matrices A and B is only defined if A and B have both the same number of rows and the same number of columns as each other.
2. The sum $A + B$ is a matrix with the same number of rows and columns as both A and B.
3. The element in the ith row and jth column of $A + B$ is the sum of two elements, one from the ith row and jth column of A and the other from the ith row and jth column of B. As a simple example, suppose

$$A = \begin{pmatrix} 2 & -1 & 2 \\ -3 & 2 & 0 \end{pmatrix} \qquad B = \begin{pmatrix} -1 & -2 & 3 \\ 2 & 0 & 4 \end{pmatrix}$$

Since A and B both have two rows and three columns, the sum $A + B$ is defined and also has two rows and three columns. The element in the first row and the first column of the sum $A + B$ is the sum of the elements in the first row, first column of A (that element is 2) and the element in the first row, first column of B (that element is -1). Thus we get $2 + (-1) = +1$. Similarly, the element in the first row, second column is

$$(-1) + (-2) = -3$$

Continuing in this way we get

$$A + B = \begin{pmatrix} 1 & -3 & 5 \\ -1 & 2 & 4 \end{pmatrix}$$

EXERCISES FOR SECTION 9.5

1. For the following matrices

$$A = \begin{pmatrix} 3 & -2 \\ 1 & 0 \\ 1 & 2 \end{pmatrix} \qquad B = \begin{pmatrix} 2 & -1 & 4 \\ 3 & 0 & 2 \end{pmatrix}$$

$$C = \begin{pmatrix} 1 & 5 & 1 \\ 5 & 1 & 5 \end{pmatrix} \qquad D = \begin{pmatrix} 2 & -6 \\ -3 & 2 \\ 0 & 4 \end{pmatrix}$$

where possible find
  *(a) $A - D$
  (c) $B + C$
  (e) $D - C$
  (b) $A^T + B$
  (d) $C^T - B^T$
  (f) $D - B^T$

2. The populations (in thousands) of the four states listed below in 1950 were:

	White	Negro	Indian	Japanese	Chinese
New York	13877.4	918.1	10.6	3.9	20.2
California	9960.9	462.2	19.9	84.9	58.3
Alabama	2081.0	979.6	0.9	0.0	0.2
Illinois	8049.0	646.0	1.4	11.6	4.2

*(a) Compute the increase in the population of each race in 1960 compared to 1950 (see Figure 9.4).

 (b) Compute the increase in the population of each race in 1970 compared to 1950 (see Figure 9.3).

*3. In the text we computed the increase in the population of various races in four states between 1970 and 1960 (see Figure 9.5). In Exercise 2a you were asked to compute the same increases between 1960 and 1950. Using these results, compute the change in the increases over these two 10-year periods (i.e., find the rate at which the increases themselves are changing).

4. Let P_{1970} be the population matrix for 1970 $\left[\text{i.e., } (9.19)\right]$. Let P_{1960} be the similar matrix for 1960 $\left[\text{see } (9.20)\right]$. Similarly, P_{1950} is the 1950 matrix as given in Exercise 2 above. If C is the change matrix,

$$C_{1970} = P_{1970} - P_{1960}$$

$$C_{1960} = P_{1960} - P_{1950}$$

Let A be the rate at which C changes so that

$$A_{1970} = C_{1970} - C_{1960}$$

Find A_{1970} in terms of P_{1950}, P_{1960}, and P_{1970}. (See also Exercise 3.)

5. Find x and y if

(a) $\begin{pmatrix} x & 2 \\ 1 & -1 \end{pmatrix} + \begin{pmatrix} 2 & 1 \\ -1 & y \end{pmatrix} = \begin{pmatrix} 5 & 3 \\ 0 & 1 \end{pmatrix}$

(b) $\begin{pmatrix} 2 & -1 \\ 1 & 2 \end{pmatrix} + \begin{pmatrix} x & y \\ 3 & 1 \end{pmatrix} = \begin{pmatrix} 6 & 2 \\ x & y \end{pmatrix}$

(c) $\begin{pmatrix} x & x^2 \\ -xy & y \end{pmatrix} + \begin{pmatrix} y & -y \\ 1 & 2x \end{pmatrix} = \begin{pmatrix} 3 & -3 \\ 5 & 2 \end{pmatrix}$

(d) $\begin{pmatrix} 2 & x \\ y & 0 \\ xy & 3 \end{pmatrix} + \begin{pmatrix} x & 1 \\ 2 & 1 \\ 3 & -2 \end{pmatrix} = \begin{pmatrix} 7 & 6 \\ 2 & 1 \\ 3 & -1 \end{pmatrix}$

(e) $\begin{pmatrix} 2 & 2y & -2 \\ y & 0 & y^2 \end{pmatrix} - \begin{pmatrix} 1 & x & -5 \\ x & y & x^2 \end{pmatrix} = \begin{pmatrix} 1 & 6 & 3 \\ 3 & -3 & 9 \end{pmatrix}$

6. The following table shows the enrollment at a moderate-size university for 1975 and 1976. Students are classified according to their class year and their place of residence.

(a) Write two 3×4 matrices M_{1975} and M_{1976} showing these two populations for the two years.

(b) Find a 3×4 matrix C that shows the change (increase) in the populations in the four classes from 1975 to 1976.

		Place of Residence	
	Dormitory	Fraternity or Sorority	Off-Campus
Freshman	1175	250	1150
Sophomore	325	275	825
1975 Junior	210	280	610
Senior	45	280	575
Freshman	1020	250	1155
Sophomore	520	280	750
1976 Junior	190	275	685
Senior	55	250	570

(c) What vector should the matrix C in part b be multiplied by to produce the total change in the occupancy of the three types of residency from 1975 to 1976? What are those changes?

(d) What vector should the matrix C in part b be multiplied by to produce the total change in the number of students in each class year? What are those changes?

(e) What is the change in the total population of the university from 1975 to 1976?

9.6 MATRIX PRODUCTS

Recall that in order to multiply an $n \times p$ matrix named A by a column vector $\mathbf{v}$, it is necessary that the column vector have p components. That is, if the left-hand multiplicand A is $n \times p$ then the right-hand multiplicand $\mathbf{v}$ must be $p \times 1$.

Now suppose the right-hand multiplicand, instead of being a vector $\mathbf{v}$, is a matrix B which, of course, can be thought of as a generalization of a column vector. Indeed, we may think of any matrix as a group of column vectors each with the same number of rows and written side by side. Suppose that each column of B has p rows. That is, the right-hand matrix B will need to be $p \times q$.

When we multiplied an $n \times p$ matrix by a $p \times 1$ column vector, the result was a column vector with n components (i.e., an $n \times 1$ matrix). In other words, $A\mathbf{v} = \mathbf{w}$ and $\mathbf{w}$ has n components. Each of the n components resulted from the multiplication of a row of the matrix A by the sole column $\mathbf{v}$. On the other hand, we may think of the matrix product

$$A \cdot B$$

as A multiplied by q column vectors (the q columns of B). Since each multiplication of A by a column vector produces n numbers, the result of $A \cdot B$ is $n \times q$ numbers. Each of these is an element of AB. Hence AB must be an $n \times q$ matrix.

In line with these observations the rule we will use for multiplying two matrices is: the element of the ith row and the jth column of the product is the result of multiplying the ith row of the left matrix by the jth column of the right matrix. Schematically we

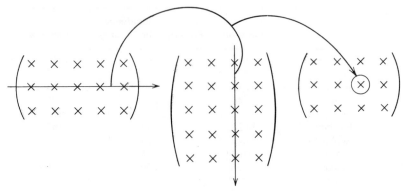

FIGURE 9.6 Schematic representation of matrix multiplication.

can represent this as shown in Figure 9.6. In this example a 3×5 matrix is multiplied by a 5×4 matrix. The arrows indicate that if we multiply the second row of the first matrix by the third column of the second matrix, we obtain the element in the second row and the third column of the product. When performing such a multiplication, each row is considered as a row vector and each column as a column vector. The multiplication of row by column is then the multiplication of a row vector by a column vector. The result of such a multiplication is a scalar and hence one element of the product.

Let us now look at a numerical example. Consider the product of

$$A = \begin{pmatrix} 4 & 0 & 5 \\ 0 & 1 & -6 \\ 3 & 0 & 4 \end{pmatrix} \quad \text{and} \quad B = \begin{pmatrix} 4 & 0 & -5 \\ -18 & 1 & 24 \\ -3 & 0 & 4 \end{pmatrix} \quad (9.22)$$

These are both 3×3 matrices. Thus the product will also be a 3×3 matrix. The element or component in the first row and first column of the product is obtained by multiplying the first row of A (considered as a row vector) by the first column of B (considered as a column vector). Thus we multiply

$$(4, 0, 5) \quad \text{by} \quad \begin{pmatrix} 4 \\ -18 \\ -3 \end{pmatrix}$$

and obtain $(4 \times 4) + [0 \times (-18)] + [5 \times (-3)] = 16 + 0 - 15 = 1$.

The component in the first row and the second column is obtained by multiplying the first row of A by the second column of B, or

$$(4, 0, 5) \quad \text{by} \quad \begin{pmatrix} 0 \\ 1 \\ 0 \end{pmatrix}$$

or 0. Similarly, we obtain the other seven elements of the product so that

$$A \cdot B = \begin{pmatrix} 1 & 0 & 0 \\ 0 & 1 & 0 \\ 0 & 0 & 1 \end{pmatrix}$$

This last matrix is a special matrix called the *identity matrix*. It has one's along its diagonal and zeroes elsewhere. The identity matrix always is *square* (i.e., has the same number of rows as columns).

Let us consider one other example before turning to matrices in BASIC. Suppose

$$C = \begin{pmatrix} 2 & 3 & 1 \\ 1 & -2 & 3 \end{pmatrix} \qquad D = \begin{pmatrix} 0 & 4 \\ 2 & -1 \\ 1 & 1 \end{pmatrix}$$

Now C is a 2×3 matrix and D is a 3×2 matrix. Therefore the product $C \cdot D$ is defined, since the number of columns of C is three, which is the same as the number of rows of D. On the other hand, the product $D \cdot C$ also is defined, since D has two columns and C has two rows. The product $C \cdot D$ will be a square matrix with two rows and two columns. The product $D \cdot C$, on the other hand, will be 3×3.

To obtain the element in the first row, first column of CD, we multiply the first row of C by the first column of D

$$(2 \quad 3 \quad 1) \qquad \text{by} \qquad \begin{pmatrix} 0 \\ 2 \\ 1 \end{pmatrix}$$

to obtain $0 + 6 + 1 = 7$.

To obtain the first row, second column element, we multiply the first row by the second column or

$$(2 \quad 3 \quad 1) \qquad \text{by} \qquad \begin{pmatrix} 4 \\ -1 \\ 1 \end{pmatrix}$$

which produces $8 - 3 + 1 = 6$. Similarly, the component of the product in the second row, first column is the second row of C multiplied by the first row of D, or

$$(1 \quad -2 \quad 3) \qquad \text{by} \qquad \begin{pmatrix} 0 \\ 2 \\ 1 \end{pmatrix}$$

or $0 - 4 + 3 = -1$. Finally, the second row, second column element results from multiplying the second row of C by the second column of D, or

$$(1 \quad -2 \quad 3) \quad \text{by} \quad \begin{pmatrix} 4 \\ -1 \\ 1 \end{pmatrix}$$

which produces $4 + 2 + 3 = 9$.

Putting all of these results together

$$C \cdot D = \begin{pmatrix} 7 & 6 \\ -1 & 9 \end{pmatrix}$$

However,

$$D \cdot C = \begin{pmatrix} 4 & -8 & 12 \\ 3 & 8 & -1 \\ 3 & 1 & 4 \end{pmatrix}$$

Clearly,

$$C \cdot D \neq D \cdot C$$

Indeed, CD is 2×2 and DC is 3×3, so the two could not possibly be equal.

Even when the products CD and DC have the same number of rows and columns, they generally are not equal.[4] This loss of the so-called commutative law is no accident. The definition we have used for matrix multiplication is a useful one, as we have seen in Section 9.4, and as we will see again in later sections. The price we pay for this usefulness is the fact that

$$C \cdot D \neq D \cdot C$$

There is nothing mystical about commutative or associative laws, even though most of the algebra and arithmetic we have encountered follow such laws. The purpose of mathematics is to help us solve problems. We use whatever laws are the most helpful. For matrices the multiplication law used here is by far the most helpful.

EXERCISES FOR SECTION 9.6

1. For the following matrices

$$A = \begin{pmatrix} 2 & 1 & 6 \\ -1 & 2 & 3 \end{pmatrix} \qquad B = \begin{pmatrix} 1 & 2 \\ -2 & 1 \end{pmatrix}$$

$$C = \begin{pmatrix} 1 & 6 \\ 1 & 2 \\ -3 & 1 \end{pmatrix} \qquad D = \begin{pmatrix} 1 & 0 & 2 \\ 2 & -2 & 3 \\ -1 & 2 & 1 \end{pmatrix}$$

compute where possible

---

[4]The reader should not conclude that for any two matrices $CD \neq DC$. See, for example, Exercise 9. However, we cannot indiscriminately interchange the order of matrices in a product.

*(a) $A \cdot C$ (b) $C \cdot A$
(c) $A \cdot B$ (d) $C \cdot D$
(e) $C \cdot B$ (f) $B \cdot C$
(g) $A \cdot D$ (h) $D^T \cdot C$

2. For the matrices in Exercise 1 compute where possible
    *(a) $A \cdot C \cdot B$ (b) $C \cdot A \cdot D$
    (c) $A \cdot D \cdot B$ (d) $B \cdot B \cdot B$
    (e) $A \cdot B \cdot A$ (f) $A \cdot A \cdot D \cdot D$

3. For the matrices in Exercise 1 compute where possible
    *(a) $AC - B$ (b) $CB + C$
    (c) $CA + D$ (d) $C - D^T C$
    (e) $AD + C^T$ (f) $B^2 + B$

4. For each of the following matrices M find M^2, M^4, and M^8.
    *(a) $M = \begin{pmatrix} 2 & 1 \\ 1 & 2 \end{pmatrix}$ (b) $M = \begin{pmatrix} 1 & 0 \\ 0 & 1 \end{pmatrix}$

    (c) $M = \begin{pmatrix} 0 & 1 \\ 1 & 0 \end{pmatrix}$ (d) $M = \begin{pmatrix} 1 & 0 & 0 \\ 0 & 1 & 0 \\ 0 & 0 & 1 \end{pmatrix}$

    (e) $M = \begin{pmatrix} 2 & -1 \\ -1 & 2 \end{pmatrix}$

5. Suppose A is a 2×3 matrix, B is a 3×4 matrix, C is a 3×3 matrix, and D is a 4×4 matrix. What are the number of rows and columns in each of the following?
    *(a) AB (b) DB^T
    (c) BD^T (d) CA^T
    (e) ACB (f) BD^2
    (g) BDB^T (h) BB^T

6. Let

$$X = \begin{pmatrix} 1 & 1 \\ 1 & 0 \end{pmatrix}$$

*(a) Find a matrix Y so that

$$XY = \begin{pmatrix} 1 & 1 \\ 0 & 1 \end{pmatrix}$$

(b) Find a matrix Z so that

$$ZX = \begin{pmatrix} 3 & 2 \\ 3 & 2 \end{pmatrix}$$

(c) Find a matrix R so that

$$RX = \begin{pmatrix} 0 & 0 \\ -1 & 1 \end{pmatrix}$$

7. If the only nonzero elements in a square matrix are along the upper left to lower right diagonal, the matrix is said to be *diagonal*. The matrices

$$A = \begin{pmatrix} 1 & 0 & 0 \\ 0 & 2 & 0 \\ 0 & 0 & 1 \end{pmatrix} \qquad B = \begin{pmatrix} 2 & 0 & 0 \\ 0 & -1 & 0 \\ 0 & 0 & 3 \end{pmatrix}$$

$$C = \begin{pmatrix} 3 & 0 & 0 \\ 0 & -2 & 0 \\ 0 & 0 & -3 \end{pmatrix}$$

are diagonal matrices.

*(a) Compute AB.   (b) Compute A^2.

(c) Compute CB.   (d) Compute AC.

(e) Describe a simple rule for multiplying two diagonal matrices.

8. Form the following products

*(a) $\begin{pmatrix} 4 & 0 & 5 \\ 0 & 1 & -6 \\ 3 & 0 & 4 \end{pmatrix} \begin{pmatrix} 4 & 0 & -5 \\ -18 & 1 & 24 \\ -3 & 0 & 4 \end{pmatrix}$

(b) $\begin{pmatrix} 2 & -1 & 1 \\ 8 & -5 & 2 \\ -11 & 7 & -3 \end{pmatrix} \begin{pmatrix} 1 & 4 & 3 \\ 2 & 5 & 4 \\ 1 & -3 & -2 \end{pmatrix}$

(c) $\begin{pmatrix} 1 & 0 & 0 \\ -13 & 1 & -5 \\ 2 & 0 & 1 \end{pmatrix} \begin{pmatrix} 1 & 0 & 0 \\ 3 & 1 & 5 \\ -2 & 0 & 1 \end{pmatrix}$

9. From the following matrices

$$X = \begin{pmatrix} 2 & 1 \\ 1 & 2 \end{pmatrix} \qquad Y = \begin{pmatrix} 4 & -1 \\ -1 & 4 \end{pmatrix}$$

(a) Compute XY.

(b) Compute YX.

(c) Does $XY \neq YX$?

9.7 MATRICES IN BASIC

On the surface it would appear that BASIC programs that use vectors or matrices could become quite complicated and tedious. However, BASIC provides us with special statements that make calculations with matrices just as simple as computation with scalars. For example, to add two matrices A and B, we merely write

    MAT C = A+B

The word "MAT" signals addition of matrices where "LET," as in

    LET C = A+B

signals the addition of scalars.

Before describing the various operations with matrices, we note that the names of matrices in BASIC must be a single letter; nothing more, nothing less. Thus A, Z, and X are legitimate names for matrices. However, A1, X2, and P3 are not. Recall that this latter set of names are valid names for regular (scalar) variables in BASIC.

To read the 3×3 matrix A given in (9.22), we write

```
 50 DIM A(3,3)
100 MAT READ A
150 DATA 4, 0, 5
160 DATA 0, 1, -6
170 DATA 3, 0, 4
```

The first statement records the fact that A is 3×3. The word "DIM" is an abbreviation of "Dimension." Notice that the data is listed *row by row*.

MAT statements also can be used to add and multiply matrices. For example,

```
MAT S = A+B
```

adds the matrices A and B and places the sum in the matrix S. Of course, A and B must have the same number of rows and columns as each other. Similarly,

```
MAT P = A*B
```

computes the product of the matrices A and B and assigns the values of this product to the matrix P. It is our problem, as users of BASIC, to make sure the product is defined.

Consider the BASIC program in Figure 9.7. It produces the sum and difference of two 2×3 matrices A and B. The results are shown in Figure 9.8. Notice that the MAT PRINT statement prints the entire matrix row by row.

```
100    DIM A[2,3],B[2,3],S[2,3],D[2,3]
110    MAT   READ A,B
120    MAT S=A+B
130    PRINT "SUM OF A AND B IS"
140    MAT   PRINT S
150    PRINT
160    MAT D=A-B
170    PRINT "DIFFERENCE OF A AND B IS"
180    MAT   PRINT D
200    DATA 4,0,-1
210    DATA -2,5,1
220    DATA 1,-2,3
230    DATA 2,-1,-1
300    END
```

FIGURE 9.7 BASIC program to find the sum and difference of two matrices.

Similarly, the BASIC program in Figure 9.9 (shown together with the program's printed output) multiplies the two matrices A and B given in (9.22) of the previous section.

```
SUM ØF A AND B IS
 5               -2              2

 0                4              0
```

```
DIFFERENCE ØF A AND B IS
 3                2             -4
-4                6              2
```

FIGURE 9.8 Output of BASIC program (Figure 9.7) to find the sum and the difference of two matrices.

Notice that DIM statements were needed in Figures 9.7 and 9.9 for the computed matrices S, D, and P in addition to the original matrices A and B. Some versions of BASIC only require the dimensions of the original matrices be given but, even in those BASIC versions, using the extra DIM statements, as we did here, does no harm. It is like using unneeded parentheses in an expression. A good rule is: when in doubt, use parentheses and use extra DIMs.

```
100   DIM A[3,3],B[3,3],P[3,3]
110   MAT   READ A,B
120   MAT P=A*B
130   PRINT "PRØDUCT ØF A AND B IS"
140   MAT   PRINT P
150   DATA 4,0,5
160   DATA 0,1,-6
170   DATA 3,0,4
180   DATA 4,0,-5
190   DATA -18,1,24
200   DATA -3,0,4
300   END

RUN
```

```
PRØDUCT ØF A AND B IS
 1                0              0

 0                1              0

 0                0              1
```

FIGURE 9.9 BASIC program to compute the product of two matrices.

In Section 9.3 we noted that if a matrix is multiplied by a scalar, each element of the matrix is multiplied by the scalar. For example,

$$\text{MAT C = (3.2)*M}$$

multiplies the matrix M by the scalar 3.2. Similarly, if X1 is a simple variable in BASIC (i.e., a scalar) and M is a matrix, then

$$\text{MAT C = (X1)*M}$$

produces the product of the scalar X1 and the matrix M. Notice that the constant, or scalar variable, must be enclosed in parentheses.

Recall that with simple variables (scalars), BASIC statements such as

```
LET X = X + Y
```

were allowed, as were statements such as

```
LET X = X*Y
```

If A and B are matrices with the appropriate numbers of rows and columns,

```
MAT A = A + B
```

is allowed, but

$$\text{MAT A = A*B} \tag{9.23}$$

is not allowed.

If we want to compute the product A*B and replace A by that new product matrix, we must do so in two steps:

```
MAT C = A*B
MAT A = C
```

where the name C is, of course, quite arbitrary.

We note also that BASIC allows only one operation in any one MAT statement. For example,

```
MAT A = B*C + D
```

is not allowed. Instead, two statements are required.

```
MAT A = B*C
MAT A = A + D
```

In most cases this does not involve any substantial hardship to the user of BASIC. In contrast to the usual algebra, long expressions do not occur often in matrix algebra.

If we wish to multiply a matrix by a column vector, we treat the column vector as an $n \times 1$ matrix. For example, to compute $A\mathbf{x}$ where

$$A = \begin{pmatrix} 6 & -1 & 2 \\ 1 & 2 & -1 \end{pmatrix} \qquad \mathbf{x} = \begin{pmatrix} 2 \\ 1 \\ -3 \end{pmatrix}$$

we write

```
100   DIM A[2,3],X[3,1],Y[2,1]
110   MAT   READ A,X
120   MAT Y=A*X
130   MAT   PRINT Y
140   DATA 6,-1,2,1,2,-1
150   DATA 2,1,-3
160   END
```

The result is a two-component column vector.

$$y = \begin{pmatrix} 5 \\ 7 \end{pmatrix}$$

Notice that in the DIM statement the second subscript for both X and Y is 1.

Similarly, a row vector $\mathbf{v}^T$ with n components is represented as a $1 \times n$ matrix and must appear in a DIM statement as V(1,N).

We already have discussed MAT statements to add, subtract, and multiply matrices. The statement

$$\text{MAT A = TRN(B)} \tag{9.24}$$

forms the transpose of B and assigns its values to A; that is

$$A = B^T$$

There also are MAT statements to reduce the number of DATA statements needed in many programs. For example, if we wish a 5×4 matrix C to be all zeroes, then

```
10 DIM C(5,4)
20 MAT C = ZER
```
(9.25)

accomplishes this. The ZER stands for "zero." On the other hand, if we want C to be all ones, we use

```
10 DIM C(5,4)
20 MAT C = CON
```
(9.26)

The CON stands for "constant," in particular a constant equal to 1. Finally, it often is convenient to have a square matrix (same number of rows and columns) that has ones along the diagonal and zeroes elsewhere; for example,

$$\begin{pmatrix} 1 & 0 & 0 & 0 \\ 0 & 1 & 0 & 0 \\ 0 & 0 & 1 & 0 \\ 0 & 0 & 0 & 1 \end{pmatrix}$$

To produce this matrix, we can use

```
10 DIM D(4,4)
20 MAT D = IDN
```
(9.27)

The IDN stands for "identity," since such a matrix is called an *identity matrix*.

Often it is useful to print only certain elements of a matrix instead of printing the entire matrix. For example, suppose we have a 3×5 matrix A and wish to print the element in the second row, fourth column (i.e., A_{24}). To do so, we write

PRINT A(2,4)

Notice that the word "MAT" is not used. Notice also that the subscripts appear separated by commas and enclosed in parentheses—row first, column second.

There are two occasions in which the name of a matrix may appear followed by two integers enclosed in parentheses. The first is in a DIM statement where the total number of rows and columns are specified. The second is in any statement where we wish to pick out one special element of the matrix, as we did above in the PRINT statement.

We close with one more example. In Figure 9.9 we produced a program that multiplied two matrices A and B given in (9.22). Instead of printing the entire product matrix, suppose we only wish to print the diagonal elements of the product (i.e., P_{11}, P_{22}, and P_{33}). To do so, we replace statements 130 and 140 with

```
130 FOR K=1 TO 3
135 PRINT P(K,K)
140 NEXT K
```

The resulting printed output will be

1
1
1

EXERCISES FOR SECTION 9.7

1. For the following matrices

$$A = \begin{pmatrix} 1.6 & -2.3 & 6.1 \\ -3.6 & 0.9 & 1.1 \\ 2.3 & 0.6 & -6.4 \end{pmatrix} \quad B = \begin{pmatrix} 7.9 & 0 & 6.4 \\ 1.2 & -1.3 & 3.2 \\ 0.2 & 2.1 & 0 \end{pmatrix}$$

$$C = \begin{pmatrix} 0 & 1.0 & 2.0 \\ -2.1 & 7.3 & -7.3 \\ 0 & 1.9 & 0 \end{pmatrix}$$

write BASIC programs to compute and print
*(a) $AB - C$ (b) $A^2 - B^2$
 (c) C^{16} (d) $B - I$ where I is the 3×3 identity matrix
 (e) ABC (f) A^2, A^4, A^8

*2. Write a BASIC program to create and print a 10×10 matrix with 3's along the diagonal and -1's elsewhere.

3. Write a BASIC program to create and print a 7×7 matrix with 0's along the diagonal and 2's elsewhere.

4. For the following matrix and vector

$$A = \begin{pmatrix} 3 & -2 & 1 \\ 0 & 1 & 2 \\ 6 & -1 & 3 \end{pmatrix} \qquad x = \begin{pmatrix} 1 \\ 4 \\ 2 \end{pmatrix}$$

write BASIC programs to compute and print

*(a) $A\mathbf{x}$
 (b) $\mathbf{x}^T A$
 (c) $\mathbf{x}^T A^T$
 (d) $A^T\mathbf{x}$

5. Write a BASIC program to compute and print M^2, M^4, M^8, and M^{16} where

$$M = \begin{pmatrix} 1/2 & 0 & 0 & 1/4 \\ 1/12 & 1/2 & 0 & 0 \\ 1/6 & 1/4 & 3/4 & 1/4 \\ 1/4 & 1/4 & 1/4 & 1/2 \end{pmatrix}$$

9.8 CASE STUDY 9: A DEMOGRAPHIC MODEL

Over any given period of time people and families migrate from one part of the United States to another. Some people move from east to west and some move from west to east. There is a flow of people in both directions. The question is: If we know the pattern of movement, and if this pattern remains unchanged for a substantial period of time, what will be the eventual distribution of the population throughout the nation?

We start with a particularly simple case in order to establish the principles behind the migration process and the methods of solution for the problem. Later in this section we will enlarge our view and expand the problem.

Consider a nation divided into two parts: east and west. We will assume that in any given time period, say five years, there is some probability that a person will move from one part of the nation to the other. If we select an individual citizen at random, what is the probability that he resides at that moment in the east?

If he does live in the east, he could have lived there during the previous time period and not moved, or he could have lived in the west and moved. If we let (1) E_0 represent the event "lives in the east in period 0," (2) W_0 represent "lives in the west in period 0," (3) EE represent "stays in the east," and (4) WE represent "moves from west to east during the time period," then the probability that an individual lives in the east in period 1 is

$$P(E_1) = P\big[(E_0 \cap EE) \cup (W_0 \cap WE)\big]$$

But the events

$$E_0 \cap EE = \{\text{lives in the east } and \text{ stays there}\}$$

and

$$W_0 \cap WE = \{\text{lives in the west } and \text{ moves east}\}$$

are mutually exclusive so

$$P(E_1) = P(E_0 \cap EE) + P(W_0 \cap WE)$$

We take the events E_0 and EE to be independent so

$$P(E_0 \cap EE) = P(E_0) \cdot P(EE)$$

What we are saying is that the tendency to stay in the east (EE) remains the same regardless of the number of people in the east.[5] Similarly,

$$P(W_0 \cap WE) = P(W_0) \cdot P(WE)$$

Therefore,

$$P(E_1) = P(E_0) \cdot P(EE) + P(W_0) \cdot P(WE) \qquad (9.28)$$

Similarly, the probability of a randomly selected individual being resident in the west in period 1 is

$$P(W_1) = P(E_0) \cdot P(EW) + P(W_0) \cdot P(WW) \qquad (9.29)$$

Suppose we form the following 2×2 matrix.

$$T = \begin{pmatrix} P(EE) & P(WE) \\ P(EW) & P(WW) \end{pmatrix} \qquad (9.30)$$

Then, if we also form the column vector,

$$\mathbf{x}_0 = \begin{pmatrix} P(E_0) \\ P(W_0) \end{pmatrix} \qquad (9.31)$$

where the subscripts of 0 all refer to period 0,[6] then from our rules of matrix multiplication, (9.28) and (9.29) may be written

$$\mathbf{x}_1 = \mathbf{T}\mathbf{x}_0 \qquad (9.32)$$

---

[5] Of course, this assumption may not be valid in an actual situation. It may be that overcrowding, large $P(E_0)$, results in a tendency for people to move, small $P(EE)$.

[6] Notice that the subscript 0 for the vector $\mathbf{x}_0$ does not represent the zeroth component. $\mathbf{x}_0$ has two components, $P(E_0)$ and $P(W_0)$.

We pause to make some observations about (9.32). It does not represent a single equation, but it represents two equations:

$$x_{1,1} = t_{11}x_{0,1} + t_{12}x_{0,2}$$
$$x_{1,2} = t_{21}x_{0,1} + t_{22}x_{0,2}$$

where $x_{1,2}$, for example, is the second component of the vector $\mathbf{x}_1$. These last two equations are essentially (9.28) and (9.29) rewritten. If T were a 4×4 matrix and if $\mathbf{x}_0$ and $\mathbf{x}_1$ were 4×1, then (9.32) would be four equations. To emphasize this point, we will refer to (9.32) as a *matrix equation* (i.e., an equation involving matrices). All equations dealing with matrices are called matrix equations. What appears to be one equation, if it is a matrix equation, is in reality several equations.

For example, suppose

$$T = \begin{pmatrix} .7 & .2 \\ .3 & .8 \end{pmatrix}$$

This means that 70% of the persons in the east remain there and 30% move west. Similarly, 80% of those in the west do not move, while 20% move east. If the population in period 0 is evenly distributed between the two sections,

$$\mathbf{x}_0 = \begin{pmatrix} .5 \\ .5 \end{pmatrix}$$

so, from (9.32),

$$\mathbf{x}_1 = \begin{pmatrix} .7 & .2 \\ .3 & .8 \end{pmatrix}\begin{pmatrix} .5 \\ .5 \end{pmatrix} = \begin{pmatrix} .45 \\ .55 \end{pmatrix}$$

or

$$x_{1,1} = 0.45 \quad \text{and} \quad x_{1,2} = 0.55$$

The net result is a gain of 5% in the west and a coresponding loss in the east.

We call the matrix T the *transition matrix* or the movement matrix. It represents the probabilities of all movements or transitions. Notice that each column of T sums to one. The vectors $\mathbf{x}_0$ and $\mathbf{x}_1$ are called *population vectors*. They represent the probability of finding a particular person in each of the regions at some given point in time. Another way of viewing the population vectors is that they indicate the fraction of the total population that lives in each region at the time in question.

With this as background we now consider the nation to be separated into four parts: east, midwest, mountains, and far west. The transition matrix is now a 4×4 matrix. In (9.30) the first column described what happened to people in the east (i.e., the probabilities that they stayed there or moved west). Now column 1 will contain the probabilities of staying in the east or moving to one of the three other regions. Since everyone in the east in period 0 must reside in one and only one of these four regions in period 1, the movements into the four regions are a complete set of mutually exclusive events. Therefore the sum of the probabilities in the first column must be one.

Suppose that the probability that an eastern resident remains in the east is 1/2. The other probabilities for eastern residents are: moves to the midwest with probability 1/12, moves to the mountains with probability 1/6, and moves to the far west with probability 1/4. These do sum to one. The first column of T becomes

$$\begin{pmatrix} 1/2 \\ 1/12 \\ 1/6 \\ 1/4 \end{pmatrix}$$

Similarly, we can construct columns 2, 3, and 4 of T to represent the probabilities of movement of the people from the midwest, mountains, and far west, respectively. When this is done, T becomes

$$T = \begin{pmatrix} 1/2 & 0 & 0 & 1/4 \\ 1/12 & 1/2 & 0 & 0 \\ 1/6 & 1/4 & 3/4 & 1/4 \\ 1/4 & 1/4 & 1/4 & 1/2 \end{pmatrix} \tag{9.33}$$

It is important to understand the meaning of this matrix. The third column says that the probability that a resident of the mountains stays in the mountains is 3/4 and that the probability that he leaves and moves to the far west is 1/4. In other words, three fourths of the people in the mountain region stay there for one more time period, while one fourth move west.

While each column describes movement out of a region, each row of T describes movement *into* a region. The third row, for example, says that the people now in the mountain region consist of 1/6 of those who lived in the east one period ago, 1/4 of those who lived in the midwest, 3/4 of those who already were in the mountains, and 1/4 of those who were in the far west. Notice that the rows need not sum to one.

We can raise some interesting questions regarding the movement implied by this matrix. For example:

1. Given some initial population distribution, what is the population distribution after the first period?
2. Assuming that the movement pattern described by the matrix T is the same for the next succeeding time period, what is the population distribution after the second period?
3. What will the long-run population distribution be if this pattern of movement continues indefinitely?

4. Is there some population distribution that will be unchanged by the movements described by the matrix T?

The reader should try to think of other questions that could provide policymakers with useful information. Indeed, the reader should pause and try to understand why even these questions might be of interest to a sociologist, a congressman, a businessman, and so forth.

We proceed now to develop a means for answering the questions we have posed above. Hopefully the development will lead the reader to further questions and to the means of answering them.

Suppose first that 3/4 of the population is in the east and the remaining 1/4 is in the midwest (i.e., the mountains and the far west are uninhabited). This was roughly the case in the early years of this nation's development. This means that the probability of finding any given individual in the east in period zero is 3/4. The probability of finding someone in the midwest is 1/4. This being the case, the population vector is

$$\mathbf{x}_0 = \begin{pmatrix} 3/4 \\ 1/4 \\ 0 \\ 0 \end{pmatrix} \tag{9.34}$$

We can find the population probabilities after one period by multiplying the matrix T by this column vector; that is,

$$\mathbf{x}_1 = T\mathbf{x}_0$$

$$\mathbf{x}_1 = \begin{pmatrix} 1/2 & 0 & 0 & 1/4 \\ 1/12 & 1/2 & 0 & 0 \\ 1/6 & 1/4 & 3/4 & 1/4 \\ 1/4 & 1/4 & 1/4 & 1/2 \end{pmatrix} \begin{pmatrix} 3/4 \\ 1/4 \\ 0 \\ 0 \end{pmatrix} = \begin{pmatrix} 3/8 \\ 3/16 \\ 3/16 \\ 1/4 \end{pmatrix} \tag{9.35}$$

This then answers question (1) above. Notice that the population has shifted westward as one might have expected. The probability of finding someone in the far west is now 1/4 compared with zero the previous period. The far west has increased more than the mountains, and the east has suffered a greater loss than the midwest. Do you think trends such as these will continue?

To determine the population vector in the second time period

$$\mathbf{x}_2 = T\mathbf{x}_1$$

Using T from (9.33) and $\mathbf{x}_1$ from (9.35),

$$\mathbf{x}_2 = \begin{pmatrix} 1/4 \\ 1/8 \\ 5/16 \\ 5/16 \end{pmatrix} \tag{9.36}$$

After two time periods, 1/4 of the population resides in the east, 1/8 in the midwest, and the remaining 5/8 are divided equally between the mountains and the far west. We have answered question (2) above.

We could, of course, continue to compute successive population vectors as follows:

$$\mathbf{x}_3 = T\mathbf{x}_2$$

$$\mathbf{x}_4 = T\mathbf{x}_3$$

and so on. In general,

$$\mathbf{x}_k = T\mathbf{x}_{k-1} \qquad k = 1, 2, 3, \ldots \tag{9.37}$$

This is a difference equation, but it differs from our earlier difference equations. The variables $\mathbf{x}_k$ and T are no longer scalars; they are matrices. Therefore we call (9.37) a *matrix difference equation*. It represents an unending or infinite set of matrix equations. Each of these matrix equations represents four algebraic equations. In addition to (9.37) we need to specify some initial or beginning population probabilities (e.g., $\mathbf{x}_0$). Having done so, there will be one and only one solution of (9.37).

Hence (9.37) is a first-order, linear matrix difference equation. Its solution depends on the difference equation and the initial condition, as it did for ordinary difference equations.

Suppose we choose a different initial population vector to accompany (9.33).

$$\mathbf{x}_0 = \begin{pmatrix} 1/6 \\ 1/36 \\ 17/36 \\ 1/3 \end{pmatrix} \tag{9.38}$$

Then

$$\mathbf{x}_1 = T\mathbf{x}_0 = \mathbf{x}_0$$

and

$$\mathbf{x}_2 = \mathbf{x}_1$$

$$\mathbf{x}_3 = \mathbf{x}_2$$

and so on. That is, the distribution of the population does not change. One sixth of the population is always in the east, 1/36th is in the midwest, and so on. This does not mean that no individuals ever move. Some people do move each time period, but the *net* movement into and out of each region is zero. Thus, while one sixth of the population always is in the east, it is a different one sixth from time period to time period.

The population vector given in (9.38) is an *equilibrium vector* where the term ''equilibrium'' is being used in the same sense that we used it with ordinary difference equations. That is, if the population probabilities ever reach (9.38), they will stay there. This answers question (4) above.

Does every transition matrix have an equilibrium vector? How can we find the equilibrium vector? Is it stable (i.e., if we change the population vector so that it is not quite in equilibrium, will the population distribution return to its equilibrium values)? In the remainder of this section we will attempt to answer these questions.

First, we return to our numerical example given by (9.33) and (9.34) and attempt to answer question (3) (i.e., what is the long run population distribution?). We will use a computer to assist us. The BASIC program in Figure 9.10 computes $\mathbf{x}_1, \mathbf{x}_2, \ldots, \mathbf{x}_{30}$ but prints only $\mathbf{x}_{10}$, $\mathbf{x}_{20}$, and $\mathbf{x}_{30}$. Notice that the population vector is not changing and is equal to

$$\begin{pmatrix} 0.166667 \\ 0.0277778 \\ 0.472222 \\ 0.333333 \end{pmatrix} = \begin{pmatrix} 1/6 \\ 1/36 \\ 17/36 \\ 1/3 \end{pmatrix}$$

But this is the population vector $\mathbf{x}_0$ given in (9.38) (i.e., the equilibrium vector). It appears, therefore, that the long-run population distribution is given by the equilibrium vector. Moreover, it appears that the equilibrium vector may be stable. Of course, we have tried only one starting point given by (9.34), but others are examined in the exercises.

Can we "solve" the linear matrix difference equation (9.37) as we did the ordinary linear difference equations earlier? For $k = 1$ (9.37) produces

$$\mathbf{x}_1 = T\mathbf{x}_0$$

and for $k = 2$

$$\mathbf{x}_2 = T\mathbf{x}_1$$

Using the first of these to replace $\mathbf{x}_1$, in the second,

$$\mathbf{x}_2 = T^2\mathbf{x}_0$$

where T^2 represents the product of the matrix T with itself. Similarly,

$$\mathbf{x}_3 = T^3\mathbf{x}_0$$

and so on. In general,

$$\mathbf{x}_k = T^k\mathbf{x}_0 \tag{9.39}$$

and this is the solution of (9.37)

```
10   DIM T[4,4],X[4,1],Y[4,1]
20   MAT  READ T[4,4]
30   MAT  READ X[4,1]
40   FØR J=1 TØ 3
50   FØR I=1 TØ 10
60   MAT Y=T*X
70   MAT X=Y
80   NEXT I
90   PRINT
100  MAT  PRINT X
110  NEXT J
120  DATA .5,0,0,.25
130  DATA 8.33333E-02,.5,0,0
140  DATA .166667,.25,.75,.25
150  DATA .25,.25,.25,.5
160  DATA .75,.25,0,0
170  END
```

.166911

2.85101E-02

.471246

.333333

.166667

2.77789E-02

.472222

.333334

.166667

2.77778E-02

.472224

.333334

FIGURE 9.10 BASIC program with its output to compute population vectors after 10, 20, and 30 time periods.

Notice the similarity between (9.39) and the corresponding result, (3.31) in Section 3.3, for ordinary first-order difference equations.

If we compute successive powers of the matrix T we will find that the resulting matrices tend toward the following matrix.

$$T_E = \begin{pmatrix} 1/6 & 1/6 & 1/6 & 1/6 \\ 1/36 & 1/36 & 1/36 & 1/36 \\ 17/36 & 17/36 & 17/36 & 17/36 \\ 1/3 & 1/3 & 1/3 & 1/3 \end{pmatrix} \tag{9.40}$$

Notice first that each column of T_E is identical with every other column. Moreover, notice that each of these columns is identical with the population vector given in (9.38). That population vector, (9.38), produced a new population vector that was identical to the starting population vector. Each column of T_E is equal to the equilibrium vector.

We make one further observation. If T_E were the transition matrix (instead of T), then for *any* initial population distribution x_0, all succeeding distributions will be 1/6, 1/36, 17/36, and 1/3. This is so because the sum of the components of any x_0 must be 1. The first component of $T_E x_0$ will be 1/6 times the sum of the components of x_0. Since the sum of the components of x_0 is 1, the first component of $T_E x_0$ is 1/6. Similarly, each of the other components will be 1/36, 17/36, and 1/3, respectively, times the sum of the components of x_0 (i.e., times one).

We now ask: Under what conditions will the powers of T tend, in the long run, to some fixed matrix T_E?

In the example we used, the powers of the matrix T given in (9.33) did tend toward the matrix T_E given in (9.40). Will this always be so? If so, will the columns of the matrix T_E tell us what the equilibrium population is? Will such an equilibrium be stable?

The answers to these questions are contained in the following statements. If any power of T has all *positive* elements, then T^k tends, in the long run, to a fixed matrix T_E. Moreover, each column of T_E is identical to every other column of T_E. Each column represents the long-run population distribution that will be attained regardless of the initial population distribution.

Each column of T_E, therefore, represents an equilibrium population or an equilibrium vector of the difference equation (9.37). Moreover, the equilibrium value is stable, since any population will eventually, for all practical purposes, become equal to the equilibrium population.

Thus we need merely compute successive powers of T until we find a power that has all positive elements. Notice that T itself has nonnegative elements; thus T^k must also contain nonnegative elements. It need not contain all positive elements, since some elements may be zero. After we have studied solutions of linear systems of equations, we will develop a method for computing what this long-run solution is without computing all of the powers of T itself.

But what if we continue to compute powers of T and do not find a matrix with all positive elements? How long must we continue to compute before we can stop?

Before attempting to answer these questions, we will look at two more examples. First, we demonstrate that if some power of T does not contain all positive components, there may be no long-run solution at all. To that end, consider again the case where the country is divided into two parts. Suppose that all of the people in the east move to the west, and vice versa. Then the matrix T becomes

$$T = \begin{pmatrix} 0 & 1 \\ 1 & 0 \end{pmatrix} \tag{9.41}$$

Moreover,

$$T^2 = \begin{pmatrix} 1 & 0 \\ 0 & 1 \end{pmatrix}$$

and

$$T^3 = \begin{pmatrix} 0 & 1 \\ 1 & 0 \end{pmatrix}$$

which is just T again. The succeeding powers of T will then alternate between T and T^2. That is, successive powers of T oscillate between T and T^2, and no long-run solution exists independent of the initial population distribution. Of course, for a given distribution of the population, the long-run population can be computed, but it will never settle down.

On the other hand, if the initial distribution is 1/2 in the east and 1/2 in the west, then the distribution will remain just that. In this special case a long-run solution does exist. It is important to note, however, that the existence of a long-run solution depends on the initial population distribution in this case. That is, the equilibrium vector

$$\mathbf{x} = \begin{pmatrix} 1/2 \\ 1/2 \end{pmatrix}$$

is unstable.

Let us now consider another example where there are three regions of the country: east, middle, and west. For a person living in the east the probability is 1/2 that he will stay there and 1/2 that he will move to the west. For the people in the middle it is certain that they will stay in the middle. Of those in the west the probability is 1/2 that a given individual will move to the middle and 1/2 that he will stay in the west.

The transition matrix T is

$$T = \begin{pmatrix} 1/2 & 0 & 0 \\ 0 & 1 & 1/2 \\ 1/2 & 0 & 1/2 \end{pmatrix} \tag{9.42}$$

where the first column and row refer to the east, the second to the middle, and the third to the west.

We wish to know if any power of T contains all positive elements. If any power of T contains all positive elements, all succeeding powers of T will also. Thus we need only compute powers of T until we reach a matrix with all positive elements. We will not compute all powers of T, since it is only T raised to large powers that are of interest to us. When we have computed T^2, we can compute T^4 by multiplying T^2 by T^2. Thus we obtain T^4 without even computing T^3. Similarly, we can obtain T^8 by squaring T^4, and we will never compute T^5, T^6, or T^7. We continue in this way computing T^{16}, T^{32}, and so on.

In the example of (9.42) T^2 becomes

$$T^2 = \begin{pmatrix} 1/4 & 0 & 0 \\ 1/4 & 1 & 3/4 \\ 1/2 & 0 & 1/4 \end{pmatrix}$$

and

$$T^4 = \begin{pmatrix} 1/16 & 0 & 0 \\ 11/16 & 1 & 15/16 \\ 4/16 & 0 & 1/16 \end{pmatrix}$$

and finally

$$T^8 = \begin{pmatrix} 1/256 & 0 & 0 \\ 247/256 & 1 & 255/256 \\ 8/256 & 0 & 1/256 \end{pmatrix} \tag{9.43}$$

It appears from these calculations that the second column will contain 0, 1, 0 indefinitely. If this is so then no power of T will contain all positive elements. Are we correct in making this supposition?

To answer this question, suppose that T is an $n \times n$ matrix each of whose columns has a sum of 1. All of the matrices we have considered in this section fulfill this requirement. The matrix T given in (9.33) is a 4×4, and all columns have a sum of 1. Similarly, the matrices given in (9.41) and (9.42) are 2×2 and 3×3, respectively, and have column sums of 1.

It can be shown that in order to determine whether any power of T has all positive elements, the process of successive squaring need only be carried out n times. Thus, if $n = 4$ (i.e., T is 4×4), we need only compute T^{16}. If $n = 3$ so that T is 3×3, we need only compute T^8. If this last matrix contains one or more zero elements, all powers of T will contain at least one zero element.

Equation 9.43 is a result of squaring the 3×3 matrix given in (9.42) three times. Since T^8 does not have all positive elements (it has three zeroes), no power of T will have all positive elements.

Earlier we noted that if any power of T has all positive elements, T^k tends toward a fixed matrix T_E as k becomes large. From the above results, we are led to conclude that for T given in (9.42), a long-run solution exists only for special initial population vectors. Unfortunately, this is not true. The long-run population is 0, 1, 0 (i.e., all of the population gravitates to the middle of the country). Moreover, this population distribution is stable (i.e., any deviation from such a distribution is wiped out as time goes on).

What, then, can we conclude? We can conclude that if some power of T has all positive components, a stable equilibrium solution exists. By a stable equilibrium solution we mean that regardless of the initial population distribution, the distribution will eventually reach an equilibrium, and succeeding small deviations from the equilibrium will disappear in time. On the other hand, if no power of T has all positive components, then there may or may not be a stable equilibrium solution.[7]

Mathematicians would say that a power of T with positive components is a *sufficient* condition to assure a stable equilibrium solution. It is not a *necessary* condition, as the examples of (9.41) and (9.42) exhibit. The distinction between necessary and sufficient conditions was encountered toward the close of Section 5.3 in the discussion of stability of a single species population.

---

[7]We already have examples of both cases. The matrix T in (9.41) had no power with positive components and had no stable equilibrium solution. The matrix T in (9.42) also had no power with positive components, but it did have a stable equilibrium solution.

The next question that naturally arises is: How can one compute the stable equilibrium solution without computing successive powers of T? The answer lies in an algorithm for computing the solution of a matrix equation. In particular, if $\mathbf{x}_k$ is the population vector after the kth time period and $\mathbf{x}_{k-1}$ is the population vector after the $(k-1)$st period,

$$\mathbf{x}_k = T\mathbf{x}_{k-1}$$

If a stable equilibrium solution exists, then $\mathbf{x}_k = \mathbf{x}_{k-1}$ for k very large. Thus we let $\mathbf{x}_k = \mathbf{x}_{k-1} = \mathbf{x}$ and get

$$\mathbf{x} = T\mathbf{x} \tag{9.44}$$

Given T, we wish to solve this equation for $\mathbf{x}$, a column vector. An efficient algorithm for doing so is called Gaussian elimination. We will discuss this algorithm in the next chapter.

Now, however, we turn to a discussion of a computer program for squaring an $n \times n$ matrix n times in order to see if the sufficient condition for a stable equilibrium is satisfied. This result will be useful to us, since Gaussian elimination will produce an equilibrium solution that may or may not be stable. If we find that T squared n times has all positive components, we can conclude that the equilibrium solution is stable.

Thus, if the program we develop here produces a matrix with all positive components, we are guaranteed that a stable equilibrium solution exists. On the other hand, if the program produces a matrix with some zero elements, we will not know whether or not an equilibrium solution is stable. In this latter case all we know is that no power of T has all positive elements.

```
10   DIM T[4,4],Q[4,4]
20   MAT   READ T
30   FØR I=1 TØ 4
40   MAT Q=T*T
50   MAT T=Q
60   NEXT I
70   MAT   PRINT T
100  DATA .5,0,0,.25
110  DATA 8.33333E-02,.5,0,0
120  DATA .166667,.25,.75,.25
130  DATA .25,.25,.25,.5
200  END
```

.166675	.166659	.166659	.166674
2.77994E-02	.027774	2.77587E-02	2.77943E-02
.472194	.472234	.472249	.472199
.333334	.333334	.333334	.333334

FIGURE 9.11 BASIC program to square a 4×4 matrix four times.

The BASIC program in Figure 9.11 squares the 4×4 matrix (9.33) four times and prints the results. In the FOR-NEXT loop from statement 30 to statement 60, Q is T^2. We then replace T by Q so that each time through the loop, T is replaced by T^2.

If we wish to change the size of the matrix we merely change the 4's in the DIM statement (statement 10) and in the FOR statement (statement 30). Of course, the DATA statements also would need to be changed.

EXERCISES FOR SECTION 9.8

1. Find the population vector after one and two time periods for the matrix T given in (9.33) for the following initial population vectors.

*(a)
$$\mathbf{x}_0 = \begin{pmatrix} 1 \\ 0 \\ 0 \\ 0 \end{pmatrix}$$

(b)
$$\mathbf{x}_0 = \begin{pmatrix} 0 \\ 1 \\ 0 \\ 0 \end{pmatrix}$$

(c)
$$\mathbf{x}_0 = \begin{pmatrix} 0 \\ 0 \\ 1 \\ 0 \end{pmatrix}$$

(d)
$$\mathbf{x}_0 = \begin{pmatrix} 0 \\ 0 \\ 0 \\ 1 \end{pmatrix}$$

2. Find the long-run population vectors for the matrix T given in (9.33) and for the initial population vectors given in Exercise 1.

3. Find the population vector after one and two time periods for the transition matrix

$$T = \begin{pmatrix} 1/2 & 0 & 1/4 & 1/4 \\ 1/12 & 1/2 & 0 & 0 \\ 1/6 & 1/4 & 1/2 & 1/4 \\ 1/4 & 1/4 & 1/4 & 1/2 \end{pmatrix}$$

and the following initial population vectors.

*(a)
$$\mathbf{x}_0 = \begin{pmatrix} 3/4 \\ 1/4 \\ 0 \\ 0 \end{pmatrix}$$

(b)
$$\mathbf{x}_0 = \begin{pmatrix} 1 \\ 0 \\ 0 \\ 0 \end{pmatrix}$$

(c)
$$\mathbf{x}_0 = \begin{pmatrix} 1/2 \\ 1/2 \\ 0 \\ 0 \end{pmatrix}$$

(d)
$$\mathbf{x}_0 = \begin{pmatrix} 2/3 \\ 1/3 \\ 0 \\ 0 \end{pmatrix}$$

4. Find the long-run population distribution for the matrix T and the initial population vectors given in Exercise 3.

5. Which of the following transition matrices have some power with all positive elements?

(a) $\begin{pmatrix} 1/3 & 0 \\ 2/3 & 1 \end{pmatrix}$

(b) $\begin{pmatrix} 1/2 & 1/2 & 1/2 \\ 1/4 & 0 & 1/2 \\ 1/4 & 1/2 & 0 \end{pmatrix}$

(c) $\begin{pmatrix} 0 & 1/2 & 1 \\ 1 & 0 & 0 \\ 0 & 1/2 & 0 \end{pmatrix}$

(d) $\begin{pmatrix} 1/2 & 0 & 5/6 \\ 1/2 & 1/3 & 1/6 \\ 0 & 2/3 & 0 \end{pmatrix}$

*6. A group of rats are being trained to run a certain maze. Records are kept of their last two performances. There are four possible records for each rat.

  i. Two successful runs, denoted SS.

  ii. The next-to-last run was successful, but the last run was unsuccessful, denoted by SU.

  iii. The next-to-last run was unsuccessful, but the last run was successful, denoted by US.

  iv. Two unsuccessful runs denoted by UU.

If a given rat has two successful runs, the probability that the next run also will be successful is 0.9. If the last run was unsuccessful but the next-to-last run was successful, the probability of the next run being successful is 0.4. If the last run was successful but its predecessor was not, the probability of a successful run on the next run is 0.7. Finally, if the last two runs were both unsuccessful, the probability of the next run being successful is 0.2.

(a) Construct a 4×4 transition matrix where each column and row is one of the four events listed above. Notice that in one time period SS can change to SU or SS but not to any other combination. That is, if the last run was successful, after the next run, the next-to-last run must be listed as successful. Similarly, SU can change to US or UU; US can change to SU or SS; and UU can change to US or UU.

(b) If we start with 480 rats, half of which have two consecutive successful runs and half of which have two consecutive unsuccessful runs, what is the distribution of rats into the four categories: SS, SU, US, and UU after one additional run?

(c) If we start with 480 rats evenly distributed into the four categories (i.e., 120 in each), what is the distribution among the four categories after one additional run?

(d) What is the long-run distribution of rats in the four categories?

7. Same as Exercise 6 except: (a) if there were two consecutive successful runs, the probability that the next run is successful is 0.8; (b) if the last run was unsuccessful but the next-to-last was successful, the probability that the next run is successful is 0.5; (c) if the last run was successful but the next-to-last was not successful, the probability that the next run is successful is 0.6; (d) if the last two runs were both unsuccessful, the probability that the next run will be successful is 0.1.

*8. Land in an urban area is divided into four usage categories: residential, commercial, parking, and vacant. Suppose the percentage usage changes from 1960 to 1970 are given by:

1970 Use	Usage in 1960 (in percent)			
	Residential	Commercial	Parking	Vacant
Residential	70	15	0	10
Commercial	20	60	0	10
Parking	10	15	80	70
Vacant	0	10	20	10

That is, of all of the land used for residential purposes in 1960, 70% remains residential in 1970, 20% has been converted for commercial use, and 10% has been converted to parking lots (see column 1).

(a) If the city was 50% commercial, 30% residential, and 10% each parking and vacant in 1960, what was the land usage distribution in the city in 1970?

(b) If the pattern of change continues as described in the above table for another 10 years, what will the land usage distribution be in the city in 1980?

(c) If the pattern change remains the same for an indefinite period of time, what will be the eventual long-run land usage distribution in the city?

(d) What would be the effect on the long-run land usage distribution of an ordinance instituted in 1960 and maintained indefinitely requiring that no vacant land could be converted directly to parking? Assume that vacant land that would be converted to parking remains vacant.

9. Same as Exercise 8 except that the percentage usage changes are:

| 1970 Use | Usage in 1960 (in percent) | | | |
	Residential	Commercial	Parking	Vacant
Residential	70	20	0	15
Commercial	10	50	0	15
Parking	20	20	70	30
Vacant	0	10	30	40

10

SYSTEMS
OF LINEAR
EQUATIONS

10.1 INTRODUCTION

Suppose we wish to construct a diet that provides us with a certain amount of protein and ascorbic acid (Vitamin C). Suppose also that we have decided that the diet should include roast beef and a lettuce salad with French dressing. The menu may include other foods, but we wish to obtain one half of our daily requirements for protein and ascorbic acid from these two foods.

Figure 10.1 summarizes the nutrient content of the two foods and the daily requirements of the two nutrients. For example this table indicates that 100 grams of salad contains one gram of protein and 15 milligrams of ascorbic acid (second row).

Let B be the amount of roast beef in the diet in units of 100 grams (i.e., $B = 1$ represents 100 grams of roast beef). Similarly, let L be the amount of lettuce in the menu. The total amount of protein in the diet is

	Protein (grams)	Ascorbic Acid (milligrams)
Roast beef (100 grams)	20	0
Salad with French dressing (100 grams)	1	15
Daily requirement	70	60

FIGURE 10.1 Nutrient content of foods and daily requirements.

$$20B + L$$

Since the daily requirement for protein is 70 grams, and we wish to obtain half of that from the beef and lettuce,

$$20B + L = 35 \qquad (10.1)$$

Similarly, the ascorbic acid requirement can be written

$$0 \cdot B + 15L = 30 \qquad (10.2)$$

because there is no ascorbic acid in the beef and each 100 grams of salad contains 15 milligrams of vitamin C.

Equations 10.1 and 10.2 are two linear equations for two unknowns, B and L. The solution is $B = 1.65$ and $L = 2$. If we substitute these two numerical values into the two equations, (10.1) and (10.2), we see that the equations are true for these particular values.

The solution tells us we should eat 200 grams of salad (about half a head of lettuce) and 165 grams of beef (two small slices).

Suppose we decide that we should be concerned with more than just the protein and ascorbic acid in our diet. In particular we wish to be sure we obtain half of our calorie

	Protein (grams)	Ascorbic Acid (milligrams)	Calories	Calcium
Roast beef (100 grams)	20	0	325	10
Salad with French dressing (100 grams)	1	15	80	40
Milk (100 grams)	3.3	1.5	70	120
Spinach (100 grams)	2.3	50	22	70
Daily requirement	70	60	1400	1000

FIGURE 10.2 Nutrient content and daily requirements of four foods.

and calcium requirements. Moreover, we wish to enlarge the food selection to include milk and spinach as well as roast beef and salad. Figure 10.2 summarizes the nutrient content of the four foods as well as the four nutrient requirements.

If B = the amount of roast beef in the diet, L = the amount of salad, M = the amount of milk, and S = the amount of spinach, all in 100 gram units, then the number of grams of protein is

$$20B + L + 3.3M + 2.3S$$

The satisfy half the daily protein requirement,

$$20B + L + 3.3M + 2.3S = 35 \qquad (10.3)$$

Similarly, the ascorbic acid requirement is

$$15L + 1.5M + 50S = 30 \qquad (10.4)$$

The calorie requirement is

$$325B + 80L + 70M + 22S = 700 \qquad (10.5)$$

and the calcium requirement is

$$10B + 40L + 120M + 70S = 500 \qquad (10.6)$$

These are four equations in the four unknowns: B, L, M, and S. Our concern in this chapter is with methods—both algebraic and computer—for solving such systems of equations.

Matrix notation provides us with a convenient and concise way to write such equations. Suppose we let the matrix A represent the table in Figure 10.2; that is,

$$A = \begin{pmatrix} 20 & 0 & 325 & 10 \\ 1 & 15 & 80 & 40 \\ 3.3 & 1.5 & 70 & 120 \\ 2.3 & 50 & 22 & 70 \end{pmatrix}$$

We let the row vector $\mathbf{r}^T$ be one half the daily requirements.

$$\mathbf{r}^T = (35, 30, 700, 500)$$

We let $\mathbf{d}^T$ be a row vector whose components are B, L, M, and S; that is,

$$\mathbf{d}^T = (\text{B, L, M, S})$$

Then

$$\mathbf{d}^T A = \mathbf{r}^T \qquad (10.7)$$

is a matrix equation that represents all four algebraic equations: (10.3), (10.4), (10.5), and (10.6). We could equally well replace these four equations with the matrix equation

$$A^T \mathbf{d} = \mathbf{r} \qquad (10.8)$$

The reader should assure himself that (10.7) and (10.8) represent identical sets of four algebraic equations. One way to do so is to write out all four equations in each case, (10.7) and (10.8), using the components of A, $\mathbf{d}$, and $\mathbf{r}$.

We close this section by returning to the problem of finding the equilibrium population in the population movement problem of Section 9.8. If T is the transition matrix and $\mathbf{x}$ is the equilibrium vector then, from (9.44),

$$\mathbf{x} = T\mathbf{x} \qquad (10.9)$$

If we use the particular transition matrix given in (9.33), (10.9) becomes

$$
\begin{aligned}
x_1 &= (1/2)x_1 && + (1/4)x_4 \\
x_2 &= (1/12)x_1 + (1/2)x_2 \\
x_3 &= (1/6)x_1 + (1/4)x_2 + (3/4)x_3 + (1/4)x_4 \\
x_4 &= (1/4)x_1 + (1/4)x_2 + (1/4)x_3 + (1/2)x_4
\end{aligned}
$$

Once again these are four equations in the four unknowns: x_1, x_2, x_3, and x_4. In what follows we will develop the techniques to solve all of the systems of equations discussed in this section and many other systems of equations as well.

EXERCISES FOR SECTION 10.1

(Note: Solutions are given for those exercises marked with an asterisk.)

*1. Suppose we want a diet of roast beef and salad that yields 165 milligrams of ascorbic acid and 10 grams of protein for someone on a special low-protein, high-Vitamin C diet. How much of each food should he eat? Discuss the results.

2. Find a two-food diet (roast beef and salad) that provides 36 grams of protein and 90 milligrams of ascorbic acid.

3. Find a two-food diet (roast beef and salad) that provides 70 grams of protein and 60 milligrams of ascorbic acid.

4. Find a two-food diet (milk and spinach) that provides 10 grams of protein and 165 milligrams of ascorbic acid.

5. Find a two-food diet (salad and milk) that provides 35 grams of protein and 1280 milligrams of calcium.

6. Find a two-food diet (roast beef and milk) that provides 75 milligrams of ascorbic acid and 4150 calories.

10.2 TWO EQUATIONS IN TWO UNKNOWNS

Consider a particularly simple system of linear equations. That is, consider the system of two equations in two variables.

$$2x_1 - 3x_2 = 1 \tag{10.11}$$

$$4x_1 - 4x_2 = 4 \tag{10.12}$$

Suppose we multiply the first equation by 2. We obtain

$$4x_1 - 6x_2 = 2 \tag{10.13}$$

Next, we subtract (10.13) from (10.12) and obtain

$$(4x_1 - 4x_2) - (4x_1 - 6x_2) = 4 - 2$$

or

$$2x_2 = 2 \tag{10.14}$$

Notice that x_1 is missing from this last equation.

The last equation, (10.14), implies that

$$x_2 = 1$$

Knowing this, we can replace x_2 by 1 in any one of the preceding equations. In particular let us replace x_2 by 1 in (10.11) as follows.

$$2x_1 - 3(1) = 1$$

or

$$2x_1 = 4$$

This implies that

$$x_1 = 2$$

The question that arises is: Are $x_1 = 2$ and $x_2 = 1$ a solution of the original system of equations given by (10.11) and (10.12)?

We say two numbers, a and b, are a *solution* of the system of (10.11) and (10.12) if, when we replace x_1 by a and x_2 by b, *both* equations are true.

From our definition of solution the answer to the question "Are $x_1 = 2$ and $x_2 = 1$ a solution?" will be yes if, when we replace x_1 by 2 and x_2 by 1, both are equations are true. Making these replacements in (10.11), we get

$$2(2) - 3(1) = 1$$

and this is true. Then, making the replacements in (10.12),

$$4(2) - 4(1) = 4$$

which also is true. Therefore $x_1 = 2$ and $x_2 = 1$ are a solution of the system of equations.

Notice that we obtained the solution by multiplying one of the original equations by a constant and subtracting the result from the second equation. The multiplicative constant was chosen so that one of the variables (in this case x_1) was missing from the result of the subtraction. Thus we "eliminated" x_1 from one equation. We will make a habit of using this strategy in solving any system of linear equations. That is, we will systematically eliminate variables from equations until we obtain an equation that looks like (10.14). Equations such as (10.14) are particularly easy to solve, since they contain only one unknown.

Before proceeding to systems of three or more equations, however, we pause to consider when a system of equations even has a solution. In other words, when does a

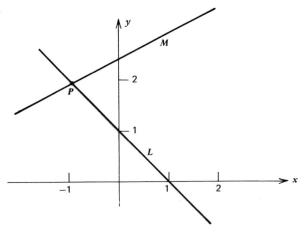

FIGURE 10.3 Graph of (10.15) and (10.16)—only one solution.

solution exist? Clearly, a solution exists for the system (10.11) and (10.12), since we found one such solution: $x_1 = 2$ and $x_2 = 1$. But is it the only solution?

To assist us in determining if any solution exists and if so, how many exist, we turn to geometry and graphs.

Consider first the system of equations

$$x + y = 1 \tag{10.15}$$

$$-x + 2y = 5 \tag{10.16}$$

The graph of the first of these is a line L in Figure 10.3. All pairs of values x and y that satisfy (10.15) must lie on line L. For example, $x = 1, y = 0$ satisfies (10.15) and lies on line L. So do $x = 0, y = 1$ and $x = -1, y = 2$.

Line M in Figure 10.3 represents the second equation, (10.16). The values of x and y that satisfy both equations must lie on both lines. But there is only one point that lies on both lines—point P. At P, $x = -1$ and $y = 2$, so that is the solution, and the only solution, of the system of two equations (10.15) and (10.16). To verify this, we replace x by -1 and y by 2 to obtain

$$-1 + 2 = 1$$

$$-(-1) + 2(2) = 5$$

Both of these are true.

In this case, then, there is one, and only one, solution. However, consider the equations

$$2x + 3y = 6 \tag{10.17}$$

$$4x + 6y = 6 \tag{10.18}$$

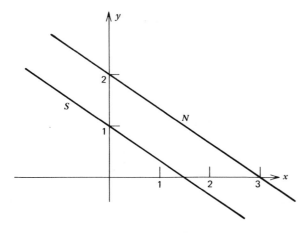

FIGURE 10.4 Graph of (10.17) and (10.18)—no solution.

The graphs of these equations are shown in Figure 10.4. Line N represents (10.17), and line S represents (10.18). Any values of x and y that satisfy both equations must lie on both lines. But the two lines are parallel, so no point lies on both lines. The system of two equations (10.17) and (10.18) has no solution.

What happens to (10.17) and (10.18) if we try our strategy of multiplying the first equation by a constant and subtracting the result from the second equation in a way that eliminates x? We multiply (10.17) by 2

$$4x + 6y = 12$$

and subtract this from (10.18) to obtain

$$0 + 0 = -6$$

Clearly, this can not be true. Algebraically this conveys the same information as Figure 10.4 does, (i.e., no solution exists).

But suppose we change the right-hand side of (10.18) to 12. Then the two equations become

$$2x + 3y = 6 \tag{10.19}$$

$$4x + 6y = 12 \tag{10.20}$$

The graphs now become those shown in Figure 10.5. The two lines representing (10.19) and (10.20) are identical. Any point on line N produces values of x and y that satisfy both equations. The pair $x = 3$, $y = 0$ satisfies both equations, as does $x = 1$, $y = 4/3$ and $x = 0$, $y = 2$, and so on.

Algebraically, if we multiply (10.19) by 2, we obtain

$$4x + 6y = 12$$

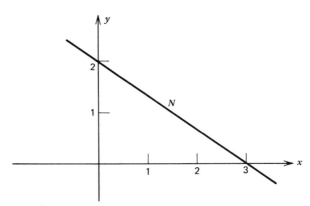

FIGURE 10.5   Graph of (10.19) and (10.20)—infinite number of solutions.

and, subtracting this from (10.20),

$$0 = 0$$

which is certainly true. In such a case the system of equations has an infinite number of solutions.

These are the only three possibilities.

1. One and only one solution exists.
2. No solution exists.
3. There are an infinite number of solutions.

While we have only demonstrated these conclusions for systems of two equations, they are the only three alternatives possible, no matter how many equations are present, provided all of the equations are linear.

We primarily will be interested in possibility 1, but we should be aware of the other two possibilities. Our strategy of systematically eliminating variables will tell us which possibility in fact applies in any given example. If we arrive at a statement such as

$$x_9 = 7$$

possibility 1 applies. If we arrive at a contradiction such as

$$3 = 2$$

possibility 2 applies. If we arrive at an identity (an equation which is always true) such as

$$2 = 2$$

possibility 3 applies.

EXERCISES FOR SECTION 10.2

1. Find solutions to the following systems of equations.

   *(a) $2x + 3y = 1$ (b) $2x + 3y = 5$
           $x + 2y = 1$ $x + 2y = 3$

   (c) $2x + 3y = 12$ (d) $2x + 3y = 19$
           $x + 2y = 7$ $x + 2y = 6$

   (e) $2x + 3y = 4$ (f) $2x + 3y = 2$
           $x + 2y = 0$ $x + 2y = 2$

2. For each of the following systems of equations decide if the system has one solution, no solutions, or an infinite number of solutions.

   *(a) $x - 2y = 1$ (b) $2x + 3y = 0$
          $-2x + 4y = 2$ $x + 2y = 0$

(c) $x + y = 8$
$ x - y = 0$
(e) $x = 1$
$ y = 2$

(d) $x - 2y = 1$
$ -2x + 4y = -2$
(f) $x - 2y = 1$
$ -2x + 4y = 1$

3. Where possible find at least one solution for each of the following systems of equations.

*(a) $x - 2y = 0$
$ -2x + 4y = 0$
(c) $2x + 3y = 0$
$ x + 2y = 0$
(e) $x + y = 8$
$ x - y = 0$

(b) $x - 2y = 1$
$ -2x + 4y = -2$
(d) $x = 1$
$ y = 2$
(f) $x - 2y = 1$
$ -2x + 4y = 1$

4. A plant manufactures two products: metal tables and metal desk chairs. Each table uses 12 bolts and 8 set screws. Each desk chair uses 14 bolts and 6 set screws. There are 2000 bolts and 1000 screws in stock.

*(a) Let t = number of tables and let c = number of chairs to be constructed. If the entire supply of both bolts and screws is to be completely used, write two linear equations for t and c.

(b) Solve the linear equations in part a.

(c) Do parts a and b if the supply of bolts and screws is doubled.

5. A shop sells two different nut mixes. Each pound of the regular mix contains 12 ounces of peanuts and 4 ounces of cashews. Each pound of the party mix contains 6 ounces of peanuts and 10 ounces of cashews. The shopkeeper has 75 pounds of peanuts and 45 pounds of cashews.

(a) Let r be the number of pounds of the regular mix, and let p be the number of pounds of the party mix. Write two linear equations for r and p.

(b) Solve the linear equations in part a to find the number of pounds to be made of each mix.

(c) Do parts a and b if the supply of both types of nuts is doubled.

10.3 THREE EQUATIONS IN THREE UNKNOWNS

In the last section we solved a system of two equations in two unknowns by eliminating one of the unknowns from one of the equations. This technique of elimination can be extended to any number of equations. In this section we will do so for the case of three equations. In the succeeding sections we will turn to a computer to help us with four or more equations.

For example, consider the equations:

$$3x_1 - x_2 + x_3 = 2 \tag{10.21}$$

$$9x_1 - x_2 = 7 \tag{10.22}$$

$$6x_1 + 2x_2 - 2x_3 = 8 \tag{10.23}$$

We seek values for the three unknowns: $x_1, x_2,$ and x_3. When these values are used to replace the unknowns all three of the equations are to be satisfied simultaneously.

Our strategy is to eliminate x_1 from both of the last two equations. We will then eliminate x_2 from the last equation. The last equation will then contain only the unknown x_3. We proceed with that strategy now.

If we multiply (10.21) by 3, the x_1 term will be $9x_1$. If we subtract this new equation from (10.22), we will obtain an equation from which x_1 will be eliminated. Multiplying (10.21) by 3, we obtain

$$9x_1 - 3x_2 + 3x_3 = 6$$

Subtracting this from (10.22),

$$2x_2 - 3x_3 = 1 \tag{10.24}$$

This equation will replace (10.22), since it is (10.22) with equal quantities subtracted from both sides.

If we multiply (10.21) by 2, we obtain

$$6x_1 - 2x_2 + 2x_3 = 4$$

Here the coefficient of x_1 is identical with the coefficient of x_1 in (10.23). Subtracting this last equation from (10.23),

$$4x_2 - 4x_3 = 4 \tag{10.25}$$

and this replaces (10.23). Therefore the original system of three equations has been replaced by

$$3x_1 - x_2 + x_3 = 2 \tag{10.21}$$
$$2x_2 - 3x_3 = 1 \tag{10.24}$$
$$4x_2 - 4x_3 = 4 \tag{10.25}$$

As promised, x_1 has been eliminated from the last two equations. It is important to realize that any solution of the last three equations is also a solution of (10.21), (10.22), and (10.23), and vice versa.

We now proceed to eliminate x_2 from the last equation above $\left[$i.e., (10.25)$\right]$. To do so, we multiply (10.24) by 2 to obtain

$$4x_2 - 6x_3 = 2$$

The multiplier 2 was chosen to make the coefficient of x_2 in this newly formed equation identical with the coefficient of x_2 in (10.25). Subtracting the last equation from (10.25),

$$2x_3 = 2 \tag{10.26}$$

This equation replaces (10.25), so the system of equations now becomes

$$3x_1 - x_2 + x_3 = 2 \qquad (10.21)$$

$$2x_2 - 3x_3 = 1 \qquad (10.24)$$

$$2x_3 = 2 \qquad (10.26)$$

Such a system of equations is called *triangular,* since they form a triangle. We can easily find the value of x_3 from (10.26). Having that value and using it in (10.24), we can find the value of x_2. Finally, we can use the values of both x_2 and x_3 in (10.21) to find the value of x_1. This process is called *back substitution*, since we "substitute" the values of the variables working our way "backward" through the system of equations.

In this particular case, back substitution produces, from (10.26),

$$x_3 = 1$$

Using this in (10.24), we get

$$2x_2 - 3(1) = 1$$

or

$$x_2 = 2$$

Finally, using these values for x_2 and x_3 in (10.21), we get

$$3x_1 - 2 + 1 = 2$$

or

$$x_1 = 1$$

It is left to the reader to substitute these values of x_1, x_2, and x_3 in the original system of equations given by (10.21), (10.22), and (10.23) to verify that they do, indeed, constitute a solution to that original system.

The strategy for finding a solution of any system of linear equations should now be clear. We will first multiply the first equation by the appropriate quantities so that when the results are subtracted from each of the other equations, the x_1 terms are eliminated. We next multiply the new second equation by constants that will eliminate x_2 from the equations from the third one on. We continue in this way until the system of equations is in triangular form. That is, the last equation contains only one variable, the next to last contains two variables, and so on. We then use back substitution to find the last variable, the next to last variable, and we continue in this way until all variables have been computed.

Of course, things are not always as simple as we have indicated. Suppose the system of three equations is:

$$- x_2 + x_3 = -1 \qquad (10.27)$$

$$9x_1 - x_2 \qquad = 7 \qquad (10.28)$$

$$6x_1 + 2x_2 - 2x_3 = 8 \qquad (10.29)$$

There is nothing by which we can multiply the first equation, (10.27), to eliminate x_1 from the last two equations. In such cases there are several courses of action open to us. We could eliminate x_2 or x_3 instead of x_1 from the last two equations. But that strategy is fraught with bookkeeping difficulties. A better solution is simply to reorder the equations (i.e., interchange the first two equations). This produces

$$9x_1 - x_2 \qquad = 7$$

$$-x_2 + x_3 = -1$$

$$6x_1 + 2x_2 - 2x_3 = 8$$

Since these are identical with (10.27), (10.28), and (10.29), they have the same solution. Now we can eliminate x_1 from the last two equations. Indeed, x_1 already has been eliminated from the second equation. To eliminate x_1 from the third equation, we multiply the first by 6/9 and subtract.

The difficulty raised by the zero coefficient in (10.27) may not be obvious at the outset as it was here. Consider, for example, the system of equations:

$$3x_1 - x_2 + x_3 = 12 \qquad (10.30)$$

$$9x_1 - 3x_2 \qquad = 30 \qquad (10.31)$$

$$6x_1 + 2x_2 - 2x_3 = 12 \qquad (10.32)$$

We proceed as before and multiply (10.30) by 3 and subtract the result from (10.31) to get

$$- 3x_3 = -6$$

We then multiply (10.30) by 2 and subtract this result from (10.32).

$$4x_2 - 4x_3 = -12$$

These two equations replace (10.31) and (10.32), so the system is replaced by

$$3x_1 - x_2 + x_3 = 12 \qquad (10.30)$$

$$- 3x_3 = -6 \qquad (10.33)$$

$$4x_2 - 4x_3 = -12 \qquad (10.34)$$

If we try to eliminate x_2 from (10.34) using our multiply-and-subtract rule, we will encounter a difficulty. We cannot achieve this elimination by multiplying (10.33) by anything, since (10.33) does not contain x_2. On the other hand, if we multiply (10.30) by -4 and subtract the result from (10.34), we will eliminate x_2, but x_1 will reappear.

Once more the resolution of this dilemma is to interchange the last two equations so that the system is:

$$3x_1 - x_2 + x_3 = 12 \tag{10.30}$$

$$4x_2 - 4x_3 = -12 \tag{10.34}$$

$$-3x_3 = -6 \tag{10.33}$$

This system is in triangular form, and we can use back substitution to obtain $x_3 = 2$, $x_2 = -1$, and $x_1 = 3$.

Our strategy for eliminating x_1 from the last two equations and then eliminating x_2 from the last equation may require that we interchange two equations from time to time.

The scheme of elimination followed by back substitution usually is called *Gaussian elimination*.

EXERCISES FOR SECTION 10.3

1. Use elimination and back substitution to solve each of the following systems of equations.

*(a)
$$2x + 3y - 6z = -1$$
$$x - y + 2z = 2$$
$$-2x + y - z = -2$$

(b)
$$8x + 2y - z = -14$$
$$x + z = 3$$
$$-y + 2z = 7$$

(c)
$$8x + 2y - z = 4$$
$$x + z = 3$$
$$-y + 2z = 3$$

(d)
$$2x + 3y + 6z = 5$$
$$x - y + 2z = 2$$
$$-2x + y - z = -4$$

2. Use elimination and back substitution to solve the following systems of equations.

*(a)
$$x_1 + x_2 = 2$$
$$x_2 + x_3 = 2$$
$$x_3 + x_4 = 2$$
$$x_1 + x_2 - x_3 = 1$$

(b)
$$2x_1 + x_2 = 2$$
$$2x_1 + x_4 = 2$$
$$2x_1 + x_3 = 2$$
$$2x_1 = 2$$

(c) $\begin{aligned} x_1 - x_2 + x_3 - 2x_4 &= 3 \\ x_1 + x_2 + x_3 + x_4 &= 4 \\ x_1 + x_2 - x_3 + x_4 &= 0 \\ 2x_1 - x_2 + 2x_3 - x_4 &= 8 \end{aligned}$

(d) $\begin{aligned} x_1 - x_2 + x_3 - 2x_4 &= -5 \\ x_1 + x_2 + x_3 + x_4 &= 2 \\ x_1 + x_2 - x_3 + x_4 &= 0 \\ 2x_1 - x_2 + 2x_3 - x_4 &= -5 \end{aligned}$

*3. Solve the four food diet problem given by (10.3), (10.4), (10.5), and (10.6).

4. Suppose in the four-food diet problem that all nutrient requirements remain the same except for ascorbic acid. Find the amounts of all four foods for the following ascorbic acid requirements.
   (a) 26 milligrams.
   (b) 28 milligrams.
   (c) 32 milligrams.
   (d) 34 milligrams.
   [*Note*. The values above are one half the daily requirements for ascorbic acid and, there-fore, should be the right side of (10.4).]

5. A plant manufactures three products: metal tables, metal desk chairs, and metal desks. Each table requires five bolts, eight set screws, and four angle braces. Each chair uses eight bolts, five screws, and two angle braces. Each desk uses six bolts, nine screws, and six angle braces. There are 4000 bolts, 4000 screws, and 2000 angle braces in stock.
   (a) Let t be the number of tables, c be the number of chairs, and d be the number of desks to be constructed. If the entire supply of bolts, screws, and angle braces is to be completely used, write three linear equations for t, c, and d.
   (b) Solve the linear equations in part a.
   (c) Do parts a and b if the supply of all three parts is doubled.

6. A shop sells three different nut mixes. Each pound of the regular mix contains nine ounces of peanuts, four ounces of cashews, and three ounces of hazelnuts. Each pound of the party mix contains five ounces of peanuts, eight ounces of cashews, and three ounces of hazel-nuts. Finally, each pound of the deluxe mix contains eight ounces of cashews and eight ounces of hazelnuts. The shopkeeper has 38 pounds of peanuts, 64 pounds of cashews, and 42 pounds of hazelnuts, and wishes to use all of his supplies.
   (a) Let r be the number of pounds of regular mix, p be the number of pounds of party mix, and d be the number of pounds of deluxe mix. Write three equations for r, p, and d.
   (b) Solve the equations in part a.
   (c) Do parts a and b if the supply of all three nuts is doubled.

10.4 MATRIX REPRESENTATION

As we implied in the introduction (Section 10.1), we can use matrices to write down systems of linear equations in a concise form. Suppose we let

$$A = \begin{pmatrix} 3 & -1 & 1 \\ 9 & -1 & 0 \\ 6 & 2 & -2 \end{pmatrix} \tag{10.34}$$

$$\mathbf{b} = \begin{pmatrix} 2 \\ 7 \\ 8 \end{pmatrix} \quad \text{and} \quad \mathbf{x} = \begin{pmatrix} x_1 \\ x_2 \\ x_3 \end{pmatrix} \tag{10.35}$$

Then (10.21), (10.22), and (10.23) can be expressed as

$$A\mathbf{x} = \mathbf{b} \tag{10.36}$$

Indeed, if A is a square $n \times n$ matrix and if $\mathbf{x}$ and $\mathbf{b}$ are both n-component column vectors, the (10.36) represents a system of n linear equations in n unknowns.

The standard problem is: Given the $n \times n$ matrix A and the n-component vector $\mathbf{b}$, find the n-component vector $\mathbf{x}$.

We have used a multiply-and-subtract rule to eliminate unknowns until A is a triangular matrix (zeroes below the diagonal). We then used back substitution to find a solution. Of course, if n is very large, the arithmetic involved in elimination and back substitution becomes tedious and error-prone, so we turn to the computer for help.

EXERCISES FOR SECTION 10.4

1. If

$$A\mathbf{x} = \mathbf{b}$$

write out the matrix A for each of the following systems of equations.

*(a) $\quad 2x_1 - 3x_2 = 0$
$\quad\quad 3x_1 + x_2 = 1$

(b) $\quad\quad\quad 2x_2 + x_3 = 1$
$\quad\quad x_1 \quad\quad - 2x_3 = 0$
$\quad\quad x_1 + x_2 + x_3 = 4$

(c) $x_1 \quad\quad\quad\quad = 4$
$\quad\quad x_2 \quad\quad = 1$
$\quad\quad\quad\quad x_3 = -1$

(d) $\quad\quad x_2 + x_3 = 1$
$\quad\quad x_1 + 2x_2 \quad\quad = 2$
$\quad\quad x_1 \quad\quad - x_3 = 0$

2. If

$$A\mathbf{x} = \mathbf{b}$$

write the system of equations for the following.

*(a) $\quad A = \begin{pmatrix} 1 & 0 & 2 \\ 1 & 1 & -2 \\ 0 & 2 & 1 \end{pmatrix} \quad \mathbf{b} = \begin{pmatrix} 1 \\ 0 \\ 1 \end{pmatrix}$

(b)
$$A = \begin{pmatrix} 0 & 1 \\ 1 & 0 \end{pmatrix} \qquad \mathbf{b} = \begin{pmatrix} 1 \\ 1 \end{pmatrix}$$

(c)
$$A = \begin{pmatrix} 1 & -1 & 0 & 2 \\ 0 & 1 & 0 & 1 \\ 2 & 1 & 1 & 2 \\ 3 & 1 & 0 & 1 \end{pmatrix} \qquad \mathbf{b} = \begin{pmatrix} 1 \\ 0 \\ 2 \\ 1 \end{pmatrix}$$

(d)
$$A = \begin{pmatrix} 5 & 1 & 0 \\ 2 & 1 & 2 \\ 1 & 2 & 1 \end{pmatrix} \qquad \mathbf{b} = \begin{pmatrix} 0 \\ 0 \\ 1 \end{pmatrix}$$

10.5 SOLUTION OF SYSTEMS OF EQUATIONS

We wish to find the n-component vector $\mathbf{x}$ so that

$$A\mathbf{x} = \mathbf{b} \tag{10.36}$$

where A, which is $n \times n$, and $\mathbf{b}$, which is $n \times 1$, are both known. Suppose we can find an $n \times n$ matrix C such that

$$CA = I \tag{10.38}$$

where I is the $n \times n$ identity matrix (1's along the diagonal and zeroes elsewhere). Multiplying (10.36) by C,

$$CA\mathbf{x} = C\mathbf{b}$$

But $CA = I$ and $I\mathbf{x} = \mathbf{x},$ so

$$\mathbf{x} = C\mathbf{b} \tag{10.39}$$

Thus we have solved our problem, since (10.39) produces $\mathbf{x}$ explicitly.

For example, consider the system of equations (10.21), (10.22), and (10.23), which are given in matrix form by (10.34), (10.35), and (10.36). For C we use

$$C = \begin{pmatrix} 1/6 & 0 & 1/12 \\ 3/2 & -1 & 3/4 \\ 2 & -1 & 1/2 \end{pmatrix}$$

Notice that

$$CA = \begin{pmatrix} 1 & 0 & 0 \\ 0 & 1 & 0 \\ 0 & 0 & 1 \end{pmatrix}$$

From (10.39),

$$\mathbf{x} = \begin{pmatrix} 1/6 & 0 & 1/12 \\ 3/2 & -1 & 3/4 \\ 2 & -1 & 1/2 \end{pmatrix} \cdot \begin{pmatrix} 2 \\ 7 \\ 8 \end{pmatrix}$$

Performing the multiplication of the matrix and the column vector,

$$\mathbf{x} = \begin{pmatrix} 1 \\ 2 \\ 1 \end{pmatrix}$$

which is the solution we found in Section 10.3 using elimination and back substitution.

The question is: How can we find the matrix C?

There are several numerical techniques for computing the components of C. In fact, Gaussian elimination as described in Section 10.3 can be used to find C. However, these numerical techniques are complex and are properly a topic in the field of *numerical analysis*. Moreover, BASIC will automatically produce this C matrix for us in one statement. We proceed, in the next section, to the computation of C in *BASIC* and its use in the solution of linear equations. First, however, we note that C is called the *inverse* of the matrix A. It usually is written A^{-1}, so that

$$A^{-1}A = I \tag{10.41}$$

Moreover,

$$\mathbf{x} = A^{-1}\mathbf{b} \tag{10.42}$$

We caution the reader, however, that for an arbitrary matrix, A, no inverse, A^{-1}, may exist. For example, the innocent-looking matrix

$$A = \begin{pmatrix} 5 & 7 & 11 \\ -1 & 4 & 2 \\ 2 & 1 & 3 \end{pmatrix}$$

has no inverse! If, using this matrix, we tried to find a solution $\mathbf{x}$ to

$$A\mathbf{x} = \mathbf{b}$$

we should find that either no solution could be found or that there would be an infinite number of solutions (see discussion in Section 10.2).

EXERCISES FOR SECTION 10.5

*1. Match the matrices on the left with their inverses on the right.

(a) $\begin{pmatrix} 1 & 1 & -3 \\ 2 & 5 & 1 \\ 1 & 3 & 2 \end{pmatrix}$

(i) $\begin{pmatrix} 7 & -8 & 5 \\ -4 & 5 & -3 \\ 1 & -1 & 1 \end{pmatrix}$

(b) $\begin{pmatrix} 3 & 1 & -5 \\ 2 & 1 & 7 \\ 8 & 3 & -2 \end{pmatrix}$

(ii) $\begin{pmatrix} 7 & -11 & 16 \\ -3 & 5 & -7 \\ 1 & -2 & 3 \end{pmatrix}$

(c) $\begin{pmatrix} 2 & 3 & -1 \\ 1 & 2 & 1 \\ -1 & -1 & 3 \end{pmatrix}$

(iii) $\begin{pmatrix} -1 & 2 & -3 \\ 2 & 1 & 0 \\ 4 & -2 & 5 \end{pmatrix}$

(d) $\begin{pmatrix} -5 & 4 & -3 \\ 10 & -7 & 6 \\ 8 & -6 & 5 \end{pmatrix}$

(iv) $\begin{pmatrix} -23 & -13 & 12 \\ 60 & 34 & -31 \\ -2 & -1 & 1 \end{pmatrix}$

2. Find x and y if

*(a) $\begin{pmatrix} 1 & 6 \\ 1 & 5 \end{pmatrix} \begin{pmatrix} -5 & x \\ y & -1 \end{pmatrix} = \begin{pmatrix} 1 & 0 \\ 0 & 1 \end{pmatrix}$

(b) $\begin{pmatrix} x & y \\ 7 & 5 \end{pmatrix} \cdot \begin{pmatrix} 5 & -2 \\ -7 & 3 \end{pmatrix} = \begin{pmatrix} 1 & 0 \\ 0 & 1 \end{pmatrix}$

(c) $\begin{pmatrix} 1 & 5 \\ 2 & x \end{pmatrix} \cdot \begin{pmatrix} -x & 5 \\ 2 & -y \end{pmatrix} = \begin{pmatrix} 1 & 0 \\ 0 & 1 \end{pmatrix}$

(d) $\begin{pmatrix} 5 & x \\ 4 & y \end{pmatrix} \cdot \begin{pmatrix} x & -y \\ -4 & 5 \end{pmatrix} = \begin{pmatrix} 1 & 0 \\ 0 & 1 \end{pmatrix}$

3. Using the fact that

$$\begin{pmatrix} 2 & -1 & 1 \\ 8 & -5 & 2 \\ -11 & 7 & -3 \end{pmatrix} \begin{pmatrix} 1 & 4 & 3 \\ 2 & 5 & 4 \\ 1 & -3 & -2 \end{pmatrix} = \begin{pmatrix} 1 & 0 & 0 \\ 0 & 1 & 0 \\ 0 & 0 & 1 \end{pmatrix}$$

find the solution of

*(a) $\quad 2x - y + z = 1$
$\quad\quad 8x - 5y + 2z = 1$
$\quad -11x + 7y - 3z = 1$

(b) $\quad 2x - y + z = 2$
$\quad\quad 8x - 5y + 2z = -1$
$\quad -11x + 7y - 3z = 3$

(c) $\quad x + 4y + 3z = 1$
$\quad 2x + 5y + 4z = 1$
$\quad\quad x - 3y - 2z = 1$

(d) $\quad x + 4y + 3z = 3$
$\quad 2x + 5y + 4z = -1$
$\quad\quad x - 3y - 2z = 2$

4. Using the fact that

$$\begin{pmatrix} 1 & 2 & 1 \\ 2 & 3 & 2 \\ 3 & 4 & 1 \end{pmatrix} \begin{pmatrix} -5/2 & 1 & 1/2 \\ 2 & -1 & 0 \\ -1/2 & 1 & -1/2 \end{pmatrix} = \begin{pmatrix} 1 & 0 & 0 \\ 0 & 1 & 0 \\ 0 & 0 & 1 \end{pmatrix}$$

find the solution of

(a) $x + 2y + z = 1$
  $2x + 3y + 2z = 1$
  $3x + 4y + z = 1$

(b) $x + 2y + z = 2$
  $2x + 3y + 2z = 3$
  $3x + 4y + z = 4$

(c) $-5x + 2y + z = 1$
  $4x - 2y \quad = 1$
  $-x + 2y - z = 1$

(d) $-5x + 2y + z = 2$
  $4x - 2y \quad = 3$
  $-x + 2y - z = 4$

5. Find the inverse of

$$\begin{pmatrix} 1 & 2 \\ 2 & 3 \end{pmatrix}$$

by finding a, b, c, and d so that

$$\begin{pmatrix} 1 & 2 \\ 2 & 3 \end{pmatrix} \begin{pmatrix} a & b \\ c & d \end{pmatrix} = \begin{pmatrix} 1 & 0 \\ 0 & 1 \end{pmatrix}$$

6. Find the inverse of

$$\begin{pmatrix} -2 & 1 \\ -1 & 2 \end{pmatrix}$$

by finding a, b, c, and d so that

$$\begin{pmatrix} -2 & 1 \\ -1 & 2 \end{pmatrix} \begin{pmatrix} a & b \\ c & d \end{pmatrix} = \begin{pmatrix} 1 & 0 \\ 0 & 1 \end{pmatrix}$$

10.6 SOLUTION OF LINEAR EQUATIONS IN BASIC

Given a square matrix A and a column vector $\mathbf{b}$ we wish to find a column vector $\mathbf{x}$ so that

$$A\mathbf{x} = \mathbf{b} \tag{10.43}$$

But, from (10.42), the solution of these equations is given by

$$\mathbf{x} = A^{-1}\mathbf{b} \tag{10.42}$$

In BASIC, to find the inverse of a square matrix A, we simply write

```
MAT B = INV(A)
```

then

$$B = A^{-1}$$

The word "INV" is an abbreviation of "inverse." It is, therefore, quite simple to find the solution of a system of linear equations. To illustrate this, we return to the system of three equations that we solved in Section 10.3: (10.21), (10.22), and (10.23). A BASIC program is given in Figure 10.6.

```
10   DIM A[3,3],I[3,3],X[3,1],B[3,1]
20   MAT   READ A,B
30   MAT I=INV(A)
40   MAT X=I*B
50   MAT   PRINT X
100    DATA 3,-1,1
110    DATA 9,-1,0
120    DATA 6,2,-2
130    DATA 2,7,8
200  END
```

1.

2.   **FIGURE 10.6** BASIC program to solve
                                                     system of three linear
1.   equations.

In the program the matrix I is the inverse of A; that is,

$$I = A^{-1}$$

The program should be self-explanatory. If no solution exists to (10.43) or if there are an infinite number of solutions, the program will print an error message.

Notice that to solve a system of n equations in n unknowns, only the DIM statement and the DATA statements need be changed.

Recall, however, that some matrices, for example,

$$A = \begin{pmatrix} 5 & 7 & 11 \\ -1 & 4 & 2 \\ 2 & 1 & 3 \end{pmatrix}$$

have no inverse and, if we try to solve

$$A\mathbf{x} = \mathbf{b}$$

that either no solution exists or there are an infinite number of solutions. What happens if we try to find the inverse of such a matrix using BASIC? We will get an error message indicating that no inverse can be found for A. Even if there are an infinite number of solutions to the system of linear equations, none will be produced. The reader is invited to use the above matrix in the BASIC program in Figure 10.6 (change the three DATA statements numbered 100, 110, and 120) and observe what happens.

EXERCISES FOR SECTION 10.6

1. Using the BASIC program in Figure 10.6, modified as necessary, solve the following systems of equations.

*(a) $\begin{aligned} x - y + z - 2w &= 3 \\ x + y + z + w &= 4 \\ x + y - z + w &= 0 \\ 2x - y + 2z - w &= 8 \end{aligned}$

(b) $\begin{aligned} 5x + 7y + 6z + 5w &= 1 \\ 7x + 10y + 8z + 7w &= 2 \\ 6x + 8y + 10z + 9w &= 1 \\ 5x + 7y + 9z + 10w &= 0 \end{aligned}$

(c) $\begin{aligned} x + y + z + w &= 4 \\ x + 2y + 2z + w &= 6 \\ -x + y + z + w &= 2 \\ y - z + 2w &= 0 \end{aligned}$

(d) $\begin{aligned} x - 2y + z + 2w &= 18 \\ 2x + y + 4z - w &= 16 \\ x + y - z + w &= 4 \\ x + 3y - 2z + w &= 1 \end{aligned}$

2. Using the BASIC program in Figure 10.6, modified as necessary, solve the following systems of equations.

*(a) $\begin{aligned} 3x - y + z + 2w + 3u &= 20 \\ 9x - y + 2u &= 20 \\ 6x + 2y - 2z - 2w + u &= 30 \\ x + 4y - z + 2w + 3u &= -10 \\ 2x + 3y - 2z + 9w + 4u &= 0 \end{aligned}$

(b) $\begin{aligned} 3x + 3y + 2z + 3w + 4u &= 20 \\ 9x + 2y + 3z + u &= 20 \\ 6x - 2y + 2u &= 30 \\ -x + 2z + 3w - 3u &= -10 \\ 6x - 2z + 6w + u &= 0 \end{aligned}$

(c) $\begin{aligned} 3x + 3y + 2z + 3w + 4u &= -20 \\ 9x + 2y + 3z + u &= 20 \\ 6x - 2y + 2u &= 300 \\ -x + 2z + 3w - 3u &= -50 \\ 6x - 2z + 6w + u &= 60 \end{aligned}$

(d) $\begin{aligned} 3.6x + 4.1y - 6.3z + u &= -5.4 \\ 6.1x + 4.3z - 6.4w + 1.1u &= 19.6 \\ 1.2z - 3.2w + 6.0u &= 23.4 \\ -3.8x - 8.4y + 2.2z + 2.2w + 6.5u &= 0 \\ -3.9x + 7.4y + 1.3z + 8.4u &= -7.5 \end{aligned}$

10.7 THE DEMOGRAPHIC MODEL REVISITED

In Case Study 9 (Section 9.8) we discussed the movement of the nation's population among four regions: east, midwest, mountains, and far west. There we found that the equilibrium population, stable or unstable, was given by the vector $\mathbf{x}$, where

$$T\mathbf{x} = \mathbf{x} \tag{10.44}$$

and T is the transition matrix. The particular transition matrix used in the case study was

$$T = \begin{pmatrix} 1/2 & 0 & 0 & 1/4 \\ 1/12 & 1/2 & 0 & 0 \\ 1/6 & 1/4 & 3/4 & 1/4 \\ 1/4 & 1/4 & 1/4 & 1/2 \end{pmatrix} \tag{10.45}$$

Using this in (10.44), the system of equations becomes

$$-(1/2)x_1 \qquad\qquad\qquad + (1/4)x_4 = 0 \tag{10.46}$$

$$(1/12)x_1 - (1/2)x_2 \qquad\qquad = 0 \tag{10.47}$$

$$(1/6)x_1 + (1/4)x_2 - (1/4)x_3 + (1/4)x_4 = 0 \tag{10.48}$$

$$(1/4)x_1 + (1/4)x_2 + (1/4)x_3 - (1/2)x_4 = 0 \tag{10.49}$$

Perhaps, surprisingly, this system of equations has an infinite number of solutions. To see why, suppose we add the first three equations together. We obtain

$$-(1/4)x_1 - (1/4)x_2 - (1/4)x_3 + (1/2)x_4 = 0$$

But this is just the last equation, (10.49), multiplied by -1. Hence the last equation in the set of four does not provide us with any additional information above and beyond the information contained in the first three equations. If we would try to solve this system of equations, we would arrive at

$$0 = 0$$

implying that there are an infinite number of solutions. If we would try to use the INV function in BASIC, we would obtain an error message and no solution at all.

But we know that the solution we seek is

$$\mathbf{x} = \begin{pmatrix} 1/6 \\ 1/36 \\ 17/36 \\ 1/3 \end{pmatrix} \tag{10.50}$$

How are we to pick this solution out from the infinite number of solutions of the system given by (10.46) to (10.49)? And how are we to use BASIC to find it?

To answer this question, we recall that the vector $\mathbf{x}$ must be a population vector (i.e., its components must sum to one). In other words,

$$x_1 + x_2 + x_3 + x_4 = 1 \tag{10.51}$$

We will use this to replace (10.49) which, as we pointed out, contains no information not contained in the first three equations. The new system then is

$$\left.\begin{array}{l} -(1/2)x_1 \qquad\qquad\qquad\quad + (1/4)x_4 = 0 \\ (1/12)x_1 - (1/2)x_2 \qquad\qquad\qquad = 0 \\ (1/6)x_1 + (1/4)x_2 - (1/4)x_3 + (1/4)x_4 = 0 \\ x_1 + \quad x_2 + \quad x_3 + \quad x_4 = 1 \end{array}\right\} \tag{10.52}$$

This system has one and only one solution, and that solution is given by (10.50).

The steps we used to create the final system, (10.52) are:

1. Construct the transition matrix T.
2. Subtract 1 from the diagonal elements (i.e., form $T - I$).
3. Replace the last row of $T - I$ with all 1's.
4. Construct a right side of all zeroes.
5. Replace the last zero on the right side by 1.

A BASIC program that performs these five steps and then solves the resulting system of equations is shown in Figure 10.7. Notice that the results are indeed $\mathbf{x}$, given in (10.50).

EXERCISES FOR SECTION 10.7

1. Find the equilibrium populations for the following transition matrices.

*(a) $\begin{pmatrix} 1/2 & 0 & 1/4 & 1/4 \\ 1/12 & 1/2 & 0 & 0 \\ 1/6 & 1/4 & 1/2 & 1/4 \\ 1/4 & 1/4 & 1/4 & 1/2 \end{pmatrix}$

(b) $\begin{pmatrix} 0.9 & 0 & 0.7 & 0 \\ 0.1 & 0 & 0.3 & 0 \\ 0 & 0.4 & 0 & 0.2 \\ 0 & 0.6 & 0 & 0.8 \end{pmatrix}$

(c) $\begin{pmatrix} 0.8 & 0 & 0.6 & 0 \\ 0.2 & 0 & 0.4 & 0 \\ 0 & 0.5 & 0 & 0.1 \\ 0 & 0.5 & 0 & 0.9 \end{pmatrix}$

```
1C   DIM T[4,4],I[4,4],R[4,4],X[4,1],B[4,1]
100  MAT   READ T
110  MAT   I=IDN
120  REM   **  SUBTRACT 1 FROM DIAGONAL  **
130  MAT   T=T-I
140  REM   **  PLACE ALL 1'S IN LAST ROW  **
150  FOR   K=1 TO 4
160  LET   T[4,K]=1
170  NEXT  K
180  REM   **  PUT ZEROES ON RIGHT SIDE EXCEPT FOR LAST ROW  **
190  MAT   B=ZER
200  LET   B[4,1]=1
210  REM   **  COMPUTE AND PRINT SOLUTION  **
220  MAT   R=INV(T)
230  MAT   X=R*B
240  MAT   PRINT X
300  DATA  .5,0,0,.25
310  DATA  8.33333E-02,.5,0,0
320  DATA  .166667,.25,.75,.25
330  DATA  .25,.25,.25,.5
400  END
```

.166667

2.77778E-02

.472222

.333333

FIGURE 10.7 BASIC program to find equilibrium vector for a transition matrix.

(d) $\begin{pmatrix} 0.7 & 0.15 & 0 & 0.1 \\ 0.2 & 0.6 & 0 & 0.1 \\ 0.1 & 0.15 & 0.8 & 0.7 \\ 0 & 0.1 & 0.2 & 0.1 \end{pmatrix}$

(e) $\begin{pmatrix} 0.7 & 0.2 & 0 & 0.15 \\ 0.1 & 0.5 & 0 & 0.15 \\ 0.2 & 0.2 & 0.7 & 0.3 \\ 0 & 0.1 & 0.3 & 0.4 \end{pmatrix}$

*2. The money that an individual has at his disposal during any given year is put to various uses. Suppose we separate all uses of money into four categories: (a) purchase of securities (stocks and bonds); (b) purchase of real property (land, etc.); (c) purchase of consumer goods; and (d) savings (bank accounts, retirement plans, etc.).

Suppose that during a given year the probability that a dollar now invested in securities will be withdrawn and (a) used for real property is 0.1, (b) used to purchase consumer goods is 0.2, (c) placed into savings is 0.2. Thus the probability that the dollar remains invested in securities is 0.5.

The probabilities that other money has its use changed are:

Probability Money Will Be Used For	Money now used for:		
	Real Property	Consumer Goods	Savings
Securities	0	0.2	0.1
Real property	0.8	0	0.3
Consumer goods	0	0.6	0.1
Savings	0.2	0.2	0.5

(a) What will be the long-run distribution of money in the four categories?

(b) If money is evenly distributed among the four categories, what will be its distribution in two years?

(c) In 10 years?

3. Suppose in the situation described in Exercise 2 above a sudden down-turn in the stock market discourages investment in stocks and bonds and encourages investment in real property. The following specific changes take place.

a. The probability that a dollar invested in securities is reinvested in real property is 0.3 instead of 0.1. There is a corresponding decrease in the probability that the dollar remains in securities.

b. The probability that a dollar now used for consumer goods is invested in securities is 0 and the probability that that dollar is invested in real property is 0.2.

c. The probability that a dollar in savings is used to purchase securities is zero. That money remains in savings.

(a) What will be the long run distribution of money in the four categories?

(b) If the money is evenly distributed among the four categories, what will be its distribution in two years?

(c) In 10 years?

*4. The following table shows the transition from one occupational class to another in England in 1949. The data is from S. J. Prais of Cambridge University (see Mizrahi and Sullivan, *Finite Mathematics with Applications,* Wiley, 1973, p. 326).

Occupational Class at End of 1949	Occupational Class at Start of 1949						
	Professional	Managerial	Higher-Grade Supervisory	Lower-Grade Supervisory	Skilled Labor	Semi-skilled Labor	Un-skilled Labor
Professional	.388	.107	.035	.021	.009	0	0
Managerial	.146	.267	.101	.039	.024	.013	.008
Higher-grade supervisory	.202	.227	.188	.112	.075	.041	.036
Lower-grade supervisory	.062	.120	.191	.212	.123	.088	.083
Skilled labor	.140	.206	.357	.430	.473	.391	.364
Semiskilled labor	.047	.053	.067	.124	.171	.312	.235
Unskilled labor	.015	.020	.061	.062	.125	.155	.274

These may be thought of as probabilities of any given individual changing his occupational class [e.g., the probability that a skilled laborer will become a lower grade supervisor in one year is 0.123 (see column 5, row 4)].

(a) Find the equilibrium distribution of the various occupational classes (i.e., in the long run, what is the fraction of the total population who will be professionals, managers, etc.?).

(b) Is this equilibrium distribution stable? (*Hint.* You may use the BASIC programs in Figures 9.11 and 10.7).

11
LINEAR PROGRAMMING

11.1 A SIMPLE DIET PROBLEM

In Section 10.1 we found a diet consisting of roast beef and salad that satisfied specified requirements for protein and Vitamin C. To find this diet, we solved the following two linear equations.

$$B + L = 35 \qquad\qquad (11.1)$$

$$15L = 30 \qquad\qquad (11.2)$$

where B was the amount of roast beef in units of 100 grams and L was the amount of lettuce in the same units (see also Figure 10.1). The first equation, (11.1), assured us of getting 35 grams of protein, while (11.2) guaranteed us 30 milligrams of Vitamin C.

357

Suppose now that we change our point of view and require only that we get *at least* 35 grams of protein and *at least* 30 milligrams of lettuce from our diet. That is, we replace (11.1) by

$$20B + L \geq 35 \qquad (11.3)$$

and we replace (11.2) by

$$15L \geq 30 \qquad (11.4)$$

The first of these says that the protein derived from the roast beef and salad together (i.e., $20B + L$) must be 35 or greater. Similarly, the second, (11.4), says that the ascorbic acid derived from the lettuce salad must be greater than or equal to 30.

There are many possible solutions to these two inequalities. If we eat enough beef and salad we can achieve any level of the desired nutrients. Moreover, given any suitable diet, we can always eat even a little more of each food and still satisfy both (11.3) and (11.4). However, we must be a little more careful than this last statement implies. Notice that $B = -1$ and $L = 55$ satisfy both inequalities and therefore appear to be an adequate diet. But what does $B = -1$ mean? Eat -100 grams of roast beef?

Clearly, we cannot accept negative values for either B or L. We require, in addition to (11.3) and (11.4), that

$$B \geq 0 \qquad (11.5)$$

$$L \geq 0 \qquad (11.6)$$

Of course, there are still an infinite number of solutions. For example, $B = 1.65$ and $L = 2$ is one. (This is the solution found in Section 10.1.) Another is $B = 1$ and $L = 15$. Still another is $B = 1$ and $L = 20$.

Suppose from among all these solutions that we wish to choose the one (or ones) that provide the fewest number of calories. Figure 10.2 gives the calorie content of both foods. From this figure we can see that the number of calories in our diet will be

$$325B + 80L \qquad (11.7)$$

Therefore we wish to minimize (i.e., make as small as possible) (11.7) while still satisfying the four inequalities (11.3) to (11.6).

Problems such as this are called *linear programming* problems—"linear" because all mathematical expressions contain only variables (in this case B and L) to the first power—"programming" because the result is a program of activities (i.e., eat certain foods).[1]

---

[1]The reader should not confuse the use of the word "program" with the use in the phrase BASIC programming. The latter refers to a sequence of steps for a computer. Linear programming refers to finding the "best" solution to some problem.

We close this section by noting that we could write this linear programming problem in matrix form as follows.

$$\text{Minimize } \mathbf{c}^T\mathbf{x} \qquad (11.8)$$

where

$$A\mathbf{x} \geq \mathbf{b} \qquad (11.9)$$

$$\mathbf{x} \geq \mathbf{0} \qquad (11.10)$$

The various vectors and matrices are given by

$$\mathbf{c} = \begin{pmatrix} 325 \\ 80 \end{pmatrix} \qquad \mathbf{x} = \begin{pmatrix} B \\ L \end{pmatrix} \qquad \mathbf{0} = \begin{pmatrix} 0 \\ 0 \end{pmatrix}$$

$$A = \begin{pmatrix} 20 & 1 \\ 0 & 15 \end{pmatrix} \qquad \mathbf{b} = \begin{pmatrix} 35 \\ 30 \end{pmatrix}$$

It is to be understood that if one vector is greater than or equal to another vector, *each component* of the first vector is greater than or equal to the corresponding component of the second. Thus, for example,

$$\begin{pmatrix} 3 \\ -1 \\ 2 \end{pmatrix} \geq \begin{pmatrix} 2 \\ -2 \\ 2 \end{pmatrix}$$

but

$$\begin{pmatrix} 3 \\ -1 \\ 2 \end{pmatrix} \ngeq \begin{pmatrix} 2 \\ 0 \\ 2 \end{pmatrix}$$

because -1 is not greater than or equal to 0.

EXERCISES FOR SECTION 11.1

(Note. Solutions are given for those exercises marked with an asterisk.)

1. Write equations similar to (11.3) to (11.6) if the protein and vitamin C requirements are:

	Protein	Vitamin C
*(a)	70	60
(b)	100	40
(c)	0	75

2. Find at least one solution for the system of inequalities developed in Exercise 1.

*3. Where possible write an inequality (i.e., $\geq$ or $\leq$) between each pair of vectors below.

(a) $\begin{pmatrix} 6 \\ 1 \\ 6 \end{pmatrix}$ $\begin{pmatrix} 1 \\ 1 \\ 1 \end{pmatrix}$

(b) $\begin{pmatrix} 2 \\ 1 \\ 0 \end{pmatrix}$ $\begin{pmatrix} 0 \\ 0 \\ 0 \end{pmatrix}$

(c) $\begin{pmatrix} 1 \\ 1 \\ 1 \end{pmatrix}$ $\begin{pmatrix} 1 \\ 2 \\ 3 \end{pmatrix}$

(d) $\begin{pmatrix} 1 \\ 2 \\ 3 \end{pmatrix}$ $\begin{pmatrix} 3 \\ 2 \\ 1 \end{pmatrix}$

4. Same as Exercise 3 except use

(a) $\begin{pmatrix} 0 \\ 0 \\ 0 \end{pmatrix}$ $\begin{pmatrix} 1 \\ 2 \\ 3 \end{pmatrix}$

(b) $\begin{pmatrix} -1 \\ -2 \\ -3 \end{pmatrix}$ $\begin{pmatrix} 1 \\ 2 \\ 3 \end{pmatrix}$

(c) $\begin{pmatrix} 3 \\ 2 \\ 1 \end{pmatrix}$ $\begin{pmatrix} -3 \\ -2 \\ -1 \end{pmatrix}$

(d) $\begin{pmatrix} 0 \\ 1 \\ 0 \end{pmatrix}$ $\begin{pmatrix} 1 \\ 0 \\ 1 \end{pmatrix}$

*5. Suppose that instead of minimizing the number of calories in the diet you wish to minimize the cost of the diet. If roast beef costs 25c per 100 grams and lettuce costs 10c per 100 grams, what expression replaces (11.7)?

6. Same as Exercise 5 except that roast beef costs 30c per 100 grams and lettuce costs 15c per 100 grams.

7. Instead of minimizing the number of calories in the diet suppose, you wish to minimize the calcium content. Using Figure 10.2, find the expression that replaces (11.7).

11.2 A NUT MIX PROBLEM

Suppose you have supplies of four kinds of nuts—cashews, peanuts, hazelnuts, and Brazil nuts—and you want to produce two mixtures of these nuts, a party mix and a regular mix. You would like to know how much of each mix to prepare in order to make the largest possible profit.

Given unlimited supplies and an unlimited demand for the two mixtures, you would naturally produce all you could of only one mix—the one with the largest profit. We will assume that many other stores sell nut mixtures similar to ours. Hence at a given price we can sell as many pounds of each mixture as we can produce. However, our supplies are limited. The supply limitations together with the composition of the two mixes and the costs of the various nuts are given in Figure 11.1.

	Cashews	Peanuts	Hazelnuts	Brazils
One pound of party mix contains	6 oz	0 oz	4 oz	6 oz
One pound of regular mix contains	0 oz	8 oz	6 oz	2 oz
Cost per pound	24c	16c	16c	32c
Supply available	75 lb	150 lb	125 lb	87.5 lb

FIGURE 11.1 Table summarizing supplies and costs of nuts in two mixes.

We will assume that we can sell the party mix for $1.25 per pound and the regular mix for 50c per pound. With this information we can solve the problem of maximizing (making as large as possible) our profit.

We let x_1 be the number of pounds of party mix to be made, and we let x_2 be the number of pounds of regular mix. Let

$$\mathbf{x} = \begin{pmatrix} x_1 \\ x_2 \end{pmatrix}$$

Then the number of *ounces* of cashews used will be

$$\mathbf{k}^T \mathbf{x}$$

where

$$\mathbf{k} = \begin{pmatrix} 6 \\ 0 \end{pmatrix}$$

and $\mathbf{k}$ is just the first two entries in the cashew column of Figure 11.1. This cannot exceed the supply of cashews, which is 75 pounds = 1200 ounces, so

$$\mathbf{k}^T \mathbf{x} \leqslant 1200 \qquad (11.11)$$

Similarly, from the peanut supply limitation, we obtain

$$\mathbf{p}^T \mathbf{x} \leqslant 2400 \qquad (11.12)$$

where

$$\mathbf{p} = \begin{pmatrix} 0 \\ 8 \end{pmatrix}$$

We could continue to write similar inequalities for hazelnuts and Brazil nuts. However, each vector for each nut supply will be a column of the table in Figure 11.1; each is multiplied by the same vector **x**, and the right side is obtained from the last row of Figure 11.1. Hence we can use matrix notation to abbreviate and condense our inequalities. We let the matrix A be the first two rows of Figure 11.1; that is,

$$A = \begin{pmatrix} 6 & 0 & 4 & 6 \\ 0 & 8 & 6 & 2 \end{pmatrix} \tag{11.13}$$

We let the row vector, $\mathbf{b}^T$, be the last row of Figure 11.1 multiplied by 16 to convert the supplies to ounces.

$$\mathbf{b}^T = (1200, 2400, 2000, 1400) \tag{11.14}$$

Then the four supply limitations can be written

$$A^T\mathbf{x} \leqslant \mathbf{b} \tag{11.15}$$

This matrix inequality includes both (11.11) and (11.12). Written out, (11.15) becomes

$$\left. \begin{aligned} 6x_1 && \leqslant\ 1200 \\ 8x_2 &\leqslant\ 2400 \\ 4x_1 + 6x_2 &\leqslant\ 2000 \\ 6x_1 + 2x_2 &\leqslant\ 1400 \end{aligned} \right\} \tag{11.16}$$

Of course, the amount of each mix made must be a positive number or zero, so

$$\mathbf{x} \geqslant \mathbf{0} \tag{11.17}$$

Moreover, (11.15) and (11.17) are the only restrictions we need to satisfy.

What is it that we wish to maximize? The profit on each pound of either mix is the selling price less the cost of the ingredients.[2] The costs (in cents) of one pound of each of the two mixes are given by

$$\frac{1}{16}A\mathbf{q} \tag{11.18}$$

where $\mathbf{q}^T$ is the third row (cost per pound) in Figure 11.1; that is,

---

[2] We are ignoring other costs such as packaging, advertising, store rental, and so forth in the interest of simplicity. They could be included with little change in approach to the problem, but with considerable added complexity in the details of the analysis.

$$\mathbf{q}^T = (24, 16, 16, 32)$$

The scalar $1/16$ appears in (11.18) because the entries in A are in ounces, while the costs in $\mathbf{q}$ are in pounds (16 ounces = 1 pound). In any case,

$$\frac{1}{16}A\mathbf{q} = \begin{pmatrix} 25 \\ 18 \end{pmatrix}$$

The profit per pound is

$$\mathbf{c} = \begin{pmatrix} 125 \\ 50 \end{pmatrix} - \begin{pmatrix} 25 \\ 18 \end{pmatrix}$$

or

$$\mathbf{c} = \begin{pmatrix} 100 \\ 32 \end{pmatrix} \tag{11.19}$$

We wish to

$$\text{Maximize } \mathbf{c}^T\mathbf{x} \tag{11.20}$$

subject to

$$A^T\mathbf{x} \leqslant \mathbf{b} \tag{11.15}$$

$$\mathbf{x} \geqslant \mathbf{0} \tag{11.17}$$

where $\mathbf{c}$, A, and $\mathbf{b}$ are given by (11.19), (11.13), and (11.14), respectively, and $\mathbf{0}$ is a vector all of whose components are zero. Notice the similarities (and the differences) between this problem and the diet problem given in (11.8), (11.9) and (11.10).

How are we to find the values of $\mathbf{x}$ (i.e., the values of x_1 and x_2) that maximize (11.20)? It seems likely that we should try to use as much of all of our supplies as possible, but can we? Suppose we try to use all of the cashews and all of the peanuts. Then the first two inequalities in (11.16) are equalities; that is,

$$\begin{aligned} 6x_1 \quad &= 1200 \\ 8x_2 &= 2400 \end{aligned}$$

so $x_1 = 200$ and $x_2 = 300$. But what of the other inequalities? If we examine the third inequality in (11.16), we find that

$$4x_1 + 6x_2 = 4 \times 200 + 6 \times 300 = 2600$$

and this exceeds 2000, so the third inequality is not satisfied (i.e., we would need more hazelnuts than we have).

If we decide to try to use all of the peanuts and hazelnuts,

$$8x_2 = 2400$$
$$4x_1 + 6x_2 = 2000$$

so $x_1 = 50$ and $x_2 = 300$. With these values of x_1 and x_2 all four inequalities are satisfied. The profit from (11.20) is 14600, or \$146. But is this the best we can do?

We could continue to try exhausting other combinations of supplies, but that would be tedious and, moreover, how would we ever know if we could do better? In the next section we will use graphs of inequalities to determine the solution of a general linear programming problem. In Section 11.4 we will return to the nut mix problem and use these graphical methods to maximize our profit.

EXERCISES FOR SECTION 11.2

1. Determine the values of x_1 and x_2 by exhausting the supplies of:
   *(a) Cashews and hazels.
   (b) Cashews and Brazils.
   (c) Peanuts and Brazils.
   (d) Hazels and Brazils.

2. Which, if any, of the supplies of nuts are exceeded in the four parts of Exercise 1?

3. What are the profits in the four parts of Exercise 1?

4. Suppose we replace Figure 11.1 with:

	Cashews	Peanuts	Hazels	Brazils
One pound of party mix contains	8	0	3	5
One pound of regular mix contains	0	10	5	1
Cost per pound	48c	16c	32c	48c
Supply available	40 lb	70 lb	90 lb	35 lb

Suppose also that the party mix sells for \$1.40 per pound and the regular mix sells for 65 cents per pound. Find the matrix A and the vectors c^T and b for (11.15) and (11.20).

5. You are a manufacturer producing two products. Each unit of the first product requires 2 hours of labor and \$2 worth of raw materials. Each unit of the second only requires 1 hour of labor but uses \$3 worth of raw materials. Both products sell for \$10 per unit. The labor supply is 100 hours, and there is a supply of \$180 of raw material. You wish to maximize the gross income (i.e., total sales dollars). Express this as a linear programming problem. In particular, find A, c^T, and b for (11.15) and (11.20).

6. A truck driver is to deliver some heavy cartons for a customer. His truck can carry 1200 cubic feet of material and up to 15,000 pounds of weight. The customer has two sizes of cartons.

The first weighs 20 pounds and is 1 cubic foot. The second weighs 60 pounds and is 5 cubic feet. He is paid 40 cents for each carton of the first type which he delivers, and he is paid 60 cents for each carton of the second type. He wishes to maximize his income. Express this as a linear programming problem $\left[\text{i.e., find } A, \mathbf{b}, \text{ and } \mathbf{c}^T \text{ for (11.15) and (11.20)}\right]$.

11.3 GRAPHICAL SOLUTION OF A LINEAR PROGRAMMING PROBLEM

Consider the following linear programming problem.

$$\text{Maximize } x_1 + x_2 \qquad (11.21)$$

where

$$2x_1 + x_2 \leqslant 2 \qquad (11.22)$$

$$x_1 + 3x_2 \leqslant 3 \qquad (11.23)$$

$$x_1 \geqslant 0 \qquad (11.24)$$

$$x_2 \geqslant 0 \qquad (11.25)$$

In matrix form this is

$$\text{Maximize } c^T \mathbf{x}$$

where

$$A\mathbf{x} \leqslant \mathbf{b}$$

$$\mathbf{x} \geqslant \mathbf{0}$$

and

$$c = \begin{pmatrix} 1 \\ 1 \end{pmatrix} \qquad \mathbf{b} = \begin{pmatrix} 2 \\ 3 \end{pmatrix} \qquad \mathbf{x} = \begin{pmatrix} x_1 \\ x_2 \end{pmatrix}$$

$$A = \begin{pmatrix} 2 & 1 \\ 1 & 3 \end{pmatrix}$$

Suppose we try to graph the inequality (11.22). We first replace the inequality by the equality

$$2x_1 + x_2 = 2$$

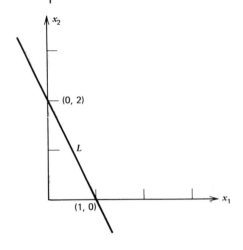

FIGURE 11.2 Graph showing line L: $2x_1 + x_2 = 2$.

and graph the equation. It is line L shown in Figure 11.2. All values of x_1 and x_2 satisfying this equation lie on line L. But where do the values of x_1 and x_2 that satisfy the inequality (11.22) lie?

Some of them lie on Line L, since those values on L certainly satisfy (11.22), but there are more. For example, the point $x_1 = 0$ and $x_2 = 0$ satisfies (11.22), as does $x_1 = 0$ and $x_2 = 1$. Indeed, it is easy to verify that *any* point lying to the *left* of line L produces values of x_1 and x_2 that satisfy (11.22). Therefore the points that satisfy (11.22) lie not on a line but in an area that includes line L and all of the points to its left. That area is shown as the shaded area in Figure 11.3. Certainly, then, any values of x_1 and x_2 that are allowable values in our linear programming problem lie in that shaded area in Figure 11.3. But we need to observe some other restrictions as well. In

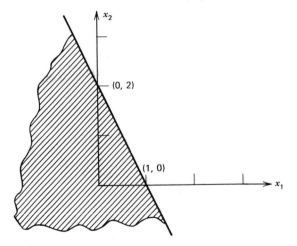

FIGURE 11.3 Shaded area that corresponds to inequality (11.22).

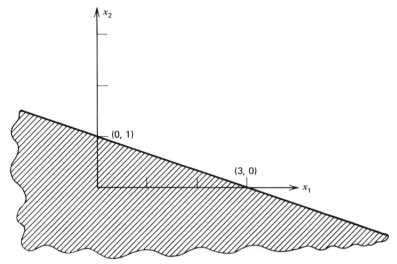

FIGURE 11.4 Shaded area that corresponds to inequality (11.23).

particular we need to observe the inequality (11.23). The values of x_1 and x_2 that satisfy (11.23) correspond to the shaded area in Figure 11.4. Since both (11.22) and (11.23) must be satisfied, we can allow only those points in both of the shaded areas (Figures 11.3 and 11.4). Those points are shown as a shaded area in Figure 11.5. The area in Figure 11.5 is the intersection of the areas in the two previous figures. All of the points that satisfy (11.22) and (11.23) as well lie in this shaded area.

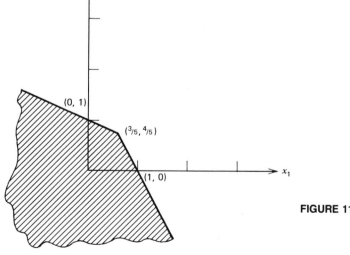

FIGURE 11.5 Shaded area that corresponds to both (11.22) and (11.23).

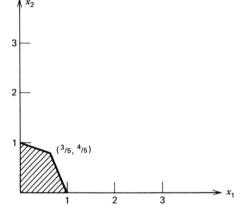

FIGURE 11.6 Shaded area that corresponds to (11.22) to (11.25).

There also are two other inequalities that must be satisfied [i.e., (11.24) and (11.25)]. In Figure 11.6 the shaded area corresponds to those values of x_1 and x_2 that satisfy all four inequalities: (11.22), (11.23), (11.24) and (11.25).

If we select any point within the shaded area in Figure 11.6, the corresponding values of x_1 and x_2 are allowable in our linear programming problem (i.e., they satisfy all of the inequalities). There are an infinite number of such points. Out of this infinite number we wish to select the one or ones that maximize $x_1 + x_2$ [see (11.21)]. We proceed to that now.

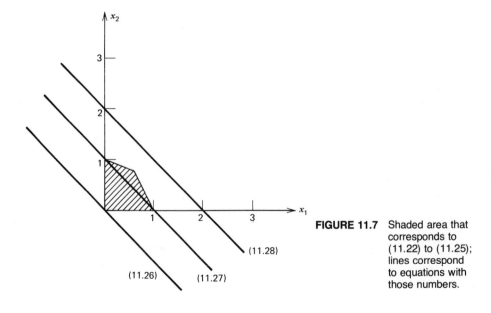

FIGURE 11.7 Shaded area that corresponds to (11.22) to (11.25); lines correspond to equations with those numbers.

Consider the three lines:

$$x_1 + x_2 = 0 \qquad (11.26)$$

$$x_1 + x_2 = 1 \qquad (11.27)$$

$$x_1 + x_2 = 2 \qquad (11.28)$$

They are shown in Figure 11.7, where the shaded area in Figure 11.6 also is reproduced. The lines are labeled with the corresponding equation numbers. Notice that the three lines are parallel to each other and, that as the right side of the equation increases from 0 to 1 to 2, the line moves to the right.

Any point on the left-most line corresponds to values of x_1 and x_2, that sum to zero. Similarly, any point on the middle line has x_1 and x_2 values that sum to one. In terms of maximizing $x_1 + x_2$, line (11.27) is better than (11.26), but (11.28) is better than either of the other two. If we want $x_1 + x_2$ to be equal to 2, we can choose any point on the right-most line.

Since we want $x_1 + x_2$ to be as large as possible, we want the right side, a, of

$$x_1 + x_2 = a$$

to be as large as possible. Hence we wish to draw a line parallel to those in Figure 11.7 and to position the line as far to the right as possible.

What is to keep us from moving the line to the right indefinitely? We must have values of x_1 and x_2 that satisfy the four inequalities (11.22) to (11.25), and this will force us to stop moving to the right.

Line (11.27) has many points lying within the shaded area. Some are $x_1 = 1, x_2 = 0$ and $x_1 = 1/2, x_2 = 1/2$ and $x_1 = 0, x_2 = 1$. Any one of these satisfies the four inequalities and any one produces a value of one for $x_1 + x_2$. On the other hand, (11.28) has no points lying within the shaded area. Since all points for which $x_1 + x_2 = 2$ lie on (11.28), there are no points within the shaded area that produce a value of two for $x_1 + x_2$. Therefore the largest value of $x_1 + x_2$ in our linear programming problem must be less than two.

Of all of the lines parallel to those in Figure 11.7, the only allowable ones must have at least one point lying within the shaded area. Since we want $x_1 + x_2$ to be as large as possible, we seek the line that is as far to the right as possible but that still touches the shaded area. That is line M in Figure 11.8. It touches the shaded area at only one point, the corner where $x_1 = 3/5$ and $x_2 = 4/5$. This is the solution to the linear programming problem.

In general we can solve a linear programming problem with only two unknown quantities as follows.

1. Draw the lines that result when each of the inequalities is written as an equation.
2. Find the area (either to the left or right)[3] of each line that satisfies the corresponding inequality.

[3]If the line is horizontal the area will be above or below the line.

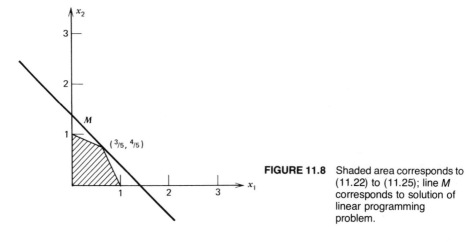

FIGURE 11.8 Shaded area corresponds to (11.22) to (11.25); line *M* corresponds to solution of linear programming problem.

3. Find the area common to all of the areas found in step 2.
4. Assume an arbitrary value of the expression to be maximized and draw the line that results from this assumption.
5. Determine whether moving the line in step 4 to the right or left (or up or down) increases the value of the expression.
6. Move the line in step 4 parallel to itself in the direction indicated in step 5 until it barely touches the area found in step 3.
7. The point (or points) on the line found in step 6 and that lie in the area in step 3 is (are) a solution to the linear programming problem.

It should be clear that we will only stop moving the line as described in step 6 at a corner of the area. Hence solutions of linear programming problems always occur at corners of the areas.[4]

Of course, the procedure described above can only be carried out if there are only two unknown quantities. If there were more, we would not be able to draw the graphs. However, the general results at which we arrived still hold when there are three, four, or more unknowns. That is, we need only examine the corners in looking for the "best" solution. Later we will use a BASIC program to find solutions to problems with more than two unknowns.

EXERCISES FOR SECTION 11.3

1. Draw a graph showing the area representing the values of x_1 and x_2 that satisfy the following inequalities.

   *(a) $x_1 + x_2 \leqslant 1$
   $x_1 \geqslant 0$
   $x_2 \geqslant 0$

---

[4]They may occur in other places as well, as we will see but; if there is a solution, or solutions, at least one of the solutions occurs at a corner.

(b) $x_1 - 10x_2 \leqslant 0$
$\quad x_1 - 5x_2 \leqslant 5$
$\quad x_1 - x_2 \leqslant 13$
$\quad x_1 \leqslant 20$
$\quad\quad x_1 \geqslant 0$
$\quad\quad x_2 \geqslant 0$

(c) $x_1 - x_2 \leqslant 1$
$\quad -x_1 + x_2 \leqslant 1$
$\quad x_1 \leqslant 2$
$\quad 2x_1 + 2x_2 \leqslant 7$
$\quad\quad x_1 \geqslant 0$
$\quad\quad x_2 \geqslant 0$

(d) $x_1 + x_2 \leqslant 3$
$\quad x_1 + 5x_2 \leqslant 10$
$\quad 2x_1 \leqslant 3$
$\quad 16x_1 + 5x_2 \leqslant 40$
$\quad\quad x_1 \geqslant 0$
$\quad\quad x_2 \geqslant 0$

2. Use the inequalities in the corresponding parts of Exercise 1, find the values of x_1 and x_2 that:
  *(a) Maximize $x_1 + 2x_2$.
  (b) Maximize $x_1 + x_2$.
  (c) Maximize $3x_1 - x_2$.
  (d) Maximize $2x_1 + x_2$.

*3. (a) Draw a graph showing the allowable values for B and L in the diet problem (11.3) to (11.6).
  (b) Find the values of B and L that minimize the calories in the diet $\left[\text{see }(11.7)\right]$.

4. (a) Draw a graph showing the allowable values for x_1 and x_2 in the nut mix problem $\left[\text{see } (11.16) \text{ and } (11.17)\right]$.
  (b) Find the values of x_1 and x_2 that maximize the profit $\left[\text{see } (11.19) \text{ and } (11.20)\right]$.

5. Using the information in Exercise 5 of Section 11.2:
  (a) Draw a graph showing the allowable values for x_1 and x_2.
  (b) Find the values of x_1 and x_2 that maximize the gross income.

6. Using the information in Exercise 6 of Section 11.2:
  (a) Draw a graph showing the allowable values for x_1 and x_2.
  (b) Find the values of x_1 and x_2 that maximize the trucker's income.

11.4 THE NUT MIX PROBLEM REVISITED

We return now to the nut mix problem discussed in Section 11.2 and find its solution. The first inequality in (11.16) requires that

$$x_1 \leqslant 200$$

The second inequality requires that

$$x_2 \leq 300$$

Proceeding in this way, the six inequalities given in (11.16) and (11.17) require that the allowable values of x_1 and x_2 lie in the shaded region in Figure 11.9.

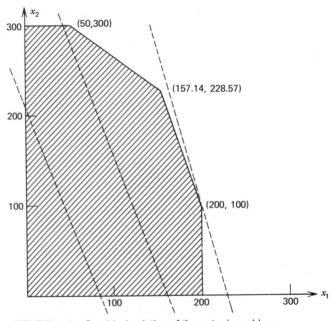

FIGURE 11.9 Graphical solution of the nut mix problem.

The dashed lines represent three different values of the profit. The profit increases as the lines move to the right. The right-most line touches the shaded area at one and only one point: $x_1 = 200$ and $x_2 = 100$. Hence those values represent the amount of the two mixes that produce the largest profit. If we make 200 pounds of Mix 1 (the party mix) and 100 pounds of Mix 2 (the regular mix), the profit is

$$100 \times 200 + 32 \times 100$$

or $232. This is the largest possible profit.

EXERCISES FOR SECTION 11.4

1. How much of each mix should be made and what is the maximum profit if the selling price of the party mix is changed to:
   *(a) $0.75?
   (b) $1?
   (c) $1.50?

2. Same as Exercise 1 except the selling price of the regular mix is changed to:
   *(a) $0.40
   (b) $0.60
   (c) $1.00

3. Same as Exercise 1 except the supply of cashews is:
   *(a) 40 lb
   (b) 60 lb
   (c) 80 lb

4. Same as Exercise 1 except that there is an arbitrarily large supply of:
   *(a) Cashews.
   (b) Brazil nuts.
   (c) Hazelnuts.

11.5 TYPES OF SOLUTIONS AND SOME DEFINITIONS

When we studied systems of linear equations, we found that there were only three different types of solutions.

1. One and only one solution.
2. No solution.
3. An infinite number of solutions.

This is true of solutions of linear programming problems as well. However, in the case of an infinite number of solutions, we may have an infinite maximum value or we may have a finite one. We will find it useful to distinguish these two cases.

First, we introduce some terminology that is quite standard in linear programming and that you are likely to encounter if you read further about this subject.

The expression that we wish to maximize (or minimize) is called the *objective*. For example, in the nut mix problem, the objective was the profit; that is,

$$100x_1 + 32x_2$$

The inequalities that must be satisfied are called the *constraints*. In the nut mix problem the constraints are given by (11.16) and (11.17).

In Section 11.3 we considered the problem of maximizing the objective,

$$x_1 + x_2 \tag{11.21}$$

subject to the four constraints:

$$2x_1 + x_2 \leq 2 \tag{11.22}$$

$$x_1 + 3x_2 \leq 3 \tag{11.23}$$

$$x_1 \geq 0 \tag{11.24}$$

$$x_2 \geq 0 \tag{11.25}$$

The maximum value of the objective was found to be 7/5, and it occurred when $x_1 = 3/5$ and $x_2 = 4/5$. There were no other values of x_1 and x_2 that produced a value of 7/5 for the objective. Hence there was one and only one solution.

Consider now the problem

$$\text{Maximize } x_1 + 3x_2 \qquad (11.29)$$

where

$$x_1 + x_2 \leqslant -1 \qquad (11.30)$$

$$x_1 \geqslant 0 \qquad (11.31)$$

$$x_2 \geqslant 0 \qquad (11.32)$$

There are no values of x_1 and x_2 that satisfy all three constraints. The last two tell us that the smallest x_1 and x_2 can be is zero. But then $x_1 + x_2 = 0$, which exceeds -1 [see (11.30)]. Here, then, is an example of a linear programming problem that has no solution.

We next turn our attention to

$$\text{Maximize } x_1 + 3x_2 \qquad (11.29)$$

subject to the constraints (11.22) to (11.25) in the first problem in this section. The allowable values of x_1 and x_2 are shown in Figure 11.10, where the dashed lines indicate different values of the objective. The objective increases as the dashed lines move up. The upper-most line represents the maximum solution. Notice that that line touches not one corner but two. Moreover, it touches the entire line segment joining the two corners. The maximum value of the objective (11.29) is 3. That value is achieved if $x_1 = 0$ and $x_2 = 1$. It also is achieved if $x_1 = 3/5$ and $x_2 = 4/5$. These are the two

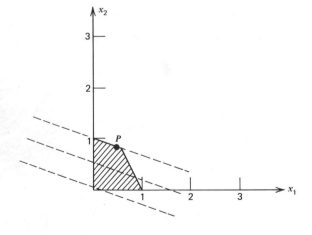

FIGURE 11.10 Linear programming problem with infinite number of solutions, bounded objective.

corners. However, the objective also equals 3 when $x_1 = 1/2$ and $x_2 = 5/6$. This is point P in Figure 11.10, and it lies on the line segment joining the two corners. This, then, is an example of a linear programming problem with an infinite number of solutions. However, the objective is finite, in this case 3. Hence we refer to this as a *bounded* solution, since the objective cannot get larger than some bound.

It is not always the case that the objective is bounded. Consider

$$\text{Maximize } 4x_1 + x_2$$

where

$$x_1 + x_2 \geqslant 1$$
$$x_1 \geqslant 0$$
$$x_2 \geqslant 0$$

The constraints define the shaded region shown in Figure 11.11. Once again the dashed lines represent the objective, and the values increase as the lines move to the right. But now we can continue moving the lines to the right indefinitely, since the constraints do not stop us. The value of the objective can be made arbitrarily large, so the solution is said to be *unbounded*. Lest the reader think that an unbounded constraint region means an unbounded solution, we consider these same constraints, but we

$$\text{Minimize } 4x_1 + x_2$$

The same dashed lines in Figure 11.11 can be used. The objective gets smaller as the lines move to the left. Hence the left-most line represents the smallest value of the objective, which is 1. It occurs when $x_1 = 0$ and $x_2 = 1$.

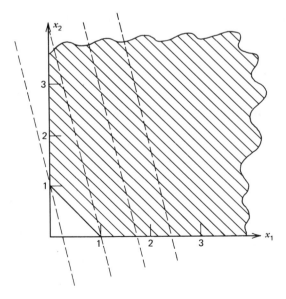

FIGURE 11.11 Linear programming problem with an unbounded solution.

EXERCISES FOR SECTION 11.5

1. Classify each of the following linear programming problems as having one and only one solution, no solution, an infinite number of solutions with a bounded objective, or an unbounded objective.

*(a) Maximize $x_1 + x_2$

$x_1 \leqslant 1$

$x_2 \leqslant 1$

(b) Minimize $x_1 + x_2$

$x_1 \leqslant 1$

$x_2 \leqslant 0$

(c) Minimize $x_1 + x_2$

$0 \leqslant x_1 \leqslant 1$

$0 \leqslant x_2 \leqslant 1$

*(d) Maximize $x_1 + x_2$

$x_1 + x_2 \geqslant 1$

$-x_1 - x_2 \geqslant -1$

$x_1 \geqslant 0$

$x_2 \geqslant 0$

(e) Minimize $x_1 + x_2$

$x_1 + x_2 \geqslant 1$

$-x_1 - x_2 \geqslant 0$

$x_1 \geqslant 0$

$x_2 \geqslant 0$

(f) Minimize $x_1 + x_2$

$x_1 + x_2 \leqslant 0$

$x_1 \geqslant 0$

$x_2 \geqslant 0$

*2. Suppose that in the nut mix problem of Sections 11.2 and 11.4 we must deliver 250 pounds of the party mix to a preferred customer. What type of solution does the problem have?

3. Suppose in the nut mix problem of Sections 11.2 and 11.4 that the selling price of the party mix is $1.21. What type of solution does the problem have?

11.6 A BASIC PROGRAM FOR LINEAR PROGRAMMING

As we noted at the close of Section 11.3, the use of graphs to solve linear programming problems is limited to those problems with only two unknowns. If, for example, we had a nut mix problem where there were four different mixes, we would find our graphical methods of no use. Fortunately, there are algebraic methods for solving linear programming problems, and these algebraic methods lend themselves to computer programs. Both the algebraic methods and the corresponding programs are complex, so we will not go into their details here. Instead, we present a BASIC program to solve linear programming problems and will give no explanation of its details. We merely will describe how you can use the BASIC program to solve linear programming problems. This is much in the same spirit as we used in INV statement in Section 10.6 to find the inverse of a matrix.

This use of a computer program without knowledge of its inner workings often is referred to as the use of a "canned program." It is as if someone has prepared and packaged a program for our use. A great deal of actual computer use involves such canned programs. Why, after all, should we go to the trouble to write a program if someone else has already done the job for us? Every computer center has a battery of canned programs to do most standard tasks—statistical tests, sorting names into alphabetical order, and so on. It is important, indeed essential, that such canned programs are accompanied by written documentation that: (1) clearly states the problem (or problems) that the program solves, (2) describes exactly how the input data is to be provided to the program, and (3) describes how to read and interpret the printed output.

The BASIC program for linear programming problems is given in Appendix D. It finds a solution (if one exists) to

$$\text{Maximize } \mathbf{c}^T\mathbf{x} \qquad (11.33)$$

where

$$A\mathbf{x} \leq \mathbf{b} \qquad (11.34)$$

$$\mathbf{x} \geq 0 \qquad (11.35)$$

and where

$$\mathbf{b} \geq 0 \qquad (11.36)$$

The vectors $\mathbf{c}$ and $\mathbf{b}$ and the matrix A must be specified by the user of the program. Notice that *all* components of $\mathbf{b}$ must be nonnegative, see (11.36). The program produces a vector $\mathbf{x}$ that maximizes (11.33).

The nut mix problem falls into this category of problems [see (11.15), (11.17) and (11.20)]. However, the diet problem given in Section 11.1 does not fit these criteria. It is possible to write a BASIC program that solves more general problems, including the diet problem, but we will not do so. Most computing centers have canned programs that solve the most general linear programming problem.

One advantage of the restriction (11.36) is that it guarantees that there is always at least one solution. Indeed,

$$\mathbf{x} = 0$$

satisfies all of the constraints [i.e., (11.34) and (11.35)]. Therefore the case of "no solution" as described in Section 11.5 does not arise. If there is one and only one solution, the BASIC program finds it. If there are an infinite number of solutions, there are two possibilities. The objective could be bounded. In this case the BASIC program finds one solution from among the infinite number. On the other hand, the objective could be unbounded. In this case the BASIC program prints "SOLUTION IS UNBOUNDED".

The values of A, $\mathbf{b}$, and $\mathbf{c}$ must be given in DATA statements. In addition, the number of rows and columns in each of these are specified in a DIM statement *and* in a DATA statement. At this point the reader should refer to Appendix D. The following data must be provided in the statements whose numbers are shown on the left.

100   *Number of rows and columns in matrix A* as follows:

        A (number of rows, number of columns in A)

        B (number of rows in A)

        T (number of rows in A)

        S (number of columns in A)

        C (number of columns in A)

In Appendix D statement 100 is shown for the case where A is 4×2.

400 to   *Components of A, row by row*
700
1000 to *Components of* $\mathbf{b}$
1300
1600   *Components of* $\mathbf{c}$
1800   Number of rows in A, number of columns in A, in that order

No other data are required.

For example, to solve the nut mix problem given by (11.15), (11.17) and (11.20) where A, $\mathbf{b}$, and $\mathbf{c}$ are given by (11.13), (11.14), and (11.19), we note that A is 4×2 [this A is the transpose of the A given in (11.13)—see (11.15)], so we use

```
100 DIM A(4,2), B(4), T(4), S(2), C(2)
```

Then we use

```
400 DATA 6,0
500 DATA 0,8
600 DATA 4,6
700 DATA 6,2
```

[See (11.13) and (11.15).] And

```
1000 DATA 1200
1100 DATA 2400
1200 DATA 2000
1300 DATA 1400
```

[See (11.14).] And

```
1600 DATA 100,32
```

$\left[\text{See (11.19)}.\right]$ And, finally,

$$1900 \text{ DATA } 4,2$$

since A is 4×2.

These DIM and DATA statements are precisely those shown in Appendix D. If we run this program with these data the printed output is

$$X(\ 1\ ) = 200$$
$$X(\ 2\ ) = 100$$

indicating that

$$x_1 = 200$$

$$x_2 = 100$$

The printed output will include the names of all of the unknowns and their values— provided the value is not zero. If the name of an unknown is missing, the value of that unknown is zero. Of course, if the objective is unbounded, no values are printed at all. Instead the words "SOLUTION IS UNBOUNDED" are printed.

In the next section we will use this BASIC program to solve an enlarged nut mix problem.

EXERCISES FOR SECTION 11.6

1. Use the BASIC program in Appendix D to solve the nut mix problems in Exercise 1, Section 11.4.

2. Same as Exercise 1 except use Exercise 2, Section 11.4.

3. Same as Exercise 1 except use Exercise 3, Section 11.4.

4. Same as Exercise 1 except use Exercise 4, Section 11.4.

11.7 CASE STUDY 10: THE NUT MIX PROBLEM ENLARGED

Suppose we wish to make four different nut mixes from five kinds of nuts: cashews, peanuts, hazelnuts, Brazil nuts, and almonds. The makeup of each mix, each mix's selling price, the cost of each type of nut, and the supply of each type of nut are given in Figure 11.12.

If we let $\mathbf{q}$ be a vector whose components are the cost per pound of each nut; that is,

$$\mathbf{q}^T = (24,\ 12,\ 16,\ 30,\ 20)$$

and if A is the 4×5 matrix of the constituents of each mix; that is,

Mix	Selling Price	Number of ounces per pound				
		Cashews	Peanuts	Hazels	Brazils	Almonds
1	$2 per pound	6	2	0	6	2
2	$1.50 per pound	3	4	2	5	2
3	$1.25 per pound	1	8	2	4	1
4	$1 per pound	0	10	3	3	0
Cost per pound		24¢	12¢	16¢	30¢	20¢
Supply		75 lb	150 lb	100 lb	80 lb	75 lb

FIGURE 11.12 Data for four-mix nut mix problem.

$$A = \begin{pmatrix} 6 & 2 & 0 & 6 & 2 \\ 3 & 4 & 2 & 5 & 2 \\ 1 & 8 & 2 & 4 & 1 \\ 0 & 10 & 3 & 3 & 0 \end{pmatrix}$$

then the four component vector given by

$$\frac{1}{16}A\mathbf{q}$$

has components that are the costs of the ingredients in each of the four mixes [see also (11.18) in Section 11.2]. Therefore

$$\mathbf{c} = \begin{pmatrix} 200 \\ 150 \\ 125 \\ 100 \end{pmatrix} - \frac{1}{16}A\mathbf{q} = \begin{pmatrix} 175.75 \\ 128.625 \\ 106.75 \\ 83.875 \end{pmatrix}$$

We wish to

$$\text{Maximize } \mathbf{c}^T\mathbf{x}$$

where

$$A^T\mathbf{x} \leqslant \mathbf{b}$$

$$\mathbf{x} \geqslant 0$$

and

$$\mathbf{b} = \begin{pmatrix} 1200 \\ 2400 \\ 1600 \\ 1280 \\ 1200 \end{pmatrix}$$

We will use the BASIC program in Appendix D to find a solution. To do so, we type the following statements.

```
100 DIM A(5,4),B(5),T(5),S(4),C(4)
400 DATA 6,3,1,0
500 DATA 2,4,8,10
600 DATA 0,2,2,3
650 DATA 6,5,4,3
700 DATA 2,2,1,0
1000 DATA 1200
1100 DATA 2400
1200 DATA 1600
1250 DATA 1280
1300 DATA 1200
1600 DATA 175.75,128.625,106.75,83.875
1900 DATA 5,4
```

The printed output is

```
X( 1     ) = 200
X( 4     ) = 26.6667
```

Thus

$$x_1 = 200$$
$$x_2 = 0$$
$$x_3 = 0$$
$$x_4 = 26.6667$$

and we should make 200 pounds of Mix 1, 26⅔ pounds of Mix 4, and we should not make any of either Mix 2 or Mix 3.
The profit in cents is

$$175.75 \times 200 + 83.875 \times 26.6667$$

or $373.87.
   In the exercises you are asked to re-solve the four-mix nut problem for various other cases.

EXERCISES FOR SECTION 11.7

*1. In the four-mix nut problem, suppose the price of Mix 4 is raised from $1 per pound to $1.50 per pound. All other data remains unchanged. How much of each mix should be made? What is the total profit?

2. Same as Exercise 1 except that the price of Mix 1 is raised to $2.50 per pound, while the price of Mix 4 remains at $1 per pound.

*3. In the four-mix nut problem, suppose the supply of cashew nuts is raised from 75 to 100 pounds. All other data remains fixed. How much of each mix should be made? What is the total profit?

4. Same as Exercise 3 except that the supply of cashew nuts is lowered to 50 pounds and, at the same time, the supply of brazil nuts is increased to 100 pounds.

APPENDIX A
SOLUTIONS TO
SELECTED EXERCISES

This appendix contains solutions to the exercises marked with an asterisk (*) in the text.

SELF-ASSESSMENT EXAMINATION

Part I. Exponents and factoring

1. 6
2. −
3. x^2
4. −, −
5. 4
6. −
7. 6
8. 4
9. 5, 3
10. 4, 3

Part II. Inequalities

11. (a), (c), and (d) are true.
12. (a) <
    (b) <
13. (a), (b), and (d) are true.
14. +
15. −

Part III. Graphs

16.

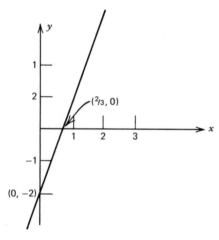

y intercept is −2
x intercept is 2/3

17.

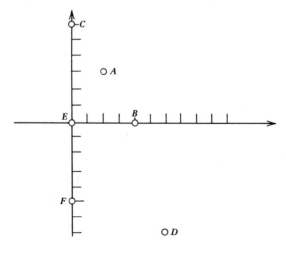

Part IV. Sets

18. $A \cup B = \{2, 3, 4, 5, 6, 7, 8\}$
19. $A \cap B = \{3, 5\}$.
20. $A \cap B \cap D = \{N\}$
21. $B \cap C \cap D = \phi$ (the empty set)

Part V. Venn Diagrams

22.

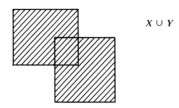

$X \cup Y$

23.

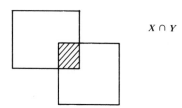

$X \cap Y$

24.

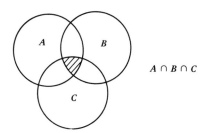

$A \cap B \cap C$

25.

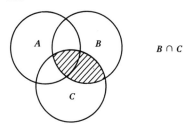

$B \cap C$

26.

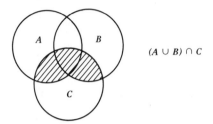

$(A \cup B) \cap C$

CHAPTER I

Section 1.2

1. (a) $L_1 = .7$ $L_6 = .999271$
     $L_2 = .91$ $L_7 = .999781$
     $L_3 = .973$ $L_8 = .999934$
     $L_4 = .9919$ $L_9 = .999980$
     $L_5 = .99757$ $L_{10} = .999994$

   (b) Increasing A increases the rate with which L_k approaches 1. For $A = 0.3$ the amount learned jumps from .92 to .97 in three sessions (see L_7 and L_{10} in Section 1.2). For $A = .7$ the same approximate increase from .91 to .97 occurs in one session (see L_2 and L_3 in part a above).

3. (a) $L_1 = .65$ $L_6 = .941176$
     $L_2 = .755$ $L_7 = .958823$
     $L_3 = .8285$ $L_8 = .971176$
     $L_4 = .87995$ $L_9 = .979823$
     $L_5 = .915965$ $L_{10} = .985876$

   (b) Increasing L_0 advances the amount learned by two sessions but does not change the fundamental pattern. Note the following correspondences.

$L_0 = 0$	$L_0 = 0.5$
$L_3 = .66$	$L_1 = .65$
$L_4 = .76$	$L_2 = .76$
$L_5 = .83$	$L_3 = .83$

and so on. In general, increasing L_0 merely advances the session at which a given amount of learning is achieved. It does not change the rate at which the value 1 is approached.

5. (a) $L_1 = .55$ $L_6 = .73428$
     $L_2 = .595$ $L_7 = .760852$
     $L_3 = .6355$ $L_8 = .784766$
     $L_4 = .67195$ $L_9 = .80629$
     $L_5 = .704755$ $L_{10} = .825661$

   (b) Decreasing A and increasing L_0 does not change the basic pattern of learning.

Section 1.3

2. $T_2 = 5.5$ $T_{11} = 3925.0$
 $T_3 = 11.25$ $T_{12} = 7934.5$
 $T_4 = 23.88$ $T_{13} = 15996.8$
 $T_5 = 50.81$ $T_{14} = 32186.3$
 $T_6 = 107.22$ $T_{15} = 64662.7$
 $T_7 = 223.83$ $T_{16} = 129762$
 $T_8 = 462.74$ $T_{17} = 260180$
 $T_9 = 949.11$ $T_{18} = 521345$
 $T_{10} = 1934.7$ $T_{19} = 1044168$
         $T_{20} = 2090554$

This is an expanding economy. The total national income continues to grow indefinitely.

4. (a) $T_2 = 4$ $T_{14} = 1.96875$
  $T_3 = 2.5$ $T_{15} = 1.992188$
  $T_4 = 1.5$ $T_{16} = 2.007813$
  $T_5 = 1.25$ $T_{17} = 2.011719$
  $T_8 = 1.5$ $T_{18} = 2.007813$
  $T_7 = 1.875$ $T_{19} = 2.001953$
  $T_8 = 2.125$ $T_{20} = 1.998047$
  $T_9 = 2.1875$ $T_{21} = 1.99707$
  $T_{10} = 2.125$ $T_{22} = 1.998047$
  $T_{11} = 2.03125$ $T_{23} = 1.999512$
  $T_{12} = 1.96875$ $T_{24} = 2.000488$
  $T_{13} = 1.953125$ $T_{25} = 2.000732$

(b) The results are similar to those when $T_0 = 2$ and $T_1 = 3$. The oscillations die out more slowly in this case.

(c) In the long run the solution approaches 2.

(d) The long-run behavior does not depend on the choice of T_0 and T_1.

6. Government spending is 1 except for the following periods.

Period	Government Spending
12	2.36192
13	33.82349
26	33.96322
27	20.51553
40	33.96322
41	20.51553

Thereafter, every fourteenth year government spending is 33.96322 and the year following it is 20.51553. The increases do not grow in time; over every 14-year period the total government expenditures will be 66.47875.

8. The national incomes are:

$T_2 = 4$ $T_5 = 4.608$

$T_3 = 4.8$ $T_6 = 2.8672$

$T_4 = 5.12$ $T_7 = -0.49152$

Thus, in period seven, we increase government spending to .49152 so that $T_7 = 0$. In the seventh period the equation is, therefore,

$$T_7 = 2.4T_6 - 1.6T_5 + 0.49152$$

T_8 will be -4.58752 if we use

$$T_8 = 2.4T_7 - 1.6T_6$$

since $T_7 = 0$. Thus we use

$$T_8 = 2.4T_7 - 1.6T_6 + 4.58752$$

which implies that government expenditures are 4.58752 and $T_8 = 0$.

In period nine, then, government spending drops back to zero, so

$$T_9 = 2.4T_8 - 1.6T_7$$

Since $T_8 = T_7 = 0$, it follows that $T_9 = 0$ and all succeeding national incomes are zero.

9. $T_k = AT_{k-1} + ABT_{k-2} - ABT_{k-3} + 1$

10. $0.3T_k + 0.7T_{k-1} = 0$

Section 1.4

1. (a) 27   (b) 60
3. (a) 7   (b) 22
5. (a) $N_k < 3000$ (b) $N_k = 3000$ (c) $N_k > 3000$
7. 12
9. $N_1 = 650$ $N_8 = 1809.003$

$N_2 = 825.5$ $N_9 = 1878.106$

$N_3 = 1019.41$ $N_{10} = 1923.892$

$N_4 = 1219.335$ $N_{11} = 1953.177$

$N_5 = 1409.713$ $N_{12} = 1971.467$

$N_6 = 1576.14$ $N_{13} = 1982.718$

$N_7 = 1709.753$ $N_{14} = 1989.571$

                           $N_{15} = 1993.721$

The population increases steadily and seems to be approaching 2000.

11. $N_1 = 2400$ $N_7 = 2013.853$

$N_2 = 2208$ $N_8 = 2008.274$

$N_3 = 2116.147$ $N_9 = 2004.95$

$N_4 = 2066.99$ $N_{10} = 2002.965$

$N_5 = 2039.297$ $N_{11} = 2001.777$

$N_6 = 2023.269$ $N_{12} = 2001.066$

$N_{13} = 2000.639$ $N_{17} = 2000.083$
$N_{14} = 2000.383$ $N_{18} = 2000.05$
$N_{15} = 2000.23$ $N_{19} = 2000.03$
$N_{16} = 2000.138$ $N_{20} = 2000.018$

The population steadily decreases and seems to be approaching 2000.

12. If $A = -1$, then $1 + A = 0$ so, from (1.36), $N_k = 0$ for $k = 1, 2, 3, \ldots$. Thus the population is exterminated after one period. If $A < -1$, $1 + A < 0$ and $(1 + A)^k < 0$. From (1.36), $N_k < 0$. In this case the population is exterminated in less than one period.

Section 1.5

1.

3.

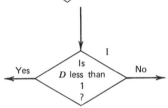

5.

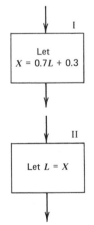

7.

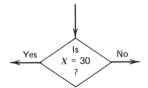

CHAPTER 2

Section 2.2

1. (a) LET K = 12
   (c) LET F9 = 2.1 + 3*12.3
   (e) LET X = 6*L
   (g) LET Z = Z-1

2. In statement 100, the word LET is missing. In statement 200, no arithmetic symbol, presumably an *, appears between the 2 and the A. There is no END statement.

4.
```
100   READ R,S
200   DATA 2,-3.4
300   LET T=R/S
400   PRINT R,S,T
500   END
```

6.
```
350 PRINT "REGULAR", "OVERTIME", "TOTAL"
```

Section 2.3

1. Ask for two numbers as input and print the larger of the two.

3. Take as input one number after another. If the input is positive, the number is printed. Otherwise nothing is printed. The program stops if and only if a value of zero is used as input.

5.

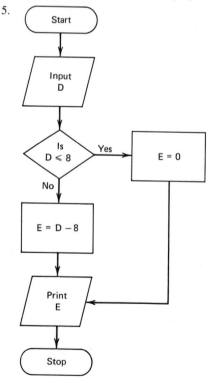

```
10   INPUT D
20   IF D <= 8 THEN 50
30   LET E=D-8
40   GOTO 60
50   LET E=0
60   PRINT E
70   END
```

7.

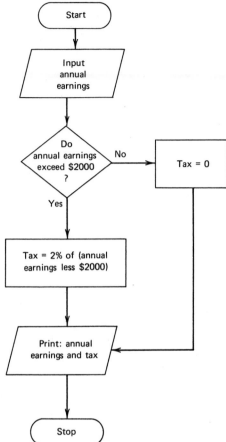

```
10   INPUT A
20   IF A<2000 THEN 50
30   LET T=.02*(A-2000)
40   GØTØ 60
50   LET T=0
60   PRINT "ANNUAL EARNINGS =";A;" TAX = ";T
70   END
```

Section 2.4

1. Print each of the integers from 1 to 15 and with each integer the sum of all of the positive integers less than or equal to that integer.

3.
```
100   FØR K=1 TØ 40
200   PRINT K
300   NEXT K
400   END
```

```
5. 10   PRINT "CENTIGRADE","FAHRENHEIT"
   20   FØR C=0 TØ 100
   30   LET F=1.8*C+32
   40   PRINT C,F
   50   NEXT C
   60   END


7. 100   FØR J=1 TØ 20
   200   PRINT J,J*J
   300   NEXT J
   400   END


9. 10    INPUT A,B
   20    INPUT N
   30    LET K=1
   40    LET X=(1+A-B*N)*N
   50    PRINT K,X
   60    LET N=X
   70    IF K=20 THEN 100
   80    LET K=K+1
   90    GØTØ 40
   100   END
```

or

```
   10    INPUT A,B
   20    INPUT N
   30    LET K=1
   40    LET N=(1+A-B*N)*N
   50    PRINT K,N
   60    IF K=20 THEN 90
   70    LET K=K+1
   80    GØTØ 40
   90    END
```

Section 2.5

```
1. 100 LET S = SQR(2)
```

Another way is 100 LET S = 2↑(1/2)

```
3. 10 FØR N = 1 TØ 40
   20 PRINT N, SQR(N)
   30 NEXT N
   40 END


5. 300 LET X = ABS(X)


7. 600 IF ABS(M-N) < 0.0001 THEN 800


9. 700 LET F = X - INT(X)


11. 600 LET A = 0.1*INT(B*10 + 0.5)
```

13.
```
10 INPUT A,B
20 INPUT N
30 LET X = (1+A-B*N)*N
40 PRINT X
50 IF ABS(X-N) < 1 THEN 80
60 LET N = X
70 GOTO 30
80 END
```

CHAPTER 3

Section 3.1

1. (a) and (c) are difference equations. (e) and (g) are not because only one value of the subscript appears.
2. (a) − (iv); (b) − (ii); (c) − (i);   (d) − (iii).
4. (a) $y_{k+1} = (2/3)y_k - 5/3$
    $M = 2/3, C = -5/3$
   (c) $r_{k+2} = (3/2)r_k$
    $N = 0, M = 3/2, C = 0$

Section 3.2

2. (a), (b), and (f) are solutions.
4. (a) − (iii); (b) − (ii); (c) − (iv); (d) − (i).
6. (b) and (c) are particular solutions.
8. $y_k = k(k + 1)/2$

Section 3.3

2. (a) $S_k = (2/3) - (1/2)^k$
   (b) The long-run solution is 2/3.
4. (a) $R_k = 2/3$.
   (b) The long-run solution is 2/3.
6. (a) $y_k = 1 + 2k$
   (c) $y_k = 2$
   (e) $y_k = \left[1 + 3(-1)^k\right]/2$

   (g) $y_k = \left[2 + 7(1/4)^k\right]/3$
   (i) $y_k = 3 \cdot 2^k - 1$
   (k) $y_k = -2$

Section 3.4

1. (a) Linearly increasing without bound (c) exponentially increasing to a bound; (e) constant; (g) Exponentially decreasing without bound.
2. See Figure A.1. This is not quite exponentially increasing to a bound, since the population increases too rapidly initially. The equation is not linear; therefore it need not follow any of the patterns established for linear equations.

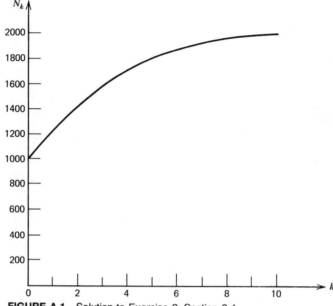

FIGURE A.1   Solution to Exercise 2, Section 3.4.

3. (a) −3
   (c) +1
   (e) 0

Section 3.5

1. (c) and (d) are stable, both with equilibrium values of 6. (k), (l), and (m) also are stable, and each has an equilibrium value of 1.
3. The stable equations and their equilibrium values are:
   (f) 3/4; (g) 1/3; (h) 1/2.
5. $1/(1 - A)$ provided $A \neq 1$. Otherwise, there is no equilibrium.

CHAPTER 4

Section 4.1

1. (a) $200   (b) $240   (c) $280
4. $B_k = A + \dfrac{Ami}{1200}k$

Section 4.2

1. (a) $250, $500, $750, $1000
   (c) $500, $1000, $1500, $2000
   (e) $750, $1500, $2250, $3000

2. 6%

4. $1048.48
6. 3 years
8. 20 years

Section 4.3

1. (a) $1276.28; $1628.89; $2078.93; $2653.30
   (c) $1610.51; $2593.74; $4177.25; $6727.50
   (e) $1280.08; $1638.62; $2097.57; $2685.06
2. 5%
4. $1100
6. 3 years
8. 15 years

Section 4.4

1. $B_{k+1} = 1 + \dfrac{mi}{1200} B_k + D; B_0 = A$

3. (a)
```
10   INPUT A,I,D
15   LET P=400*D/I
20   FØR K=4 TØ 40 STEP 4
30   LET C=(A+P)*(1+I/400)↑K-P
50   PRINT K/4,C
60   NEXT K
70   END
```

   (b)
```
?2000,5.5,50
     1          2316.46
     2          2650.68
     3          3003.67
     4          3376.47
     5          3770.21
     6          4186.05
     7          4625.24
     8          5089.09
     9          5578.99
     10         6096.38
```

   (c) $4000
   (d) $3530.51. To check this result, run the program in part a using $A = 3530.51$, $I = 5.5$, and $D = 0$.

6. (a)
```
10   INPUT A,I
20   FØR K=365 TØ 3650 STEP 365
30   LET C=A*(1+I/36500.)↑K
50   PRINT K/365,C
60   NEXT K
70   END
```

   (b)
```
?2000,5.5
     1          2113.03
     2          2232.44
     3          2358.6
     4          2491.9
     5          2632.72
     6          2781.51
     7          2938.7
     8          3104.77
     9          3280.23
     10         3465.61
```

Daily compounding produces only a modest increase (about 1/3 of 1%) over quarterly compounding.

9. 7% will produce a balance slightly larger than $2A$.

Section 4.5

1. (a) For $A = \$1000, X = \$60.06$
   For $A = \$2000, X = \$120.12$
   For $A = \$4000, X = \$240.23$
   (b) The payment X is directly proportional to the loan A.
3. (a) For $i = 5\%, X = \$60.06$
   For $i = 10\%, X = \$64.76$
   For $i = 15\%, X = \$69.67$
   (b) Increasing i also increases X; however, the increases in X are rather modest compared to the change in i. For example, doubling i results in an increase of less than 7% in X.
5. (a) For $A = \$1000$, total interest $= \$81.08$
   For $A = \$2000$, total interest $= \$162.16$
   For $A = \$4000$, total interest $= \$324.16$
   (b) The total interest is proportional to A.
7. (a) Larger, since the interest is compounded more rapidly; therefore the payment must be larger to account for the increase in interest.
   (b) The period for interest calculation is $m/2$. The interest in the first half of the $(k + 1)$st period is

$$I_1 = B_k \times \frac{i}{100} \times \frac{m}{24}$$

The balance then is $B_k + I_1$, so the interest in the second half of the period is

$$I_2 = (B_k + I_1) \times \frac{i}{100} \times \frac{m}{24}$$

and the balance (excluding any repayment) is $B_k + I_1 + I_2$. Thus

$$B_{k+1} = B_k + I_1 + I_2 - X$$

where X is the payment. Let

$$S = mi/1200$$

and

$$Q = 1 + S/2$$

Then

$$B_{k+1} = Q^2 B_k - X$$

and

$$B_0 = A$$

so

$$B_k = \left(A - \frac{X}{Q^2 - 1}\right) Q^{2k} + \frac{X}{Q^2 - 1}$$

Letting $B_n = 0$,

$$X = \frac{A(1 - Q^2)Q^{2n}}{Q^{2n} - 1}$$

When $Q^2 = 1 + S$, this reduces to (4.13).

9. In (4.13), the values of X, m, n and i are known and A is to be found. Solving (4.13) for A,

$$A = \frac{1200X}{mi} \left[\frac{\left(1 + \frac{mi}{1200}\right)^n - 1}{\left(1 + \frac{mi}{1200}\right)^n}\right]$$

11. (a) $B_{k+1} = B_k - (A - S)/n$; $B_0 = A - S$
    (b) $B_k = (1 - k/n) \cdot (A - S)$
    (c) $1000

12. (a) $B_{k+1} = B_k - \dfrac{n - k}{X}(A - S)$; $B_0 = A - S$

Year	Depreciation	Balance
1	$1667.67	$3333.33
2	133.33	2000
3	1000	1000
4	667.67	333.33
5	333.33	0

13. (a) $B_{k+1} = (1 - 2/n)B_k$; $B_0 = A - S$
    (b) $B_k = (A - S)(1 - 2/n)^k$

Year	Depreciation	Balance
1	$2000	$3000
2	1200	1800
3	720	1080
4	432	648
5	259.20	388.80

Section 4.6

3. (a) 11.1%. (b) 10.7%. (c) For the borrower it is better to have less frequent payments. For the lender more frequent payments are better.
5. (a) 11.1%. (b) 11.1%. (c) The true interest rate is independent of the amount of the loan.

7. From (4.19),

$$A = \frac{1200X}{mi}\left[1 - \left(1 + \frac{mi}{1200}\right)^{n}\right]$$

Using $X = 89.55$, $m = 1$, $i = 9$, and $n = 19$, this produces $A = \$1580.27$.

10. In (4.19) $A = 20{,}000$, $X = 120$, $i = 6$, and $m = 1$. We wish to find n. The BASIC program in Figure 4.11 can be used with the following changes.

```
50      INPUT I
60      PRINT "TYPE PAYMENT AMØUNT"
70      INPUT X
180     PRINT "TYPE GUESS FØR NØ. ØF MØNTHS"
190     INPUT N
210     IF N=0 THEN 260

DELETE 100,110,120,130,140,150,170
```

Using this program we can conclude that 359 payments are too few and 360 are the fewest possible.

13. 22.30%
14. 14.67%

CHAPTER 5

Section 5.1

1. The coefficient of the N_k term is $1 + A$, which is larger than 1. Therefore the solution increases exponentially without bound unless the initial population, N_0, is less than or equal to the constant term divided by $1/\left[1 - (1 + A)\right]$. But the constant term in (5.1) is zero.

4. (a)
```
TYPE VALUE FØR A
? 1.0
TYPE VALUE FØR B
? 0.0001
TYPE INITIAL PØPULATIØN
? 1000
TYPE NØ. ØF PREDICTIØNS
? 10
```

PERIØD	PØPULATIØN
0	1000
1	1900
2	3439
3	5695.33
4	8146.98
5	9656.63
6	9988.21
7	9999.99
8	10000
9	10000
10	10000

(b)

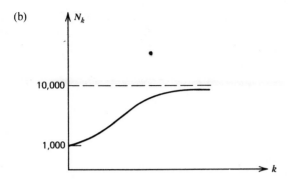

This is quite close to Figure 5.2.

6. (a)
```
TYPE VALUE FØR A
?2.0
TYPE VALUE FØR B
?0.0001
TYPE INITIAL PØPULATIØN
?1000
TYPE NØ. ØF PREDICTIØNS
?20
```

PERIØD	PØPULATIØN
0	1000
1	2900
2	7859
3	17400.6
4	21923.7
5	17706.2
6	21767.6
7	17919.9
8	21647.4
9	18081.2
10	21550.6
11	18208.9
12	21470.3
13	18313.5
14	21402.
15	18401.4
16	21343.1
17	18476.5
18	21291.4
19	18541.9
20	21245.5

(b)

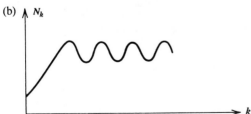

The oscillations are of nearly constant amplitude. However, each peak is slightly lower than the preceding one and each valley is slightly higher than its predecessor. Therefore, the sketch is close to Figure 5.6a.

8. (a) ```
TYPE VALUE FØR A
?3.0
TYPE VALUE FØR B
?0.0001
TYPE INITIAL PØPULATIØN
?1000
TYPE NØ. ØF PREDICTIØNS
?20
```

| PERIØD | PØPULATIØN |
|--------|------------|
| 0 | 1000 |
| 1 | 3900 |
| 2 | 14079 |
| 3 | 36494.2 |
| 4 | 12794.2 |
| 5 | 34807.6 |
| 6 | 18073.4 |
| 7 | 39628.8 |
| 8 | 1470.94 |
| 9 | 5667.4 |
| 10 | 19457.6 |
| 11 | 39970.6 |
| 12 | 117.559 |
| 13 | 468.853 |
| 14 | 1853.43 |
| 15 | 7070.2 |
| 16 | 23282. |
| 17 | 38922.8 |
| 18 | 4192.62 |
| 19 | 15012.7 |
| 20 | 37512.6 |

(b)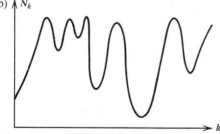

This behavior is unlike any of the others. It is perhaps closest to Figure 5.6$b$, since the population comes perilously close to zero in period 12. This raises the questions: If we allowed $k$ to increase even further, would the population eventually become extinct? If the initial population were less than 1000, would the population become extinct in period 12?

10. (a) ```
TYPE VALUE FØR A
?1.0
TYPE VALUE FØR B
?0.0002
TYPE INITIAL PØPULATIØN
?1000
TYPE NØ. ØF PREDICTIØNS
?10
```

```
PERIØD          PØPULATIØN
  0             1000
  1             1800
  2             2952
  3             4161.14
  4             4859.26
  5             4996.04
  6             5000.
  7             5000.
  8             5000
  9             5000
 10             5000
```

(b)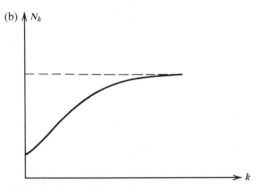

This is similar to Figure 5.2.

12. (a)
```
TYPE VALUE FØR A
?1.5
TYPE VALUE FØR B
?0.0002
TYPE INITIAL PØPULATIØN
?1000
TYPE NØ. ØF PREDICTIØNS
?20
```

```
PERIØD          PØPULATIØN
  0             1000
  1             2300
  2             4692
  3             7327.03
  4             7580.5
  5             7458.45
  6             7520.43
  7             7489.7
  8             7505.13
  9             7497.43
 10             7501.28
 11             7499.36
 12             7500.32
 13             7499.84
 14             7500.08
 15             7499.96
 16             7500.02
 17             7499.99
 18             7500.
 19             7500.
 20             7500
```

(b)

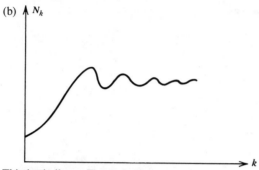

This is similar to Figure 5.6a.

14. (a) ```
TYPE VALUE FØR A
?2.5
TYPE VALUE FØR B
?0.0002
TYPE INITIAL PØPULATIØN
?1000
TYPE NØ. ØF PREDICTIØNS
?20
```

| PERIØD | PØPULATIØN |
|--------|-----------|
| 0 | 1000 |
| 1 | 3300 |
| 2 | 9372 |
| 3 | 15235.1 |
| 4 | 6901.14 |
| 5 | 14628.8 |
| 6 | 8400.34 |
| 7 | 15288. |
| 8 | 6763.29 |
| 9 | 14523.1 |
| 10 | 8646.78 |
| 11 | 15310.4 |
| 12 | 6704.81 |
| 13 | 14475.9 |
| 14 | 8755.22 |
| 15 | 15312.5 |
| 16 | 6699.23 |
| 17 | 14471.4 |
| 18 | 8765.69 |
| 19 | 15312.4 |
| 20 | 6699.35 |

(b)

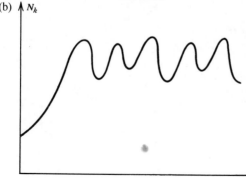
```

This is not similar to any of the figures in the text. The population oscillates with alternately larger and smaller amplitudes. Note that the peaks are successively 15,300, 14,500, 15,300, 14,500, and so on where we have rounded the results to the nearest 100. The same pattern is visible in the troughs.

16. (a)
```
TYPE VALUE FØR A
? 4.0
TYPE VALUE FØR B
? 0.0002
TYPE INITIAL PØPULATIØN
? 1000
TYPE NØ. ØF PREDICTIØNS
? 10
```

PERIØD	PØPULATIØN
0	1000
1	4800
2	19392
3	21750.1
4	14137.3
5	30713.9
6	-35099.
7	-421884.
8	-3.77066E+07
9	-2.84546E+11
10	-1.61933E+19

(b)

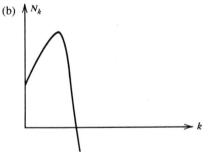

This is similar to Figure 5.6*b*.

17. (a)
```
TYPE VALUE FØR A
? 2.0
TYPE VALUE FØR B
? 0.0001
TYPE INITIAL PØPULATIØN
? 20010
TYPE NØ. ØF PREDICTIØNS
? 10
```

PERIØD	PØPULATIØN
0	20010
1	19990.
2	20010.
3	19990.
4	20010
5	19990.
6	20010.
7	19990.
8	20010
9	19990.
10	20010.

(b) For $A = 2$ the solution appears to oscillate about A/B ($= 20,000$ in this case) with constant amplitude, provided the amplitudes are not too large (see also solution to Exercise 6).

Section 5.2

1. $N = 10,000$. The numerical results in Exercise 4 of Section 5.1 appear to approach this value.

5. (a)
```
TYPE VALUE FØR A
?1.0
TYPE VALUE FØR B
?0.0001
TYPE INITIAL PØPULATIØN
?25000
TYPE NØ. ØF PREDICTIØNS
?5
```

PERIØD	PØPULATIØN
0	25000
1	-12500
2	-40625.
3	-246289.
4	-6.55841E+06
5	-4.31439E+09

(b) No. The population becomes extinct in one time period.

(c)
```
TYPE VALUE FØR A
?1.0
TYPE VALUE FØR B
?0.0001
TYPE INITIAL PØPULATIØN
?19999
TYPE NØ. ØF PREDICTIØNS
?20
```

PERIØD	PØPULATIØN
0	19999
1	1.99785
2	3.9953
3	7.989
4	15.9716
5	31.9177
6	63.7336
7	127.061
8	252.507
9	498.639
10	972.413
11	1850.27
12	3358.19
13	5588.63
14	8053.98
15	9621.3
16	9985.66
17	9999.98
18	10000
19	10000
20	10000

For $N_0 = 19,999$ the population does reach equilibrium ($A/B = 10,000$) in 18 time periods.

```
TYPE VALUE FØR A
?1.0
TYPE VALUE FØR B
?0.0001
TYPE INITIAL PØPULATIØN
?20000
TYPE NØ. ØF PREDICTIØNS
?5

PERIØD          PØPULATIØN
  0               20000
  1                 0
  2                 0
  3                 0
  4                 0
  5                 0
```

For $N_0 = 20,000$ the population becomes extinct in one time period. Hence the largest value of N_0 that allows the population to reach equilibrium is 19,999.

7. (a)
```
TYPE VALUE FØR A
?0.5
TYPE VALUE FØR B
?0.0002
TYPE INITIAL PØPULATIØN
?7500
TYPE NØ. ØF PREDICTIØNS
?5

PERIØD          PØPULATIØN
  0               7500
  1                 0
  2                 0
  3                 0
  4                 0
  5                 0
```

(b) No. The population becomes extinct in one time period.

(c) Trying N_0 just one smaller:

```
TYPE VALUE FØR A
?0.5
TYPE VALUE FØR B
?0.0002
TYPE INITIAL PØPULATIØN
?7499
TYPE NØ. ØF PREDICTIØNS
?30

PERIØD          PØPULATIØN
  0               7499
  1               1.50005
  2               2.24962
  3               3.37342
  4               5.05786
  5               7.58167
  6              11.361
  7              17.0157
  8              25.4656
  9              38.0688
 10              56.8133
```

11	84.5744
12	125.431
13	185.
14	270.655
15	391.332
16	556.369
17	772.645
18	1039.57
19	1343.21
20	1653.98
21	1933.84
22	2152.81
23	2302.3
24	2393.33
25	2444.39
26	2471.58
27	2485.63
28	2492.77
29	2496.38
30	2498.19

The largest initial population for which equilibrium is eventually reached is 7499.

9. (a) 0; 100; 1000
10. (a) 1 and 3

Section 5.3

1. (a) TYPE VALUE FØR A
   ?0.5
   TYPE VALUE FØR B
   ?0.0001
   TYPE INITIAL PØPULATIØN
   ?1
   TYPE NØ. ØF PREDICTIØNS
   ?20

PERIØD	PØPULATIØN
0	1
1	1.4999
2	2.24963
3	3.37393
4	5.05976
5	7.58708
6	11.3749
7	17.0494
8	25.545
9	38.2522
10	57.232
11	85.5204
12	127.549
13	189.697
14	280.947
15	413.527
16	603.19
17	868.402
18	1227.19
19	1690.19
20	2249.61

(b) TYPE VALUE FØR A
?0.5
TYPE VALUE FØR B
?0.0001
TYPE INITIAL PØPULATIØN
?10
TYPE NØ. ØF PREDICTIØNS
?10

PERIØD	PØPULATIØN
0	10
1	14.99
2	22.4625
3	33.6433
4	50.3518
5	75.2742
6	112.345
7	167.255
8	248.085
9	365.973
10	535.566

(c) TYPE VALUE FØR A
?0.5
TYPE VALUE FØR B
?0.0001
TYPE INITIAL PØPULATIØN
?100
TYPE NØ. ØF PREDICTIØNS
?10

PERIØD	PØPULATIØN
0	100
1	149
2	221.28
3	327.023
4	479.841
5	696.736
6	996.56
7	1395.53
8	1898.54
9	2487.37
10	3112.35

(d) The smallest value of N_0 for which equilibrium eventually is attained is 1.

4. (a) TYPE VALUE FØR A
?0.5
TYPE VALUE FØR B
?0.0001
TYPE INITIAL PØPULATIØN
?14500
TYPE NØ. ØF PREDICTIØNS
?25

PERIØD	PØPULATIØN
0	14500
1	724.999
2	1034.94
3	1445.3
4	1959.06
5	2554.79
6	3179.49

7	3758.32
8	4224.98
9	4552.43
10	4756.18
11	4872.15
12	4934.44
13	4966.79
14	4983.28
15	4991.61
16	4995.8
17	4997.9
18	4998.95
19	4999.47
20	4999.74
21	4999.87
22	4999.93
23	4999.97
24	4999.98
25	4999.99

(b) TYPE VALUE FØR A
?0.5
TYPE VALUE FØR B
?0.0001
TYPE INITIAL PØPULATIØN
?14900
TYPE NØ. ØF PREDICTIØNS
?20

PERIØD	PØPULATIØN
0	14900
1	149.
2	221.28
3	327.023
4	479.84
5	696.736
6	996.559
7	1395.53
8	1898.54
9	2487.36
10	3112.35
11	3699.85
12	4180.89
13	4523.35
14	4738.95
15	4862.66
16	4929.45
17	4964.22
18	4981.98
19	4990.96
20	4995.47

(c) TYPE VALUE FØR A
?0.5
TYPE VALUE FØR B
?0.0001
TYPE INITIAL PØPULATIØN
?15100
TYPE NØ. ØF PREDICTIØNS
?5

```
PERIØD          PØPULATIØN
   0              15100
   1             -151.
   2             -228.78
   3             -348.404
   4             -534.744
   5             -830.712


TYPE VALUE FØR A
?0.5
TYPE VALUE FØR B
?0.0001
TYPE INITIAL PØPULATIØN
?15500
TYPE NØ. ØF PREDICTIØNS
?5

PERIØD          PØPULATIØN
   0              15500
   1             -775.003
   2             -1222.57
   3             -1983.32
   4             -3368.33
   5             -6187.07
```

(e) The largest value for N_0 for which the solution will be stable is somewhere between 14,900 and 15,100. The largest value of n_0 if stability is to be achieved is between 9900 and 10,100. Thus we try $N_0 = 14,999$ and $N_0 = 15,000$.

```
TYPE VALUE FØR A
?0.5
TYPE VALUE FØR B
?0.0001
TYPE INITIAL PØPULATIØN
?14999
TYPE NØ. ØF PREDICTIØNS
?20

PERIØD          PØPULATIØN
   0              14999
   1              1.49836
   2              2.24732
   3              3.37047
   4              5.05457
   5              7.5793
   6              11.3632
   7              17.0319
   8              25.5188
   9              38.2131
  10              57.1737
  11              85.4336
  12              127.421
  13              189.507
  14              280.669
  15              413.127
  16              602.623
  17              867.619
  18              1226.15
  19              1688.88
  20              2248.09
```

```
TYPE VALUE FOR A
?0.5
TYPE VALUE FOR B
?0.0001
TYPE INITIAL POPULATION
?15000
TYPE NO. OF PREDICTIONS
?5
```

PERIOD	POPULATION
0	15000
1	0
2	0
3	0
4	0
5	0

From these results we conclude that the largest values of N_0 and n_0, respectively, are 14,999 and 9999.

6.
```
TYPE VALUE FOR A
?2.5
TYPE VALUE FOR B
?0.0001
TYPE INITIAL POPULATION
?10000
TYPE NO. OF PREDICTIONS
?5
```

PERIOD	POPULATION
0	10000
1	25000
2	25000
3	25000
4	25000
5	25000

The particular choice of N_0 produced equilibrium in one time period. Had we chosen N_0 slightly larger or smaller, equilibrium would not have been attained, since $A > 2$.

8. $N_0 = A/B$ or $N_0 = 1/B$.

10. (a) $BN_k < A$, so $A - BN_k > 0$ and $1 + A - BN_k > 1$. Therefore,

$$N_{k+1} = (1 + A - BN_k)N_k > N_k$$

(b) Suppose the statement is not true. We will show that this supposition leads to a contradiction; therefore the statement must be true. Suppose

$$N_{k+1} \geq A/B + 1/B \qquad \text{(i)}$$

Then, since,

$$N_{k+1} = (1 + A - BN_k)N_k$$

it follows that

$$N_k^2 B^2 - (1 + A)B + (1 + A) \leq 0 \qquad \text{(ii)}$$

This last inequality is not true if B is zero, since the left side is then $1 + A$. It also is not true when B is very large, since $N_k{}^2 B^2$ is very large then. Therefore, if the inequality is to be true at all, it must be true in between the zeros of

$$N_k{}^2 B^2 - (1 + A)B + (1 + A)$$

If this expression is to be zero for real values of B, the discriminant must be nonnegative; that is,

$$D = (1 + A)^2 - 4(1 + A)N_k{}^2 \geq 0$$

But $N_k > 1$, so

$$-4(1 + A)N_k{}^2 < -4(1 + A)$$

and

$$D < (1 + A)^2 - 4(1 + A)$$

The right side is just $(1 - A)^2 - 4$, so

$$D < (1 - A)^2 - 4$$

But $0 < A < 3$, so $\left| 1 - A \right| < 2$ and $(1 - A)^2 < 4$. Therefore,

$$D < 0$$

But this implies the function F has no real zeros. Hence the inequality (ii) can never be true. Hence (i) is never true, and

$$N_{k+1} < \frac{A}{B} + \frac{1}{B}$$

(c) We can conclude that if $0 < A < 2$ and if N_0 is positive and less than the equilibrium value, A/B, then the solution is stable. After all, the solution never exceeds $1/B$ more than the equilibrium value (i.e., $n_k < 1/B$) and this guarantees stability.

11. (a) 1 and 3
   (b) 1 is stable, 3 is unstable.

Section 5.4

1. (a) $A = 1.5$, $B = 0.05$
   (b) $N_3 = 30$, $N_4 = 30$
   (c) 30
4. (a) ? 62.948,75.995,91.972

A = .19293	B = -2.27738E-04	A/B = -847.159
0	62.948	
1	75.9949	
2	91.9719	
3	111.642	
4	136.02	
5	166.476	
6	204.906	
7	254.	
8	317.697	
9	401.976	
10	516.328	

(b) The predicted population in 1970 is 317.6975, while the actual population is 199.208. The predictions appear much too high, and the population does not seem to stabilize at all. One difficulty is that B is negative, as is the equilibrium population. Since (5.2) assumed that both A and B were positive, it does not seem that (5.2) is appropriate in this instance.

Section 5.5

1. $C = 1/10$

4. $R_{k+1} = R_k + HI_k$
$S_{k+1} = S_k - CS_kI_k$
$I_{k+1} = (1 + CS_k - H)I_k$

5.
```
100   PRINT "TYPE CONTACT RATE"
200   INPUT C
300   PRINT "TYPE REMOVAL RATE "
400   INPUT H
500   PRINT "TYPE INITIAL POPULATION"
600   INPUT N
700   PRINT "TYPE INITIAL NO. OF INFECTIVES"
800   INPUT IO
900   PRINT "TYPE NO. OF PERIODS TO BE PREDICTED"
1000  INPUT M
1100  PRINT
1200  PRINT "PERIOD","INFECTIVES","SUSCEPTIBLES","REMOVALS"
1300  PRINT
1400  LET SO=N-IO
1500  LET RO=0
1600  FOR J=1 TO M
1700  LET I1=(1+C*SO-H)*IO
1800  LET S1=(1-C*IO)*SO
1900  LET R1=RO+H*IO
2000  PRINT J,I1,S1,R1
2100  LET SO=S1
2200  LET IO=I1
2300  LET RO=R1
2400  NEXT J
2500  END
```

8. (a)
```
TYPE CONTACT RATE
?0.0005
TYPE REMOVAL RATE
?0.3
TYPE INITIAL POPULATION
?1000
TYPE INITIAL NO. OF INFECTIVES
?10
TYPE NO. OF PERIODS TO BE PREDICTED
?100
```

PERIOD	INFECTIVES	SUSCEPTIBLES	REMOVALS
10	50.7395	877.222	72.0383
20	105.046	579.467	315.487
30	52.8772	372.549	574.574
40	13.2288	317.923	668.848
50	2.78382	306.542	690.675
60	.565104	304.239	695.196
70	.113882	303.776	696.11
80	2.29161E-02	303.683	696.295
90	.00461	303.664	696.332
100	9.27331E-04	303.66	696.339

```
(b) TYPE CØNTACT RATE
    ?0.0005
    TYPE REMØVAL RATE
    ?0.3
    TYPE INITIAL PØPULATIØN
    ?1000
    TYPE INITIAL NØ. ØF INFECTIVES
    ?20
    TYPE NØ. ØF PERIØDS TØ BE PREDICTED
    ?100
```

PERIØD	INFECTIVES	SUSCEPTIBLES	REMØVALS
10	84.4918	781.45	134.059
20	96.3831	460.261	443.356
30	31.2215	330.809	637.969
40	6.70155	303.087	690.212
50	1.31898	297.679	701.002
60	.255238	296.635	703.11
70	.04923	296.433	703.518
80	9.48944E-03	296.395	703.596
90	1.82894E-03	296.387	703.611
100	3.52489E-04	296.386	703.614

9. 1200
```
   1950 IF I1>I0 THEN 2100
   2000 PRINT "MAX. NØ. ØF INFECTIVES IS";I0
   2050 GØTØ 2500
```

12. $I_{k+1} = (1 - CS_k - H)I_k$

$S_{k+1} = (1 - CI_k)S_k + M$

$R_{k+1} = R_k + HI_k$

Section 5.6

2. (a)(i)
```
    TYPE NØ. ØF INDIVIDUALS
    ?1000
    TYPE CØNTACT RATE
    ?0.0005
    TYPE INITIAL NØ. ØF INFECTIVES
    ?20

    PEAK ØF EPIDEMIC CURVE IS 124.997
```

(ii)
```
    TYPE NØ. ØF INDIVIDUALS
    ?1000
    TYPE CØNTACT RATE
    ?0.00025
    TYPE INITIAL NØ. ØF INFECTIVES
    ?20

    PEAK ØF EPIDEMIC CURVE IS 62.4687
```

(c)(iii)
```
    TYPE NØ. ØF INDIVIDUALS
    ?1000
    TYPE CØNTACT RATE
    ?0.000125
    TYPE INITIAL NØ. ØF INFECTIVES
    ?20

    PEAK ØF EPIDEMIC CURVE IS 31.2308
```

(b) Yes and $S = 250,000$.

4. $F = x^3$

6. $f = w(1 - w)$

CHAPTER 6

Section 6.2

1. (a)

2. (a) even or 1 to 1

3. (a) 3/4

4. Yes

6. 4

8. (a) 2/3

9. (a) 2/3

Section 6.3

1. (a) 52   (b) 13   (c) 26

3. (a) 20

4. (a) 33

5. (a) 4

6. (a) The union of two sets must have at least as many members as the smaller of the two sets.

7. (a) 651   (b) 349   (c) 538

9. The number of students taking at least one course is, from (6.7), $617 + 592 - 113 = 1096$, which exceeds the total number of students.

Section 6.4

1. (a) 1/4   (b) 1/2   (c) 4/13   (d) 9/26

4. (a) .33   (b) .25   (c) .37   (d) .36

6. (a) .452   (b) .312   (c) .113   (d) .651   (e) .349   (f) .538

8. 1

Section 6.5

1. (a) .3   (b) .8

3. (a) .4   (b) 2 to 3

5. (a) .42   (b) .88   (c) .12

7. (a) 1/13   (b) 1/13   (c) 1/169   (d) 25/169   (e) 144/169

10. (a) 7/12   (b) No. If they were independent the answer to part a would be 1/2.

Section 6.6

1. .352

4. (a) .028   (b) .104

Section 6.7

1. | 0+ | .44 | 0− | .066 |
    |-----|-----|-----|------|
    | A+ | .86 | EA− | .129 |
    | B+ | .54 | B− | .081 |
    | AB+ | 1.0 | AB− | .15 |

4. 11 to 14

Section 6.8

1. (a) .62   (c) .57
2. (a) Suppose the point is 11. Then if 11 appears on the second or succeeding tosses, we cannot tell whether we win or lose.
4.
```
100   LET N1=0
200   LET N2=0
300   LET N3=0
400   LET N4=0
500   PRINT "TYPE NØ. ØF GAMES TØ BE PLAYED"
600   INPUT N
700   LET D1=INT(6*RND(0)+1)
800   LET D2=INT(6*RND(0)+1)
900   LET P=D1+D2
1000  IF P=7 THEN 2400
1100  IF P=11 THEN 2400
1200  IF P=2 THEN 2100
1300  IF P=3 THEN 2100
1400  IF P=11 THEN 2100
1500  LET D1=INT(6*RND(0)+1)
1600  LET D2=INT(6*RND(0)+1)
1700  LET R=D1+D2
1800  IF R=7 THEN 2200
1900  IF R=P THEN 2500
2000  GØTØ 1500
2100  LET N1=N1+1
2200  LET N3=N3+1
2300  GØTØ 2600
2400  LET N2=N2+1
2500  LET N4=N4+1
2600  IF N3+N4<N THEN 700
2700  PRINT "NØ. ØF WINS =";N4
2800  PRINT "NØ. ØF LØSSES =";N3
2900  PRINT "ESTIMATE FØR PRØBABILITY ØF WINNING IS";N4/(N3+N4)
3000  PRINT "NØ. ØF WINS ØN FIRST TØSS =";N2
3100  PRINT "NØ. ØF LØSSES ØN FIRST TØSS =";N1
3200  PRINT "ESTIMATE FØR PRØBABILITY ØF WINNING ØN FIRST"
3300  PRINT "TØSS IS";N2/(N3+N4)
3400  PRINT "ESTIMATE FØR PRØBABILITY ØF LØSING ØN FIRST"
3500  PRINT "TØSS IS";N1/(N3+N4)
3600  END
```

(a) 0.222   (b) 0.111   (c) .0.667

6.
```
10   INPUT N,M
20   LET T=N
30   LET N=N+N
40   IF N<M THEN 60
50   LET N=N-M
60   PRINT N,
70   IF N=T THEN 90
80   GØTØ 30
90   END
```

CHAPTER 7

Section 7.2

1. (a) 2/3
2. (a) 2/5
3. $P(X \mid Y) = P(X)$
5. 0.35
8. (a) 0.415
9. (a) $(419 + 617 + 404)/2655 = 0.5424$
   (d) $419/(387 + 419 + 109) = 0.4579$

Section 7.3

1. (a) 1/5   (b) Yes, the bet is a fair one.
3. 4/9
5. (a) No. A patient can have both a rash and a fever (i.e., $A \cap B \neq \phi$)
   (b) Yes. $A \cap D = \phi$
   (c) No. A patient can have two or even three of the symptoms simultaneously.
   (d) Use two events. The first is the union of the first three classifications (i.e., $A \cup B \cup C$). The second event is D.
8. 0.0000505
10. (b) Yes. The profit with no testing is $4 million. Since the probability of a positive result, $P(X)$, is 0.41, we should expect to have to conduct 50 tests to find 20 positive results. The cost of 50 tests is $2.5 million. The results, however, will be 4 gushers and 6 moderate supplies for an income of $7 million and a net of $4.5 million.

Section 7.4

1. (a)

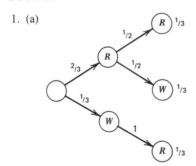

2. (a)

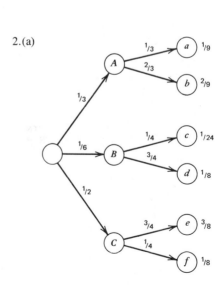

3.(a)

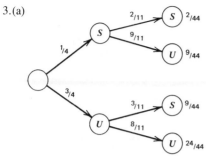

(b) 2/44   (c) 18/44

5.(a)

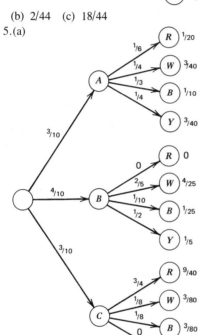

(b)

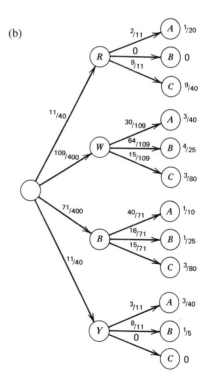

(c) 2/5   (d) 64/109

7.

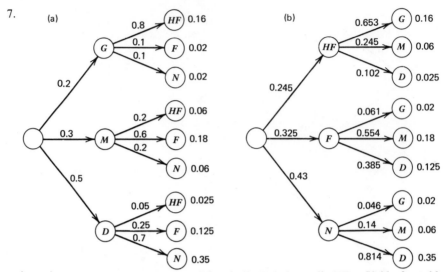

where G = gusher, M = moderate supply of oil, D = dry well, HF = highly favorable, F = favorable, N = negative

(c) 0.385   (d) 0.316

10.

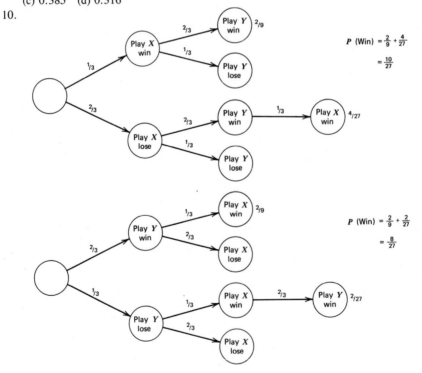

Play X first.

Section 7.5

1. NØ. ØF MUTUALLY EXCLUSIVE EVENTS IS   ?2

  PRØBABILITY ØF EVENT 1   IS ?0.0005
  PRØBABILITY ØF EVENT 2   IS ?0.9995

  PRØBABILITY ØF EVENT X GIVEN THAT EVENT 1   ØCCURS IS?0.1
  PRØBABILITY ØF EVENT X GIVEN THAT EVENT 2   ØCCURS IS?0.99

  PRØBABILITY THAT EVENT X ØCCURS IS .989555

  PRØBABILITY ØF EVENT   1   GIVEN X IS 5.05278E-05
  PRØBABILITY ØF EVENT   2   GIVEN X IS .99995

3. NØ. ØF MUTUALLY EXCLUSIVE EVENTS IS   ?5

  PRØBABILITY ØF EVENT 1   IS ?0.05
  PRØBABILITY ØF EVENT 2   IS ?0.1
  PRØBABILITY ØF EVENT 3   IS ?0.2
  PRØBABILITY ØF EVENT 4   IS ?0.15
  PRØBABILITY ØF EVENT 5   IS ?0.5

  PRØBABILITY ØF EVENT X GIVEN THAT EVENT 1   ØCCURS IS?0.9
  PRØBABILITY ØF EVENT X GIVEN THAT EVENT 2   ØCCURS IS?0.7
  PRØBABILITY ØF EVENT X GIVEN THAT EVENT 3   ØCCURS IS?0.6
  PRØBABILITY ØF EVENT X GIVEN THAT EVENT 4   ØCCURS IS?0.4
  PRØBABILITY ØF EVENT X GIVEN THAT EVENT 5   ØCCURS IS?0.1

  PRØBABILITY THAT EVENT X ØCCURS IS .345

  PRØBABILITY ØF EVENT   1   GIVEN X IS .130435
  PRØBABILITY ØF EVENT   2   GIVEN X IS .202899
  PRØBABILITY ØF EVENT   3   GIVEN X IS .347826
  PRØBABILITY ØF EVENT   4   GIVEN X IS .173913
  PRØBABILITY ØF EVENT   5   GIVEN X IS .144928

  (a) 0.345
  (b) See last five lines of computer output.
  (c) 0.307249

5. 500   -----
  550 LET S=0
  600 FØR J=1 TØ M
  700 PRINT "PRØBABILITY ØF EVENT ";J;" IS ",
  800 INPUT A(J)
  850 LET S=S + A(J)
  900 NEXT J
  910 IF S=1 THEN 1000
  920 PRINT "THE P(A(J)) DØ NØT SUM TØ 1"
  930 STØP
  1000   -----

Section 7.6

2. (a) See Figure A on next page. 0.618 or more precisely 34/55.
  (c) See Figure C on page 421. 0.583 or more precisely 116287/199584.
3. (a) 2, 3, 4, 5, 6, 7, 8
4. (a) 1/4   (c) 1/3
5. (a) 55/122

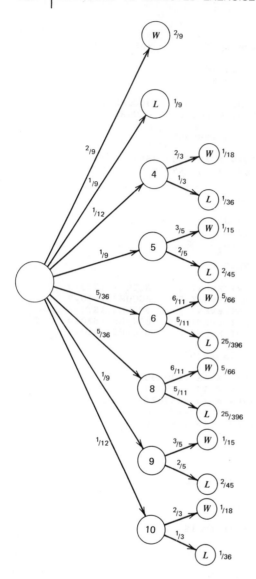

FIGURE A Solution to Exercise 2a of Section 7.6.

CHAPTER 8

Section 8.2

1. (a) 0 successes: .2373
   1 success: .3955
   2 successes: .2637
   3 successes: .0879

   4 successes: .0146
   5 successes: .0010
   (c) 0 successes: .0576
   1 success: .1977
   2 successes: .2965
   3 successes: .2541

   4 successes: .1361
   5 successes: .0467
   6 successes: .0100
   7 successes: .0012
   8 successes: .0001

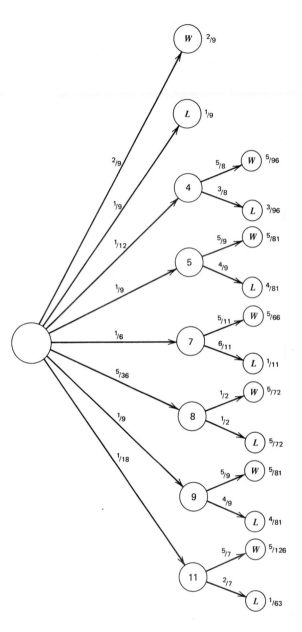

FIGURE C Solution to Exercise 2c of Section 7.6.

2. 0.7799

4. The probability of n or more successes is the probability of exactly n successes. The probability of $n - 1$ or more successes is the probability of $n - 1$ successes or n successes. Hence, between boxes 8 and 9, we insert

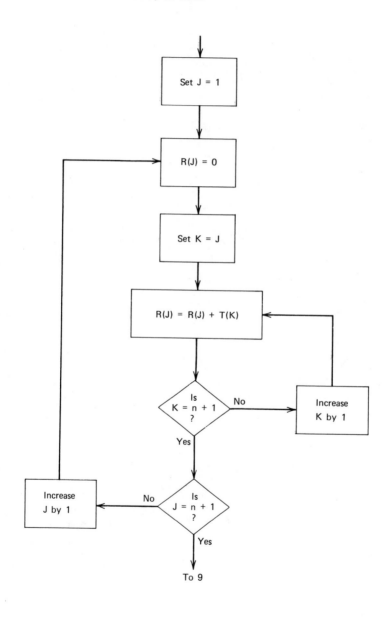

where $R(J+1)$ is the cumulative tally of experiments resulting in J successes. In Box 9 the $R(J+1)$ are used as the numerators in computing relative frequencies.

Section 8.3

1.

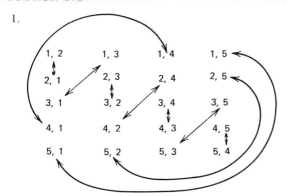

3. (b) 6
   (d) $6p^2(1 - p)^2$
4. `110 IF N >= S THEN 300`
   `120 PRINT "S EXCEEDS N"`
   `130 STØP`
6. (a) 360
7. (a) 15
8. (a) $32/625 = 0.0512$
9. (a) 0.058
10. (a) 0.544
11. (a) 0.009
12. (a) 0.03549
13. (a) 0.382
14. Yes. Against women. The probability of this happening is only 0.382.

Section 8.4

1. (a) 0.370
2. (a) 0.0015
3. (a) 0.399 or 0.601
4. 0.8334

Section 8.5

1. (a) 0.058
2. (a) 0.0027
3. 0.106

Section 8.6

1. 0.785
3. (a) 0.607
5. 0.4223. This is identical with the legal problem in Section 8.6 if (a) Tribe 1 is identified with the properly functioning machine and (b) red clay present is identified with a defective light bulb. Thus the probability of Tribe 2 is the probability of a malfunctioning machine.

CHAPTER 9

Section 9.1

1. (a) (0, 1, 5)
2. (a) Not possible.   (c) $(-2, -5, 4)$ (g) $(4, -2, 8)$
3. (a) $\begin{pmatrix} 1 \\ 1 \end{pmatrix}$ (d) Not possible.   (g) $\begin{pmatrix} 0 \\ 6 \\ 9 \end{pmatrix}$
4. (a) True.   (c) Not true.
5. No
7. (a) $\begin{pmatrix} 6 \\ -7 \\ 0 \end{pmatrix}$
8. (a) $x = 3, y = 1$
9. (a)

$$\mathbf{n} = \begin{pmatrix} 31 \\ 33\frac{1}{2} \\ 32 \\ 30 \end{pmatrix} \qquad \mathbf{u} = \begin{pmatrix} 72 \\ 69 \\ 76 \\ 78 \end{pmatrix} \qquad \mathbf{c} = \begin{pmatrix} 104 \\ 102 \\ 97 \\ 97 \end{pmatrix}$$

   (b)

$$\mathbf{t} = 50\mathbf{n} + 20\mathbf{u} + 10\mathbf{c}$$

   (c) 1973 $4030
        1974 $4075
        1975 $4090
        1976 $4030

Section 9.2

1. (a) 0
2. (a) 5
3. 1
5. (a) -2
6. Smith = 18, White = 14
8. (a) (150, 200, 100)   (b) $\begin{pmatrix} 92 \\ 87 \\ 112 \end{pmatrix}$ (c) $42,400

10. (a) (6, 0, 4, 6) (b) $\begin{pmatrix} 24 \\ 16 \\ 16 \\ 32 \end{pmatrix}$ (c) 25¢

Section 9.3

1. (a) B and D as well as C and F
2. $a = 6, b = 2, x = 1, y = 3$
4. (a) $\begin{pmatrix} 3 & 5 \\ 6 & -6 \\ -2 & 1 \end{pmatrix}$ (b) $\begin{pmatrix} 3 & 6 & -2 \\ 5 & -6 & 1 \end{pmatrix}$
6. $a = -2, b = 6, c = -1, d = 3, e = 0$

Section 9.4

1. (a) (3, 3, 0)
2. (a) $\begin{pmatrix} 1 \\ 0 \\ 0 \end{pmatrix}$ (d) (1, 0, 0)
3. (a) $\begin{pmatrix} 1 \\ 0 \\ 0 \end{pmatrix}$ (d) (1, 0, 0)
4. (a) (−1, 1)

Section 9.5

1. (a) $\begin{pmatrix} 1 & 4 \\ 4 & -2 \\ 1 & -2 \end{pmatrix}$

2. (a)

	White	Negro	Indian	Japanese	Chinese
New York	1424.6	499.4	5.9	4.8	17.4
California	4580.5	421.7	19.1	72.4	37.3
Alabama	203.4	0.6	0.4	0.5	0.1
Illinois	968.8	391.5	3.3	2.5	2.8

3.

	White	Negro	Indian	Japanese	Chinese
New York	−840.7	247.7	6.0	6.9	26.4
California	−1266.7	92.9	32.9	−16.4	37.2
Alabama	48.0	−77.8	0.7	0.1	0.2
Illinois	−379.3	−6.6	3.4	0.7	4.7

Section 9.6

1. (a) $\begin{pmatrix} -15 & 20 \\ -8 & 1 \end{pmatrix}$
2. (a) $\begin{pmatrix} -55 & -10 \\ -10 & -15 \end{pmatrix}$

3. (a) $\begin{pmatrix} -16 & 18 \\ -6 & 0 \end{pmatrix}$

4. (a) $M^2 = \begin{pmatrix} 5 & 4 \\ 4 & 5 \end{pmatrix}$ $M^4 = \begin{pmatrix} 41 & 40 \\ 40 & 41 \end{pmatrix}$ $M^8 = \begin{pmatrix} 3281 & 3280 \\ 3280 & 3281 \end{pmatrix}$

5. (a) 2×4

6. (a) $\begin{pmatrix} 0 & 1 \\ 1 & 0 \end{pmatrix}$

7. (a) $\begin{pmatrix} 2 & 0 & 0 \\ 0 & -2 & 0 \\ 0 & 0 & 3 \end{pmatrix}$

8. (a) $\begin{pmatrix} 1 & 0 & 0 \\ 0 & 1 & 0 \\ 0 & 0 & 1 \end{pmatrix}$

Section 9.7

1. (a)
```
10   DIM A[3,3],B[3,3],C[3,3],D[3,3]
20   MAT   READ A,B,C
30   MAT D=A*B
40   MAT C=D-C
50   MAT  PRINT C
100  DATA 1.6,-2.3,6.1,-3.6,.9,1.1,2.3,.6,-6.4
110  DATA 7.9,0,6.4,1.2,-1.3,3.2,.2,2.1,0
120  DATA 0,1,2,-2.1,7.3,-7.3,0,1.9,0
200  END
```

11.1	14.8	.879997
-25.04	-6.16	-12.86
17.61	-16.12	16.64

2.
```
10   DIM A[10,10],B[10,10]
20   MAT A=CØN
30   MAT A=(-1)*A
40   MAT B=IDN
50   MAT B=(4)*B
60   MAT A=A+B
70   MAT  PRINT A
80   END
```

4. (a)
```
10   DIM A[3,3],X[3,1],Y[3,1]
20   MAT   READ A,X
30   MAT Y=A*X
40   MAT  PRINT Y
50   DATA 3,-2,1,0,1,2,6,-1,3
60   DATA 1,4,2
70   END
```

RUN

-3

8

8

Section 9.8

1. (a)
$$\mathbf{x}_1 = \begin{pmatrix} 1/2 \\ 1/12 \\ 1/6 \\ 1/4 \end{pmatrix} \quad \mathbf{x}_2 = \begin{pmatrix} 5/16 \\ 1/12 \\ 7/24 \\ 5/16 \end{pmatrix}$$

3. (a)
$$\mathbf{x}_1 = \begin{pmatrix} 3/8 \\ 3/16 \\ 3/16 \\ 1/4 \end{pmatrix} \quad \mathbf{x}_2 = \begin{pmatrix} 19/64 \\ 1/8 \\ 17/64 \\ 5/16 \end{pmatrix}$$

6. (a)
$$\begin{array}{c} \\ T = \end{array} \begin{array}{cccc} SS & SU & US & UU \\ \begin{pmatrix} 0.9 & 0 & 0.7 & 0 \\ 0.1 & 0 & 0.3 & 0 \\ 0 & 0.4 & 0 & 0.2 \\ 0 & 0.6 & 0 & 0.8 \end{pmatrix} & \begin{matrix} SS \\ SU \\ US \\ UU \end{matrix} \end{array}$$

   (b) $SS = 216, SU = 24, US = 48, UU = 192$
   (c) $SS = 192, SU = 48, US = 72, UU = 168$
   (d) $SS = 280, SU = 40, US = 40, UU = 120$

8. (a) 29.5% residential, 37% commercial, 25.5% parking, 8% vacant
   (b) 27% residential, 28.9% commercial, 34.5% parking, 9.6% vacant
   (c) 9.6% residential, 8.7% commercial, 66.1% parking, 15.6% vacant
   (d) 21.2% residential, 19.2% commercial, 25% parking, 34.6% vacant

CHAPTER 10

Section 10.1

1. 1100 grams of salad and -5 grams of roast beef. However, it is not possible to eat -5 grams, so the diet is not possible.

Section 10.2

1. (a) $x = -1, y = 1$
2. (a) No solution
3. (a) Choose any value for y, then x is double that value.
4. (a) $12t + 14c = 2000$
      $8t + 6c = 1000$

Section 10.3

1. (a) $x = 1, y = 1, z = 1$
2. (a) $x_1 = 1, x_2 = 1, x_3 = 1, x_4 = 1$
3. $B = 1.0804, L = 1.1667, M = 3.606, S = 0.1418$

Section 10.4

1. (a)
$$A = \begin{pmatrix} 2 & -3 \\ 3 & 1 \end{pmatrix}$$

2. (a) $x_1 \qquad + 2x_3 = 1$
$\quad x_1 + \; x_2 - 2x_3 = 0$
$\qquad\qquad 2x_2 + x_3 = 1$

Section 10.5

1. (a) $-$(ii); (b) $-$(iv); (c) $-$(i); (d) $-$(iii).
2. (a) $x = 6, y = 1$
3. (a) $\begin{pmatrix} 1 & 4 & 3 \\ 2 & 5 & 4 \\ 1 & -3 & -2 \end{pmatrix} \begin{pmatrix} 1 \\ 1 \\ 1 \end{pmatrix} = \begin{pmatrix} 8 \\ 11 \\ -4 \end{pmatrix}$ so $x = 8, y = 11$, and $z = -4$

Section 10.6

1. (a) $x = 2, y = -1, z = 2, w = 1$
2. (a) $x = -4.36, y = -19.79, z = -30.11, w = -7.90, u = 19.73$

Section 10.7

1. (a) $\begin{pmatrix} 6/19 \\ 1/19 \\ 17/57 \\ 1/3 \end{pmatrix} = \begin{pmatrix} .3158 \\ .0526 \\ .2983 \\ .3333 \end{pmatrix}$

2.

	(a)	(b)	(c)
Securities	0.107	0.1725	0.114
Real property	0.482	0.3425	0.467
Consumer goods	0.125	0.2025	0.133
Savings	0.286	0.2825	0.286

4. (a) Professional   0.023
     Managerial   0.041
     Higher-grade supervisory   0.088
     Lower-grade supervisory   0.127
     Skilled labor   0.410
     Semi-skilled labor   0.182
     Unskilled labor   0.129
  (b) Yes

CHAPTER 11

Section 11.1

1. (a) $20B + L \geqslant 70$
$\qquad 15L \geqslant 60$
$\qquad\quad B \geqslant 0$
$\qquad\quad L \geqslant 0$
2. (a) $B = 3.3$ and $L = 4$

3. (a) $\geq$, (b) $\geq$, (c) $\leq$, (d) not possible
5. $25B + 10L$

Section 11.2

1. (a) $x_1 = 200$, $x_2 = 200$
2. (a) Supply of Brazil nuts is exceeded.
3. (a) $264

Section 11.3

1. (a)

2. (a) $x_1 = 0$, $x_2 = 1$
3. (a)

(b) $B = 1.65$, $L = 2$

Section 11.4

1. (a) $x_1 = 157.14$, $x_2 = 228.57$, profit $= 151.71
2. (a) $x_1 = 200$, $x_2 = 100$, profit $= 222
3. (a) $x_1 = 106.67$, $x_2 = 262.22$, profit $= 190.58
4. (a) $x_1 = 233.33$, $x_2 = 0$, profit $= 233.33

Section 11.5

1. (a) One and only one.
   (d) Infinite number, bounded.
2. No solution.

Section 11.6

1. (a) `1600 DATA 50,32`
2. (a) `1600 DATA 100,22`
3. (a) `1000 DATA 640`
4. (a) `1000 DATA 1.E10`

Section 11.7

1. The only change in the BASIC program is statement 1600, which should read

   `1600 DATA 175.75,128.625,106.75,133.875`

   The results of running the program are
   ```
   X( 1    ) = 103.704
   X( 4    ) = 219.259
   ```
   Thus we should make 103.7 pounds of Mix 1 and 219.26 pounds of Mix 4, yielding a profit of $475.79.
3. The only change in the BASIC program is

   `1000 DATA 1600`

   The results of running the program are
   ```
   X( 1    ) = 213.333
   ```
   Thus we should make 213 ⅓ pounds of Mix 1 and nothing else. The total profit is $374.93.

APPENDIX B
TABLE OF
BINOMIAL
PROBABILITIES

The table in this appendix gives the values of

$$P(n,s;p) = \frac{n!}{(n-s)!\,s!}\, p^s(1-p)^{n-s}$$

The table is read as follows. The number of trials is specified at the left (i.e., the portion of the table immediately below.

NUMBER OF TRIALS = 5

corresponds to 5 trials.) The number of successes is given in the column on the left labeled "SUCCESSES," and the probability of success on one trial, p, is given in the same row as the word SUCCESSES itself.

To find $P(6, 4, 0.3)$ we look down the left side of the table for "NUMBER OF TRIALS = 6." We then look down the column labeled "SUCCESSES" to the row containing a 4. We look across that row to the entry in the column headed by 0.3. The result is

$$P(6, 4, 0.3) = 0.059535$$

431

NUMBER OF TRIALS = 2

SUCCESSES				PROBABILITY OF SUCCESS ON ONE TRIAL					
	0.1	0.2	0.3	0.4	0.5	0.6	0.7	0.8	0.9
0	.81	.64	.49	.36	.25	.16	.09	.04	.01
1	.18	.32	.42	.48	.5	.48	.42	.32	.18
2	.01	.04	.09	.16	.25	.36	.49	.64	.81

NUMBER OF TRIALS = 3

SUCCESSES				PROBABILITY OF SUCCESS ON ONE TRIAL					
	0.1	0.2	0.3	0.4	0.5	0.6	0.7	0.8	0.9
0	.729	.512	.343	.216	.125	.064	.027	8.00001E-03	.001
1	.243	.384	.441	.432	.375	.288	.189	9.60001E-02	.027
2	.027	.096	.189	.288	.375	.432	.441	.384	.243
3	.001	.008	.027	.064	.125	.216	.343	.512	.729

NUMBER OF TRIALS = 4

SUCCESSES				PROBABILITY OF SUCCESS ON ONE TRIAL					
	0.1	0.2	0.3	0.4	0.5	0.6	0.7	0.8	0.9
0	.6561	.4096	.2401	.1296	.0625	.0256	.0081	.0016	.0001
1	.2916	.4096	.4116	.3456	.25	.1536	.0756	.0256	.0036
2	.0486	.1536	.2646	.3456	.375	.3456	.2646	.1536	.0486
3	.0036	.0256	.0756	.1536	.25	.3456	.4116	.4096	.2916
4	.0001	.0016	.0081	.0256	.0625	.1296	.2401	.4096	.6561

NUMBER OF TRIALS = 5

SUCCESSES				PROBABILITY OF SUCCESS ON ONE TRIAL					
	0.1	0.2	0.3	0.4	0.5	0.6	0.7	0.8	0.9
0	.59049	.32768	.16807	.07776	.03125	.01024	.00243	.00032	.00001
1	.32805	.4096	.36015	.2592	.15625	.0768	.02835	6.40001E-03	.00045
2	.0729	.2048	.3087	.3456	.3125	.2304	.1323	.0512	.0081
3	.0081	.0512	.1323	.2304	.3125	.3456	.3087	.2048	.0729
4	.00045	.0064	.02835	.0768	.15625	.2592	.36015	.4096	.32805
5	.00001	.00032	.00243	.01024	.03125	.07776	.16807	.32768	.59049

NUMBER OF TRIALS = 6

SUCCESSES				PROBABILITY OF SUCCESS ON ONE TRIAL					
	0.1	0.2	0.3	0.4	0.5	0.6	0.7	0.8	0.9
0	.531441	.262144	.117649	.046656	.015625	.004096	7.28999E-04	6.40001E-05	.000001
1	.354294	.393216	.302526	.186624	.09375	.036864	.010206	.001536	5.40001E-05
2	.098415	.24576	.324135	.31104	.234375	.13824	.059535	.01536	.001215
3	.01458	.08192	.18522	.27648	.3125	.27648	.18522	8.19201E-02	.01458
4	.001215	.01536	.059535	.13824	.234375	.31104	.324135	.24576	.098415
5	.000054	.001536	.010206	.036864	.09375	.186624	.302526	.393216	.354294
6	9.99999E-07	.000064	.000729	.004096	.015625	.046656	.117649	.262144	.531441

NUMBER OF TRIALS = 7

SUCCESSES	\ PROBABILITY OF SUCCESS ON ONE TRIAL								
	0.1	0.2	0.3	0.4	0.5	0.6	0.7	0.8	0.9
0	.478297	.209715	8.23543E-02	2.79936E-02	7.81250E-03	1.63840E-03	2.18700E-04	1.28000E-05	1.00000E-07
1	.372009	.367002	.247063	.130637	5.46875E-02	1.72032E-02	3.57210E-03	3.58401E-04	6.30001E-06
2	.124003	.275251	.317652	.261274	.164062	7.74144E-02	2.50047E-02	4.30080E-03	1.70100E-04
3	2.29635E-02	.114688	.226895	.290304	.273438	.193536	9.72405E-02	2.86720E-02	2.55150E-03
4	2.55150E-03	2.86720E-02	9.72405E-02	.193536	.273438	.290304	.226895	.114688	2.29635E-02
5	1.70100E-04	4.30080E-03	2.50047E-02	7.74144E-02	.164062	.261274	.317652	.275251	.124003
6	6.30001E-06	3.58401E-04	3.57210E-03	1.72032E-02	5.46875E-02	.130637	.247063	.367002	.372009
7	9.99999E-08	1.28000E-05	2.18700E-04	1.63840E-03	7.81250E-03	2.79936E-02	8.23543E-02	.209715	.478297

NUMBER OF TRIALS = 8

SUCCESSES	\ PROBABILITY OF SUCCESS ON ONE TRIAL								
	0.1	0.2	0.3	0.4	0.5	0.6	0.7	0.8	0.9
0	.430467	.167772	5.76480E-02	1.67962E-02	3.90625E-03	6.55360E-04	6.56100E-05	2.56001E-06	1.00000E-08
1	.382638	.335544	.197650	8.95795E-02	3.12500E-02	7.86432E-03	1.22472E-03	8.19202E-05	7.20001E-07
2	.148804	.293601	.296476	.209018	.109375	4.12877E-02	1.00019E-02	1.14688E-03	2.26800E-05
3	3.30674E-02	.146801	.254122	.278692	.218750	.123863	4.66754E-02	9.17504E-03	4.08240E-04
4	4.59270E-03	4.58752E-02	.136137	.232243	.273438	.232243	.136137	4.58752E-02	4.59271E-03
5	4.08240E-04	9.17504E-03	4.66754E-02	.123863	.218750	.278692	.254122	.146801	3.30675E-02
6	2.26800E-05	1.14688E-03	1.00019E-02	4.12877E-02	.109375	.209018	.296476	.293601	.148804
7	7.20000E-07	8.19200E-05	1.22472E-03	7.86432E-03	3.12500E-02	8.95795E-02	.197650	.335544	.382638
8	9.99999E-09	2.56000E-06	6.56100E-05	6.55360E-04	3.90625E-03	1.67962E-02	5.76480E-02	.167772	.430467

NUMBER OF TRIALS = 9

SUCCESSES	\ PROBABILITY OF SUCCESS ON ONE TRIAL								
	0.1	0.2	0.3	0.4	0.5	0.6	0.7	0.8	0.9
0	.387420	.134218	4.03536E-02	1.00777E-02	1.95312E-03	2.62144E-04	1.96830E-05	5.12001E-07	1.00000E-09
1	.387420	.301990	.155650	6.04662E-02	1.75781E-02	3.53894E-03	4.13343E-04	1.84320E-05	8.10002E-08
2	.172187	.301990	.266828	.161243	7.03125E-02	2.12337E-02	3.85787E-03	2.94912E-04	2.91601E-06
3	4.46410E-02	.176161	.266828	.250823	.164062	7.43178E-02	2.10039E-02	2.75251E-03	6.12361E-05
4	7.44017E-03	6.60603E-02	.171532	.250823	.246094	.167215	7.35139E-02	1.65151E-02	8.26686E-04
5	8.26686E-04	1.65151E-02	7.35139E-02	.167215	.246094	.250823	.171532	6.60603E-02	7.44018E-03
6	6.12360E-05	2.75251E-03	2.10039E-02	7.43178E-02	.164062	.250823	.266828	.176161	4.46418E-02
7	2.91600E-06	2.94912E-04	3.85787E-03	2.12337E-02	7.03125E-02	.161243	.266828	.301990	.172187
8	8.10000E-08	1.84320E-05	4.13343E-04	3.53894E-03	1.75781E-02	6.04662E-02	.155650	.301990	.387420
9	9.99999E-10	5.12000E-07	1.96830E-05	2.62144E-04	1.95312E-03	1.00777E-02	4.03536E-02	.134218	.387420

NUMBER OF TRIALS = 10

SUCCESSES	\ PROBABILITY OF SUCCESS ON ONE TRIAL								
	0.1	0.2	0.3	0.4	0.5	0.6	0.7	0.8	0.9
0	.348678	.107374	2.82475E-02	6.04662E-03	9.76562E-04	1.04858E-04	5.90490E-06	1.02400E-07	1.00000E-10
1	.387420	.268435	.121061	4.03108E-02	9.76562E-03	1.57286E-03	1.37781E-04	4.09600E-06	9.00002E-09
2	.193710	.301990	.233475	.120932	4.39453E-02	1.06168E-02	1.44670E-03	7.37280E-05	3.64501E-07
3	5.73956E-02	.201327	.266828	.214991	.117187	4.24673E-02	9.00170E-03	7.86432E-04	8.74802E-06
4	1.11603E-02	8.80804E-02	.200121	.250823	.205078	.111477	3.67569E-02	5.50502E-03	1.37781E-04
5	1.48803E-03	2.64241E-02	.102919	.200658	.246094	.200658	.102919	2.64241E-02	1.48803E-03
6	1.37781E-04	5.50502E-03	3.67569E-02	.111477	.205078	.250823	.200121	8.80804E-02	1.11603E-02
7	8.74799E-06	7.86432E-04	9.00170E-03	4.24673E-02	.117187	.214991	.266828	.201327	5.73957E-02
8	3.64500E-07	7.37280E-05	1.44670E-03	1.06168E-02	4.39453E-02	.120932	.233475	.301990	.193710
9	8.99999E-09	4.09600E-06	1.37781E-04	1.57286E-03	9.76562E-03	4.03108E-02	.121061	.268435	.387420
10	9.99999E-11	1.02400E-07	5.90490E-06	1.04858E-04	9.76562E-04	6.04662E-03	2.82475E-02	.107374	.348678

APPENDIX C
TABLE OF CUMULATIVE BINOMIAL PROBABILITIES

This table gives the values of

$$P*(n,s;p) = P(n,s;p) + P(n, s + 1; p) + \ldots + P(n, n; p)$$

$$= \frac{n!}{(n - s)! \, s!} \, p^s(1 - p)^{n-s} + \frac{n!}{(n - s - 1)! \, (s + 1)!} \, p^{s+1} (1 - p)^{n-s-1}$$

$$+ \ldots + \frac{n!}{0! \, n!} \, p^n(1 - p)^0$$

The table is read as follows. The number of trials, n, is found on the left. The number of successes, s, is found in the column labeled "SUCCESSES," immediately below the appropriate number of trials. This specifies the row of the table. The column is located by finding the column headed by the probability of success on one trial, p.

For example, to find $P*(6, 4, 0.3)$, we look down the left side of the table for "NUMBER OF TRIALS = 6." We then look down from there in column labeled "SUCCESSES" to the row labeled 4. We follow across that row to the column headed by 0.3. The entry found is

$$P*(6, 4, 0.3) = 0.07047$$

435

NUMBER OF TRIALS = 2

SUCCESSES	\	\	\	PROBABILITY	OF SUCCESS	ON ONE TRIAL	\	\	\
	0.1	0.2	0.3	0.4	0.5	0.6	0.7	0.8	0.9
0	1.	1.	1.	1.	1.	1.	1.	1.	1.
1	.19	.36	.51	.64	.75	.84	.91	.96	.99
2	.01	.04	.09	.16	.25	.36	.49	.64	.81

NUMBER OF TRIALS = 3

SUCCESSES	\	\	\	PROBABILITY	OF SUCCESS	ON ONE TRIAL	\	\	\
	0.1	0.2	0.3	0.4	0.5	0.6	0.7	0.8	0.9
0	1.	1.	1.	1.	1.	1.	1.	1.	1.
1	.271	.488	.657	.784	.875	.936	.973	.992	.999
2	.028	.104	.216	.352	.5	.648	.784	.896	.972
3	.001	.008	.027	.064	.125	.216	.343	.512	.729

NUMBER OF TRIALS = 4

SUCCESSES	\	\	\	PROBABILITY	OF SUCCESS	ON ONE TRIAL	\	\	\
	0.1	0.2	0.3	0.4	0.5	0.6	0.7	0.8	0.9
0	1.	1.	1.	1.	1.	1.	1.	1.	1.
1	.3439	.5904	.7599	.8704	.9375	.9744	.9919	.9984	.9999
2	.0523	.1808	.3483	.5248	.6875	.8208	.9163	.9728	.9963
3	.0037	.0272	.0837	.1792	.3125	.4752	.6517	.8192	.9477
4	.0001	.0016	.0081	.0256	.0625	.1296	.2401	.4096	.6561

NUMBER OF TRIALS = 5

SUCCESSES	\	\	\	PROBABILITY	OF SUCCESS	ON ONE TRIAL	\	\	\
	0.1	0.2	0.3	0.4	0.5	0.6	0.7	0.8	0.9
0	1.	1.	1.	1.	1.	1.	1.	1.	1.
1	.40951	.67232	.83193	.92224	.96875	.98976	.99757	.99968	.99999
2	.08146	.26272	.47178	.66304	.8125	.91296	.96922	.99328	.99954
3	.00856	.05792	.16308	.31744	.5	.68256	.83692	.94208	.99144
4	.00046	.00672	.03078	.08704	.1875	.33696	.52822	.73728	.91854
5	.00001	.00032	.00243	.01024	.03125	.07776	.16807	.32768	.59049

NUMBER OF TRIALS = 6

SUCCESSES	\	\	\	PROBABILITY	OF SUCCESS	ON ONE TRIAL	\	\	\
	0.1	0.2	0.3	0.4	0.5	0.6	0.7	0.8	0.9
0	1.	1.	1.	1.	1.	1.	1.	1.	1.
1	.468559	.737856	.882351	.953344	.984375	.995904	.999271	.999936	.999999
2	.114265	.34464	.579825	.76672	.890625	.95904	.989065	.9984	.999945
3	.01585	.09888	.25569	.45568	.65625	.8208	.92953	.98304	.99873
4	.00127	.01696	.07047	.1792	.34375	.54432	.74431	.90112	.98415
5	.000055	.0016	.019935	.04096	.109375	.23328	.420175	.65536	.885735
6	9.99999E-07	.000064	.000729	.004096	.015625	.046656	.117649	.262144	.531441

NUMBER OF TRIALS = 7

PROBABILITY OF SUCCESS ON ONE TRIAL

NUMBER OF SUCCESSES	0.1	0.2	0.3	0.4	0.5	0.6	0.7	0.8	0.9
0	1.	1.	1.	1.	1.	1.	1.	1.	1.
1	.521703	.790285	.917646	.972006	.992188	.998362	.999781	.999987	
2	.149694	.423283	.670583	.84137	.9375	.981158	.996209	.999629	.999994
3	2.56915E-02	.148032	.352931	.580096	.773438	.903744	.971205	.995328	.999824
4	.002728	.033344	.126036	.289792	.5	.710208	.873964	.966656	.997272
5	1.76500E-04	.004672	2.87955E-02	.096256	.226563	.419904	.64707	.851968	.974309
6	6.40000E-06	3.71200E-04	3.79080E-03	1.88416E-02	.0625	.15863	.329417	.576717	.850306
7	9.99999E-08	1.28000E-05	2.18700E-04	1.63840E-03	7.81250E-03	2.79936E-02	8.23543E-02	.209715	.478297

NUMBER OF TRIALS = 8

PROBABILITY OF SUCCESS ON ONE TRIAL

NUMBER OF SUCCESSES	0.1	0.2	0.3	0.4	0.5	0.6	0.7	0.8	0.9
0	1.	1.	1.	1.	1.	1.	1.	1.	1.
1	.569533	.832228	.942352	.983204	.996094	.999345	.999934	.999997	
2	.186895	.496684	.744702	.893624	.964844	.99148	.99871	.999915	.999999
3	3.80918E-02	.203082	.448226	.684605	.855469	.950193	.988708	.998769	.999977
4	5.02435E-03	5.62816E-02	.194104	.405913	.636719	.82633	.942032	.989594	.999594
5	4.31650E-04	1.04064E-02	5.79667E-02	.173637	.363281	.594086	.805896	.943718	.994976
6	2.34100E-05	1.23136E-03	1.12922E-02	4.98073E-02	.144531	.315395	.551774	.796918	.961908
7	7.30000E-07	8.44800E-05	1.29033E-03	8.51967E-03	3.51562E-02	.106376	.255298	.503316	.813105
8	9.99999E-09	2.56000E-06	6.56100E-05	6.55359E-04	3.90625E-03	1.67962E-02	.0576648	.167772	.430467

NUMBER OF TRIALS = 9

PROBABILITY OF SUCCESS ON ONE TRIAL

NUMBER OF SUCCESSES	0.1	0.2	0.3	0.4	0.5	0.6	0.7	0.8	0.9
0	1.	1.	1.	1.	1.	1.	1.	1.	1.
1	.612579	.865783	.959647	.989922	.998047	.999738	.999981		
2	.225159	.563793	.803997	.929456	.980469	.996199	.999567	.999981	
3	5.29721E-02	.261803	.537169	.768213	.910156	.974965	.99571	.999686	.999997
4	8.33109E-03	8.56417E-02	.270341	.51739	.746094	.900647	.974706	.996934	.999936
5	9.09920E-04	1.95814E-02	9.88087E-02	.266568	.5	.733432	.901192	.980419	.999109
6	6.42340E-05	3.06637E-03	2.52949E-02	9.93525E-02	.253906	.48261	.72966	.914358	.991669
7	2.98000E-06	3.13856E-04	4.32090E-03	2.52449E-02	8.98438E-02	.231787	.462831	.738197	.947028
8	8.20000E-08	1.89440E-05	4.33026E-04	3.80109E-03	1.95313E-02	7.05439E-02	.196003	.436207	.774841
9	9.99999E-10	5.12000E-07	1.96830E-05	2.62144E-04	1.95312E-03	1.00777E-02	4.03536E-02	.134218	.38742

NUMBER OF TRIALS = 10

PROBABILITY OF SUCCESS ON ONE TRIAL

NUMBER OF SUCCESSES	0.1	0.2	0.3	0.4	0.5	0.6	0.7	0.8	0.9
0	1.	1.	1.	1.	1.	1.	1.	1.	1.
1	.651321	.892626	.971753	.993953	.999023	.999895	.999994		
2	.263901	.62419	.850692	.95364	.989258	.998322	.999856	.999996	
3	7.01908E-02	.322201	.617218	.83271	.945313	.987705	.99841	.999922	
4	1.27952E-02	.120874	.350389	.617719	.828125	.945238	.989408	.999135	.999991
5	1.63494E-03	3.27935E-02	.150268	.366897	.623047	.833761	.952651	.993631	.999853
6	1.46902E-04	6.36938E-03	4.73490E-02	.166239	.376953	.633103	.849732	.967206	.998365
7	9.12160E-06	8.64358E-04	1.05921E-02	5.47619E-02	.171875	.382281	.649611	.879126	.987205
8	3.73600E-07	7.79263E-05	1.59039E-03	1.22945E-02	5.46875E-02	.16729	.382783	.677799	.929809
9	9.09999E-09	4.19840E-06	1.43680E-04	1.67772E-03	1.07422E-02	4.63574E-02	.149308	.375809	.736099
10	9.99999E-11	1.02400E-07	5.90490E-06	1.04857E-04	9.76562E-04	6.04662E-03	2.82475E-03	.107374	.348678

APPENDIX D
BASIC PROGRAM
FOR LINEAR
PROGRAMS

The BASIC program below will solve linear programming problems of the following type.

Maximize
$$c_1 x_1 + c_2 x_2 + \ldots + c_n x_n$$

where

$$a_{11}x_1 + a_{12}x_2 + \ldots + a_{1n}x_n \leq b_1$$
$$a_{21}x_1 + a_{22}x_2 + \ldots + a_{2n}x_n \leq b_2$$
$$\vdots \qquad \vdots \qquad\qquad \vdots \qquad \vdots$$
$$a_{m1}x_1 + a_{m2}x_2 + \ldots + a_{mn}x_n \leq b_m$$
$$x_1 \geq 0$$
$$x_2 \geq 0$$
$$\vdots \qquad \vdots$$
$$x_n \geq 0$$

```
LIST
APPD

100   DIM A[4,2],B[4],T[4],S[2],C[2]
200   REM  **   READ CØEFFICIENTS IN INEQUALITIES  **
300   MAT   READ A
400   DATA 6,0
500   DATA 0,8
600   DATA 4,6
700   DATA 6,2
800   REM  **  READ RIGHT-HAND SIDES ØF INEQUALITIES  **
900   MAT   READ B
1000  DATA 1200
1100  DATA 2400
1200  DATA 2000
1300  DATA 1400
1400  REM  **  READ CØEFFICIENTS IN ØBJECTIVE  **
1500  MAT   READ C
1600  DATA 100,32
1700  REM  **   SET NØ. ØF INEQUALITIES, M AND NØ. ØF VARIABLES, N  **
1800  READ M,N
1900  DATA 4,2
2000  REM  **   INITIALIZE TABLE ØF NØN-BASIC VARIABLES  **
2100  FØR J=1 TØ N
2200  LET S[J]=J
2300  NEXT J
2400  REM  **   INITIALIZE TABLE ØF BASIC VARIABLES  **
2500  FØR I=1 TØ M
2600  LET T[I]=I+N
2700  NEXT I
2800  REM  **   CHECK FØR ØPTIMAL SØLUTIØN  **
2900  FØR J=1 TØ N
3000  IF C[J]>0 THEN 4100
3100  NEXT J
3200  REM  **   ØPTIMAL SØLUTIØN FØUND  **
3300  FØR I=1 TØ M
3400  REM  **  IF VARIABLE IS ARTIFICIAL, DØ NØT PRINT  **
3500  IF T[I]>N THEN 3700
3600  PRINT " X(";T[I];") =";B[I]
3700  NEXT I
3800  STØP
3900  REM  **  JTH (SØØN TØ BECØME KTH) ENTRY IN NØN-BASIC
4000  REM       TABLE BECØMES BASIC                         **
4100  LET K=J
4200  LET N1=0
4300  LET R1=1.E+10
4400  REM  **   FIND BASIC VARIABLE TØ BECØME NØN-BASIC
4500  REM       BY MINIMIZING B(I)/A(I,K) ØVER I FØR
4600  REM       PØSITIVE VALUES ØF A(I,K)                   **
4700  FØR I=1 TØ M
```

and, in addition,

$$b_1 \geqslant 0$$
$$b_2 \geqslant 0$$
$$\cdot \quad \cdot$$
$$\cdot \quad \cdot$$
$$b_m \geqslant 0$$

```
4800    IF A[I,K] <= 0 THEN 5300
4900    LET T1=B[I]/A[I,K]
5000    IF R1<T1 THEN 5300
5100    LET R1=T1
5200    LET N1=I
5300    NEXT I
5400    REM ** CHECK FØR ALL A(I,K) NEGATIVE **
5500    IF N1>0 THEN 6100
5600    PRINT "SØLUTIØN IS UNBØUNDED"
5700    STØP
5800    REM ** N1 ENTRY IN BASIC TABLE BECØMES BASIC.
5900    REM     INTERCHANGE N1 ENTRY IN BASIC TABLE WITH
6000    REM     K ENTRY IN NØN-BASIC TABLE.            **
6100    LET T1=T[N1]
6200    LET T[N1]=S[K]
6300    LET S[K]=T1
6400    REM ** CØMPUTE NEW N1 RØW **
6500    LET A[N1,K]=1/A[N1,K]
6600    FØR J=1 TØ N
6700    REM ** SKIP CØLUMN K (SEE STATEMENT 4900) **
6800    IF J=K THEN 7000
6900    LET A[N1,J]=A[N1,J]*A[N1,K]
7000    NEXT J
7100    REM ** CØMPUTE NEW RIGHT SIDE FØR RØW N1 **
7200    LET B[N1]=B[N1]*A[N1,K]
7300    REM ** CØMPUTE NEW RØWS (EXCEPT RØW N1) **
7400    FØR I=1 TØ M
7500    REM ** SKIP RØW N1 (SEE STATEMENTS 6500 TØ 7000) **
7600    IF I=N1 THEN 8600
7700    FØR J=1 TØ N
7800    REM ** SKIP SØLUMN K (SEE STATEMENT 8500) **
7900    IF J=K THEN 8100
8000    LET A[I,J]=A[I,J]-A[I,K]*A[N1,J]
8100    NEXT J
8200    REM ** CØMPUTE RIGHT WIDE ØF RØW I **
8300    LET B[I]=B[I]-A[I,K]*B[N1]
8400    REM ** CØMPUTE CØLUMN K ENTRIES **
8500    LET A[I,K]=-A[I,K]*A[N1,K]
8600    NEXT I
8700    REM ** CØMPUTE NEW CØEFFICIENTS FØR ØBJECTIVE **
8800    FØR J=1 TØ N
8900    REM ** SKIP KTH CØEFFICIENT )SEE STATEMENT 9300) **
9000    IF J=K THEN 9200
9100    LET C[J]=C[J]-C[K]*A[N1,J]
9200    NEXT J
9300    LET C[K]=-C[K]*A[N1,K]
9400    REM ** RETURN TØ CHECK FØR ØPTIMAL SØLUTIØN **
9500    GØTØ 2900
9600    END
```

The DIM statement is

```
100 DIM A(M,N), B(M), T(M), S(N), C(N)
```

where M = the number of inequalities and N = the number of unknowns (variables). The matrix A must be given row by row in DATA statements numbered 400 through 700. The right-hand sides of the inequalities (i.e., $b_1, b_2, \ldots, b_m$) must be given in DATA statements numbered 1000 through 1300 in order. The values of the coefficients in the objective ($c_1, c_2, \ldots, c_n$) must be given in order in one DATA statement numbered 1600. Finally, the number of inequalities,

M, and the number of unknowns, N, must be given in that order in a DATA statement numbered 1900.

The program prints the names and the values of all non-zero variables. An example of the use of this program is given in Sections 11.6 and 11.7.

INDEX